Aerogel Hybrids and Nanocomposites

Aerogel Hybrids and Nanocomposites

Editors

István Lázár
Melita Menelaou

MDPI

Basel • Beijing • Wuhan • Barcelona • Belgrade • Novi Sad • Cluj • Manchester

Editors
István Lázár
Department of Inorganic
and Analytical Chemistry
University of Debrecen
Debrecen
Hungary

Melita Menelaou
Department of Chemical
Engineering
Cyprus University
of Technology
Limassol
Cyprus

Editorial Office
MDPI
St. Alban-Anlage 66
4052 Basel, Switzerland

This is a reprint of articles from the Special Issue published online in the open access journal *Gels* (ISSN 2310-2861) (available at: www.mdpi.com/journal/gels/special_issues/Aerogel_Hybrids).

For citation purposes, cite each article independently as indicated on the article page online and as indicated below:

Lastname, A.A.; Lastname, B.B. Article Title. *Journal Name* **Year**, *Volume Number*, Page Range.

ISBN 978-3-0365-9409-5 (Hbk)
ISBN 978-3-0365-9408-8 (PDF)
doi.org/10.3390/books978-3-0365-9408-8

Contents

About the Editors

István Lázár

István Lázár received his M.Sc. in Chemistry in 1984 and his Ph.D. in biologically active boron analogs of amino acids in 1988, both at Lajos Kossuth University (Debrecen, Hungary). He spent two periods as a postdoctoral research associate at The University of Texas at Dallas, USA (1989–91 and 1994). He received a C.Sc degree in the synthesis of new MRI contrast agents in 1994 (Hungarian Academy of Science, Budapest). He initiated and established the aerogel research at the University of Debrecen in 2006. Currently, he works at the Department of Inorganic and Analytical Chemistry as an Associate Professor of Chemistry. For his contribution to the education and research of inorganic and analytical chemistry, he was awarded the Knight's Cross of the Hungarian Order of Merit in 2020. His actual research focuses on two major fields: biomedical applications of new aerogel-based composite materials and the synthesis and study of catalytically or photocatalytically active aerogels.

Melita Menelaou

Melita Menelaou received her Diploma in Chemical Engineering (2004) and her Ph.D. degree (2009) from the Aristotle University of Thessaloniki (Greece). Dr. Menelaou was a member of respected research groups in Asia (Japan—Advanced Institute of Materials Research) and Europe (Czechia—Central European Institute of Technology-Brno University of Technology; Spain—University of Barcelona; Greece—School of Chemical Engineering/School of Chemistry, the Aristotle University of Thessaloniki). Currently, she is working in the Department of Chemical Engineering at Cyprus University of Technology (Cyprus), and very recently, she was elected as an Assistant Professor in the Department of Mechanical Engineering and Materials Science and Engineering at the same university. Her work focuses on the synthesis and characterization of a wide range of inorganic (nano)materials with potential technological and biomedical applications.

gels

Editorial

Editorial for the Special Issue: "Aerogel Hybrids and Nanocomposites"

István Lázár [1,*] and Melita Menelaou [2,*]

1 Department of Inorganic and Analytical Chemistry, University of Debrecen, 4032 Debrecen, Hungary
2 Department of Chemical Engineering, Cyprus University of Technology, Limassol 3036, Cyprus
* Correspondence: lazar@science.unideb.hu (I.L.); melita.menelaou@cut.ac.cy (M.M.)

Aerogel materials are porous ultralight solid materials obtained from gels, wherein a gas, commonly air, replaces the liquid component. These aerogels exhibit several distinctive characteristics, such as high porosity (exceeding 90% of the total volume), very low apparent density, very high specific surface area and high mechanical strength, when compared to the density of the material, among others. The term "aerogel" encompasses a broad range of structures and does not impose specific restrictions on materials selection or synthetic methodologies. Moreover, aerogels may vary widely in their composition and can be of inorganic or organic origin. Due to their extremely low density, huge specific surface area, and open mesoporous structure, single-component aerogels have already claimed dozens of applications, such as in catalysis, aerospace, construction industries, energy storage devices, solar-steam generation, and medical applications.

Aerogel materials exhibit a wide range of classifications based on their appearance (i.e., monoliths, powders, films), their microstructural characteristics (i.e., microporous, mesoporous, mixed porous), their composition (i.e., inorganic, organic, hybrid materials) and on their polarity and surface functionality (i.e., hydrophilic, hydrophobic, amphiphilic, oleophilic or oleophobic). Typical inorganic aerogels can derive from various oxides like silica and alumina. Organic aerogels comprise carbon-based materials like cellulose, graphene, polymers and chitosan. Hybrid aerogels are prepared by combining organic and inorganic materials at the molecular level, and harnessing the versatile and cooperating properties arising from their constituent components.

Considering these factors, this *Gels* Special Issue focuses on the synthesis, production, structure, properties and applications of such complex aerogel materials, while paying particular attention to the cooperation between the hybrid matrix components and the guest particles. We explored original contributions based on conventional and non-conventional approaches to obtain hybrid and composite aerogel materials. The aim was to delve into fundamental and applied aspects, including the technological and biomedical applications of such materials.

The recent advancements concerning the role of aerogel-based materials in bone and cartilage tissue engineering were thoroughly discussed by Lazar and colleagues. Through their literature review, the authors presented a comprehensive list and summary of various aerogel building blocks and their biological activities synthesized under different synthetic protocols, as well as how the complexity of aerogel scaffolds can influence their in vivo performance, ranging from simple single-component or hybrid aerogels to more intricate and organized structures. Lastly, the authors acknowledged the challenges of aerogel-based hard tissue engineering materials in the near and far future, and their potential connection with emerging healing techniques [1].

In addition, chronic wounds are physical traumas that significantly impair the quality of life of over 40 million patients worldwide, and aerogels can act as carriers for the local delivery of bioactive compounds at the wound site. In this regard, Remuiñán-Pose and colleagues obtained alginate aerogel particles loaded with vancomycin, an antibiotic

check for updates

Citation: Lázár, I.; Menelaou, M. Editorial for the Special Issue: "Aerogel Hybrids and Nanocomposites". *Gels* **2023**, *9*, 812. https://doi.org/10.3390/gels9100812

Received: 3 October 2023
Accepted: 7 October 2023
Published: 12 October 2023

used for the treatment of Staphylococcus aureus infections, through aerogel technology combined with gel inkjet printing and water-repellent surfaces. Alginate aerogel particles showed high porosity, large surface area, a well-defined spherical shape and a reproducible size. Several drug-loading strategies (in ink, a bath or ink-plus-bath) were tested, which, due to the increase in the specific surface area and the mesoporosity, opened the way to the preparation of more efficient drug-delivery aerogels [2].

Silica and silica-based (hybrid) aerogels represent the most extensively studied family of mesoporous materials. Within this Special Issue, Lee and colleagues published an article on the synthesis of spherical silica aerogel powder with hydrophobic surfaces from a water glass-in-hexane emulsion. The gelation was completed by heating the water glass droplets, which was then followed by the solvent exchange and surface modification steps. The authors claim that the pH of the silicic acid solution was crucial in obtaining a highly porous silica aerogel powder with a spherical morphology [3].

In another study, Pérez-Moreno and colleagues prepared chitosan (CS)-silica xerogel and aerogel hybrids using the sol–gel method, either by direct solvent evaporation at atmospheric pressure or by supercritical drying in CO_2, respectively. Both types of mesoporous materials exhibited large surface areas (821 m^2g^{-1}–858 m^2g^{-1}), outstanding bioactivity and osteoconductive properties. Through a comparative study, the authors showed that the sol–gel synthesis of CS-silica xerogels and aerogels enhanced not only their bioactive response but also their osteoconduction and cell differentiation properties. Therefore, these new biomaterials can provide adequate secretion of the osteoid for fast bone regeneration [4].

Additionally, Sai et al. described the synthesis of robust fibrous silica-bacterial cellulose (BC) composite aerogels with high performance, following a novel synthetic path. Silica sol was diffused into a fiber-like matrix, which was obtained by cutting the BC hydrogel. A secondary shaping formed a composite wet gel fiber with a nanoscale interpenetrating network structure. The tensile strength of the resulting aerogel fibers reached up to 5.4 MPa. The quantity of BC nanofibers in the unit volume of the matrix was improved significantly by the secondary shaping process. Thus, a novel method was proposed herein to prepare aerogel fibers with excellent performance to meet the requirements of wearable applications [5].

Furthermore, Győri and colleagues reported a way of functionalizing porphyrin rings of 5,10,15,20-tetrakis(4-aminophenyl)porphyrin with a silane linker, followed by complexation with selected metal ions, such as copper and iron, and binding the complexes to the silica aerogel matrix with strong covalent bonds. The as-prepared aerogel catalysts were highly compatible with the aqueous phase, in contrast to the insoluble nature of the porphyrin complexes initially synthesized. The authors studied their catalytic activities in the mineralization reaction of environmental pollutants phenol, 3-chlorophenol and 2,4-dichlorophenol with hydrogen peroxide. The as-obtained catalysts had a large specific surface area and an open mesoporous structure, essential features for heterogeneous catalysis [6].

In another study, Kaplin and colleagues obtained luminescent aerogels in a supercritical carbon dioxide medium for the first time based on sodium alginate, cross-linked with ions of rare earth elements (Eu^{3+}, Tb^{3+}, Sm^{3+}) where phenanthroline, thenoyltrifluoroacetone, dibenzoylmethane and acetylacetonate served as ligands upon SC impregnation. The intensity of the luminescence bands changed after impregnation, while the nature of the influence of the organic additives (ligands) on the luminescent properties of REE ions depended on the nature of both the ion and the ligand. Thus, the authors demonstrated that, upon SC impregnation, ligands could penetrate and act as luminescence sensitizers of rare earth ions throughout the entire thickness of aerogels [7].

The removal of polyvalent metal ions Eu(III) and Th(IV) from aqueous solutions using polyurea-crosslinked calcium alginate (X-alginate) aerogels was investigated, through batch-type experiments under ambient conditions and pH 3, by Georgiou and colleagues. Compared to other materials used for the sorption of Eu(III), such as carbon-based materials, the authors proved for the first time that X-alginate aerogels showed by far the highest sorption capacity. Regarding Th(IV) species, X-alginate aerogels showed the highest capacity per volume among the aerogels reported in the literature. Eu(III) and Th(IV) could be recovered from the beads by 65% and 70%, respectively. Thus, such characteristics, along with their stability in aqueous environments, make X-alginate aerogels attractive candidates for water treatment and metal recovery applications [8].

Carbon materials, including graphene aerogels, carbon nanotube aerogels, and carbon aerogels, have been the subjects of extensive investigation. Kubovics and colleagues fabricated composites involving reduced graphene oxide (rGO) aerogels supporting Pt/TiO_2 nanoparticles using a one-pot supercritical CO_2 gelling and drying method, where 3D monolithic aerogels with a mesa/macroporous morphology were obtained, targeted to evaluate the photocatalytic production of H_2 from methanol in aqueous media. The reaction conditions, aerogel composition and architecture were the factors that varied in optimizing the process. Using methanol as the sacrificial agent, the measured H_2 production rate for the optimized system was remarkably higher than the values found in the literature for similar $Pt/TiO_2/rGO$ catalysts and reaction media [9].

Furthermore, Cheng et al. reported a novel method of increasing the mechanical properties of poly(vinyl alcohol) (PVA) aerogels while decreasing their flammabilities, maintaining low densities and using a low-cost/toxicity additive. Sodium hydroxide (NaOH) was used as a base catalyst at flame temperatures to reduce the flammability of PVA aerogels. Low additive levels of NaOH were found to profoundly alter the aerogel properties, such as their mechanical properties and flammability, minimizing their impact on product density without using char-forming agents or halogen compounds to decrease flammability [10].

In summary, aerogel hybrids and nanocomposites represent distinctive materials with versatile applications across various technological and biomedical domains. Among these, silica aerogel stands out as the most prevalent, initially finding commercial use in thermal insulation blankets. However, research endeavors worldwide have since been dedicated to exploring various materials and structures and alternative applications for aerogel-based materials, extending their utility into sectors such as catalysis and tissue engineering. This wide-ranging applicability hinges on the chosen synthesis method, the constituents and the unique properties of the resultant aerogels. It is evident that these materials offer substantial insights, both in terms of fundamental scientific understanding and practical applications, holding the promise of continued discoveries and innovations.

Author Contributions: I.L. and M.M. have contributed equally to the conceptualization, writing and editing of this manuscript. All authors have read and agreed to the published version of the manuscript.

Acknowledgments: The Guest Editors would like to express their sincere thanks to all of the authors for their valuable contributions, and to all of the peer reviewers for their constructive comments and suggestions.

Conflicts of Interest: The authors declare no conflict of interest.

References

1. Lázár, I.; Čelko, L.; Menelaou, M. Aerogel-Based Materials in Bone and Cartilage Tissue Engineering—A Review with Future Implications. *Gels* **2023**, *9*, 746. [CrossRef] [PubMed]
2. Remuiñán-Pose, P.; López-Iglesias, C.; Iglesias-Mejuto, A.; Mano, J.F.; García-González, C.A.; Rial-Hermida, M.I. Preparation of Vancomycin-Loaded Aerogels Implementing Inkjet Printing and Superhydrophobic Surfaces. *Gels* **2022**, *8*, 417. [CrossRef] [PubMed]
3. Lee, K.-J.; Lee, J.M.; Nam, K.S.; Hwang, H. Thermal Gelation for Synthesis of Surface-Modified Silica Aerogel Powders. *Gels* **2021**, *7*, 242. [CrossRef] [PubMed]
4. Pérez-Moreno, A.; Piñero, M.; Fernández-Montesinos, R.; Pinaglia-Tobaruela, G.; Reyes-Peces, M.V.; Mesa-Díaz, M.d.M.; Vilches-Pérez, J.I.; Esquivias, L.; de la Rosa-Fox, N.; Salido, M. Chitosan-Silica Hybrid Biomaterials for Bone Tissue Engineering: A Comparative Study of Xerogels and Aerogels. *Gels* **2023**, *9*, 383. [CrossRef] [PubMed]

5. Sai, H.; Wang, M.; Miao, C.; Song, Q.; Wang, Y.; Fu, R.; Wang, Y.; Ma, L.; Hao, Y. Robust Silica-Bacterial Cellulose Composite Aerogel Fibers for Thermal Insulation Textile. *Gels* **2021**, *7*, 145. [CrossRef] [PubMed]
6. Győri, E.; Kecskeméti, Á.; Fábián, I.; Szarka, M.; Lázár, I. Environment-Friendly Catalytic Mineralization of Phenol and Chlorophenols with Cu- and FeTetrakis(4-aminophenyl)-porphyrin—Silica Hybrid Aerogels. *Gels* **2022**, *8*, 202. [CrossRef]
7. Kaplin, V.; Kopylov, A.; Koryakovtseva, A.; Minaev, N.; Epifanov, E.; Gulin, A.; Aksenova, N.; Timashev, P.; Kuryanova, A.; Shershnev, I.; et al. Features of Luminescent Properties of Alginate Aerogels with Rare Earth Elements as Photoactive Cross-Linking Agents. *Gels* **2022**, *8*, 617. [CrossRef] [PubMed]
8. Georgiou, E.; Pashalidis, I.; Raptopoulos, G.; Paraskevopoulou, P. Efficient Removal of Polyvalent Metal Ions (Eu(III) and Th(IV)) from Aqueous Solutions by Polyurea-Crosslinked Alginate Aerogels. *Gels* **2022**, *8*, 478. [CrossRef] [PubMed]
9. Kubovics, M.; Silva, C.G.; López-Periago, A.M.; Faria, J.L.; Domingo, C. Photocatalytic Hydrogen Production Using Porous 3D Graphene-Based Aerogels Supporting Pt/TiO_2 Nanoparticles. *Gels* **2022**, *8*, 719. [CrossRef] [PubMed]
10. Cheng, Z.-H.; Guo, M.-L.; Chen, X.-Y.; Wang, T.; Wang, Y.-Z.; Schiraldi, D.A. Reduction of PVA Aerogel Flammability by Incorporation of an Alkaline Catalyst. *Gels* **2021**, *7*, 57. [CrossRef]

gels

Review

Aerogel-Based Materials in Bone and Cartilage Tissue Engineering—A Review with Future Implications

István Lázár [1,*], Ladislav Čelko [2] and Melita Menelaou [3,*]

1 Department of Inorganic and Analytical Chemistry, University of Debrecen, Egyetem tér 1, 4032 Debrecen, Hungary
2 Central European Institute of Technology, Brno University of Technology, Purkynova 656/123, 612 00 Brno, Czech Republic; ladislav.celko@ceitec.vutbr.cz
3 Department of Chemical Engineering, Cyprus University of Technology, 30 Arch. Kyprianos Str., Limassol 3036, Cyprus
* Correspondence: lazar@science.unideb.hu (I.L.); melita.menelaou@cut.ac.cy (M.M.)

Abstract: Aerogels are fascinating solid materials known for their highly porous nanostructure and exceptional physical, chemical, and mechanical properties. They show great promise in various technological and biomedical applications, including tissue engineering, and bone and cartilage substitution. To evaluate the bioactivity of bone substitutes, researchers typically conduct in vitro tests using simulated body fluids and specific cell lines, while in vivo testing involves the study of materials in different animal species. In this context, our primary focus is to investigate the applications of different types of aerogels, considering their specific materials, microstructure, and porosity in the field of bone and cartilage tissue engineering. From clinically approved materials to experimental aerogels, we present a comprehensive list and summary of various aerogel building blocks and their biological activities. Additionally, we explore how the complexity of aerogel scaffolds influences their in vivo performance, ranging from simple single-component or hybrid aerogels to more intricate and organized structures. We also discuss commonly used formulation and drying methods in aerogel chemistry, including molding, freeze casting, supercritical foaming, freeze drying, subcritical, and supercritical drying techniques. These techniques play a crucial role in shaping aerogels for specific applications. Alongside the progress made, we acknowledge the challenges ahead and assess the near and far future of aerogel-based hard tissue engineering materials, as well as their potential connection with emerging healing techniques.

Keywords: aerogel; tissue engineering; artificial bone substitution; in vitro and in vivo bioactivity; biodegradation; cartilage regeneration; scaffold; osteogenesis; simulated body fluids; immortalized cell lines

check for updates

Citation: Lázár, I.; Čelko, L.; Menelaou, M. Aerogel-Based Materials in Bone and Cartilage Tissue Engineering—A Review with Future Implications. *Gels* **2023**, *9*, 746. https://doi.org/10.3390/gels9090746

Academic Editor: Shige Wang

Received: 9 August 2023
Revised: 9 September 2023
Accepted: 11 September 2023
Published: 13 September 2023

1. Introduction

Bone is a rigid tissue with essential functions in providing structural support, protecting vital organs, and enabling movement [1]. It consists of an organic matrix (20%), primarily made up of type I collagen, a mineral phase (65%) predominantly composed of hydroxyapatite [$Ca_{10}(PO_4)_6(OH)_2$, HAp], water (10%), and various bioactive factors and cells, mainly osteoblasts and osteoclasts [2]. Bone has a natural regenerative process that is regulated by biomechanical, cellular, and molecular factors [3]. Articular cartilage is a thin layer covering the ends of bones, allowing smooth gliding and facilitating proper joint function. The cartilage tissue is a sturdy, flexible avascular structure composed of collagen, proteoglycan, non-collagenous proteins, and water. A unique feature of cartilage is its close connection with the underlying hard subchondral bone, comprising three distinct components: highly mineralized subchondral, intermediate-mineralized calcified, and non-mineralized tissues, separated by a dense tidemark.

Damage to bones and cartilage often occurs due to disease or traumatic injuries. Incidences of bone- and cartilage-related disorders have been on the rise, linked to factors like aging, obesity, cancer, and sports-related injuries. These conditions can significantly impact patients' quality of life, causing pain, reduced mobility, and loss of independence. To address these challenges, the research community has been focusing on regenerative medicine approaches, including the development of biomaterial-based and tissue-engineering solutions.

Restoring bone integrity and structure is essential in cases of fractures, skeletal development, or regular physiological reshaping. It involves facilitating the transport, growth, proliferation, and differentiation of osteoprogenitor cells in the injured or defective area. To achieve successful bone repair, a well-designed system is necessary to support the three primary mechanisms of bone regeneration: (a) rapid revascularization, (b) osteogenesis induction, and (c) osteoinduction, which generate new tissue from osteogenic cells. Additionally, the process should promote osteoconduction and encourage cell growth towards the bone surface [4]. However, there are instances where bone regeneration or repair exceeds the tissue's capacity for new bone formation. Such cases may include bone deformations, neoplastic diseases, infections, avascular necrosis, and osteoporosis, among others [3].

Addressing bone defects requires orthopedic reconstruction methodologies that involve bone replacement through the implantation of natural or artificial grafts [5]. Both types of grafts must meet specific criteria, such as high biocompatibility, osteoconduction, and osseointegration [6]. Moreover, they should exhibit robust mechanical strength, be harmless to the body, and remain stable in the biological environment. Throughout their presence in the body, the grafts must not demonstrate any toxic effects [5].

Recent advancements in bone tissue engineering are centered around the development of structures that can closely mimic the behavior of natural bone in terms of both structure and performance. This involves creating materials with exceptional mechanical reinforcement and a supporting matrix, all while maintaining biocompatibility [7]. To promote successful regeneration, these materials need to be highly porous, encouraging vascularization into the damaged area and facilitating the migration of osteogenic cells. Biocompatible three-dimensional scaffolds or hydrogels often possess these desirable characteristics [8].

Scaffolds tailored for bone and cartilage tissue engineering must meet specific biological requirements. They should be biocompatible, non-toxic, and biodegradable, capable of seamlessly integrating and interacting with the surrounding environment. Porosity is a key factor, as it enables cellular infiltration and facilitates the transport of essential gases, nutrients, and regulatory factors, all crucial for cell survival. Finding the right balance is crucial, as excessively large pores reduce the surface area available for cell attachment, while overly small pores hinder cell migration and infiltration, and restrict the diffusion of vital nutrients and waste products. Research conducted by Matsiko et al. suggests that the optimal pore sizes for bone and cartilage tissue-engineering applications typically fall within the range of 100–300 μm [9]. Various fabrication methods can be employed to create these scaffolds, such as freeze drying, electrospinning, and 3D printing, as studied by Iglesias-Mejuto, García-González, Włodarczyk-Biegun, and del Campo [10,11]. Detailed insights into these methods will be provided in Section 3.

In bone and cartilage repair applications, both synthetic and natural polymers have found utility. Synthetic polymers offer versatility in their physical and chemical properties, allowing precise control over molecular weight, degradation time, and hydrophobicity. Prominent examples of synthetic materials employed in such applications, as reported by Puppi et al., include poly L-lactic acid (PLLA), polycaprolactone (PCL), polyglycolic acid (PGA), and polyethylene glycol (PEG) [12]. These synthetic materials can be used in various forms to create scaffolds of different shapes and sizes.

Conversely, natural polymers boast several advantages over their synthetic counterparts. They demonstrate biocompatibility, biodegradability with non-toxic degradation products, and possess bioactive properties that facilitate enhanced cell interactions. Some of the natural polymers used for bone and cartilage repair applications include collagen, silk, gelatin, fibrinogen, elastin, keratin, actin, and myosin. Several examples of polysaccharide-based aerogels exist, such as the crosslinked cellulose nanocrystal aerogels synthesized by Osorio et al. [13], an alginate aerogel reported by Wu et al. [14], and the development of a novel high-methoxyl pectin–xanthan aerogel coating on medical-grade stainless steel reported by Horvat et al. [15]. They can be classified as "bio-aerogels", which originated from natural, semi-synthetic, and synthetic sources, with promising biomedical applications. The processing steps of the polysaccharide-based aerogels are similar to those applied for silica and other organic counterparts and, most commonly, start with the preparation of gel from an aqueous solution (often called hydrogel or "aquagel") and the water in the pores of the aquagel is replaced with an alcohol such as methanol, to prepare an "alcogel" and to make possible the drying by supercritical carbon dioxide (scCO_2) [16]. Alternatively, the solvent exchange step can be circumvented if gelation is directly carried out in alcohol. In addition, composite materials such as PEGDA/CNF aerogel–wet hydrogel scaffold (where PEGDA: polyethylene glycol diacrylate; and CNF: cellulose nanofibril) have been proposed to overcome limitations of the single components in a concise review by Kazimierczak and Przekora [4].

Ceramic materials in the form of calcium phosphate have also been studied and proposed as potential candidates for bone regeneration and/or substitution. Hydroxyapatite (HAp), the main inorganic constituent of human bone, has high biocompatibility and osteo-conductivity, rendering it a material of particular interest for bone regeneration [3]. HAp biomaterials, however, are characterized by poor cell adhesion and difficult ingrowth, thus limiting their therapeutic effect in clinical applications. Also, it is not easy to prepare single-phase HAp porous scaffolds with both high porosity and excellent mechanical properties to be suitable candidates for bone regeneration. Thus, researchers worldwide are working to improve the properties of such materials. An example includes the work of Duan and coworkers who investigated HAp-based composite porous scaffolds instead of "HAp-only" porous scaffolds for such applications. Their experimental results suggest that they prepared a promising material in the form of HAp nanowire aerogel scaffold [17].

Other promising calcium phosphate-based ceramic materials are the a-tricalcium phosphate (α-$Ca_3(PO_4)_2$, α-TCP) and β-tricalcium phosphate (β-$Ca_3(PO_4)_2$, β-TCP). Combining the excellent biocompatibility of β-TCP and the conductivity of carbon aerogels, Tevlek et al. synthesized a β-TCP carbon–aerogel composite material. The biocompatibility of the composite material was evaluated, and their results suggested that composites may also act as promising targets for such applications [18]. Also, Lin and coworkers developed a β-TCP bioceramic platform coated with carbon aerogel as a novel approach to conquer osteosarcoma in one step [19].

In 2015, Wan et al. proposed mesoporous TiO_2 nanotube materials as a novel 3D porous network-structured scaffold for potential bone tissue engineering. The TiO_2 nanotubes were synthesized using the template-assisted sol–gel method followed by calcination. The scaffold showed an extremely large surface area of 1629 m^2 g^{-1} and a diameter of less than 100 nm [20].

Repairing cartilage defects remains a significant challenge in the field. While various clinical treatments for cartilage regeneration, such as microfracture, autologous chondrocyte implantation, Pridie perforations, and transplantation of osteochondral plugs have been developed [21], their success in fully regenerating functional cartilage tissue has been limited. To address these limitations, alternative approaches have been proposed, including the use of cell-loaded scaffold constructs. For successful cartilage regeneration through tissue engineering, an ideal scaffold must possess certain crucial characteristics, similar to those required for bone regeneration. These characteristics include a biomimetic three-dimensional (3D) architecture to facilitate cell adhesion, an appropriate porosity to support cell ingrowth, sufficient mechanical strength to maintain its shape, good biocompatibility, and biodegradability, among others. These essential features are key to developing effective strategies for cartilage repair and regeneration.

Electrospinning proves to be a highly effective technique for producing composite fibers with varying diameters and arrangements, closely resembling the morphology of the natural extracellular matrix (ECM) found in cartilage tissue, while also possessing suitable mechanical properties. For instance, Feng et al. explored a novel method involving electrospinning cartilage-derived extracellular matrix and polycaprolactone (PCL) composite nanofibrous membranes [22]. The traditional electrospinning technique primarily produces two-dimensional (2D) fiber membrane materials with minimal thickness and small pores. However, researchers have pursued the electrospinning of multi-component nanofibers to overcome the limitations associated with individual polymers and to cater to specific requirements. These requirements encompass crucial aspects like mechanical strength, biocompatibility, and degradation rate. By finely adjusting the proportion of each component in the composite fibers, these special requirements can be met, as was reviewed in detail by Chen et al. [23].

Various methods for preparing 3D electrospun nanofibrous scaffolds have been extensively researched and published. Examples include multilayering electrospinning, as reported by Zhang et al. [24] and Chainani et al. [25], as well as liquid and template-assisted electrospinning and post-treated electrospinning, as explored by Shim and colleagues [26], among others. In light of these advancements, Chen et al. [23] and Li et al. [27] successfully prepared aerogels composed of electrospun gelatin/polylactide (Gel/PLA) or gelatin/polycaprolactone (Gel/PCL) fibers, offering promising potential for cartilage regeneration. Additionally, Wang et al. developed a 3D fibrous aerogel comprising SiO_2 nanofibers with chitosan serving as bonding sites for bone regeneration [28]. These reports represent a few examples demonstrating the feasibility and potential of fibrous aerogels in the fields of cartilage and bone tissue engineering. Furthermore, inorganic components like hydroxyapatite (HAp) have been widely incorporated into implants for calcified cartilage and subchondral bone regeneration. Meanwhile, glycosaminoglycans (GAG) such as hyaluronic acid (HA) and chondroitin sulfate (CS) are frequently utilized for cartilage regeneration. These materials play crucial roles in enhancing the performance and functionality of tissue-engineering scaffolds.

The quest for robust and long-lasting bone regeneration and cartilage tissue remains an important and challenging topic. In light of this, the present review article offers an overview of (composite) aerogel materials, a remarkable category of nanoporous materials with great potential for bone and cartilage repair applications. In the following sections, detailed information on the physical and chemical properties, and the significant role that such aerogels can play in various bone-related biomedical applications will be explored.

2. Aerogel Microstructure

The discovery of aerogels dates back to 1931 when Kistler published the first article on the subject in Nature, titled "Coherent expanded aerogels and jellies" [29]. Kistler's groundbreaking work involved the successful synthesis of aerogels from silica, achieved through the condensation of sodium metasilicate. He later expanded his research to include aerogels made from alumina, tungsten, nickel tartrate, cellulose, and gelatin. In his definition, aerogels were described as "gels in which the liquid has been replaced by air, with moderate shrinkage of the solid network." For over 50 years, aerogels received little attention from the scientific community. However, in the last four decades, the interest in these materials has grown exponentially. This surge in interest can be attributed to the diverse range of applications that aerogels offer in various fields. Notably, aerogels provided solutions in catalysis, aerospace, and construction industries, for example. They have also proven valuable in energy-storage devices, solar-steam generation, and medical applications.

Aerogels are remarkable porous ultralight solid materials obtained from gels, wherein the liquid component is replaced by a gas, commonly air. These aerogels exhibit several distinctive characteristics, including (a) high porosity (exceeding 90% of the total volume),

(b) very low apparent density, (c) very high specific surface area, and (d) high mechanical strength when compared to the density of the material.

The term "aerogel" encompasses a broad description of the structure and does not impose specific restrictions on material compositions or synthetic methodologies. Hüsing and Schubert proposed that "aerogels are materials in which the typical pore structures and networks remain remarkably maintained when the pore liquid of a gel is replaced by air" [30]. This definition better captures the essential characteristic of aerogels, highlighting their porous and highly structured nature, even after the liquid component has been replaced by air.

In this regard, aerogel materials exhibit a wide range of classifications, as discussed by Karamikamkar et al. [31]. These classifications include their appearance, microstructural characteristics, composition, polarity and surface functionality.

The most commonly employed and, perhaps, the simplest method for producing aerogels is the well-established sol–gel approach, followed by the specific drying process, known as supercritical drying at or above the supercritical point [30]. The sol–gel procedure involves two main stages: the formation of a sol and the subsequent transformation into a gel. The last step involves removing the pore liquid through a specialized drying process, leading to the formation of the aerogel. This drying step plays a critical role in shaping the final physical and chemical profile of the aerogel, allowing for precise control over its characteristics and performance.

The gelation of inorganic aerogels primarily relies on hydrolysis and condensation processes, while biopolymer aerogels form through the aggregation process. Subsequent to gel formation, liquid extraction from the gel can be achieved using various techniques, resulting in materials classified as xerogels, cryogels, and aerogels [32]. The drying methods strongly affect the final properties of these materials, and are discussed in Section 3. (Figure 1)

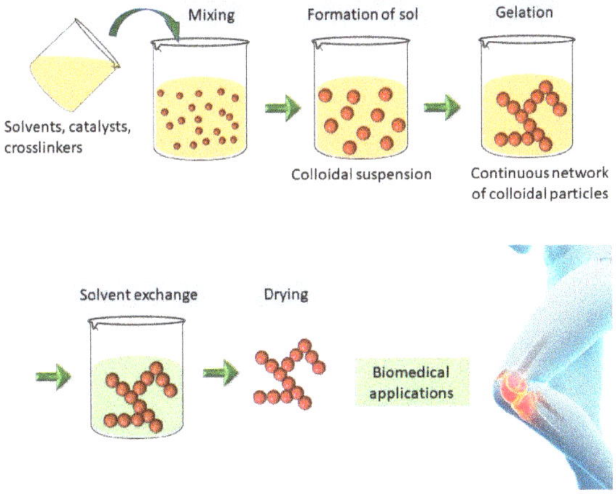

Figure 1. Graphical visualization of the general process of making aerogels in a sol–gel process followed by a specific drying technique to provide aerogel materials for biomedical applications.

To address drawbacks like mechanical issues and limited specific functionalities, various new synthetic approaches have been employed in the production of aerogels. Techniques such as ambient pressure drying and freeze drying have been utilized to tailor the physical, chemical, and biological properties of aerogels, leading to the design and synthesis of hybrid inorganic or organic–inorganic hybrid aerogels. Aerogels stand apart from conventional foams due to their nanometer-scale pores with intricate interconnectivity, resulting in their superior insulating capabilities, being 2–5 times more effective than foams, with low thermal conductivity (0.005–0.1 W/mK), and an ultra-low dielectric constant (k = 1.0–2.0). Such characteristics make aerogels highly appealing for various applications. However, one of the challenges faced by these materials lies in their mechanical properties, which can be limited. For instance, silica aerogels are known for their fragility, hygroscopic nature, and poor mechanical properties, leading to drawbacks in certain applications [31]. To expand the range of applications while preserving the unique properties of aerogels, mechanical reinforcing strategies have been devised. For silica-based aerogels, which represent the most extensively studied family, several methods have been explored in the literature to improve their mechanical properties. A common technique employed for structural reinforcement is prolonged aging time, as utilized by Hong et al., leading to the development of 3D internetworked GA@PDMS (where GA: graphene aerogel; and PDMS: poly(dimethylsiloxane)) [33]. Another widely employed approach involves surface-crosslinking of a silica backbone with a polymer. Boday et al. demonstrated the growth of silica aerogel polymer nanocomposites in the presence of poly(methyl methacrylate) (PMMA), while Leventis reported the development of silica aerogels crosslinked with isocyanate-derived polymers [34,35]. In addition, the incorporation of a secondary phase, such as an organic/inorganic phase, embedded in the structure before (or after) gelation, has proven to be an effective strategy for aerogel structural reinforcement, as demonstrated by Randall and coworkers [36]. Theoretical considerations also suggest that improving elastic recovery in silica aerogels can be achieved by including organic flexible linking groups in the silica backbone or by crosslinking the underlying structural gel with silanol groups through reactions with precursors, monomers, or polymers, as described by Lenentis et al. [37]. These methodologies have successfully enhanced the mechanical properties of aerogels while also improving their transparency.

3. Formulation and Drying Methods

3.1. Formulation Methods

3.1.1. Casting, Molding

Monolithic aerogels are crafted through a straightforward casting or molding technique, which stands as the most extensively employed procedure. The constituents for gel formation are poured in a suitable container, allowing the gelation process to reach completion while facilitating subsequent retrieval of the gelled or solidified material devoid of structural compromise. Following this, the material undergoes a series of solvent exchange steps, wherein the original solvent mixture is replaced with an organic solvent compatible with carbon dioxide, such as acetone, methanol, or ethanol. As an alternative route, water-based gels are subjected to freezing and subsequent freeze drying. A viable realization of the casting, solvent exchange, and supercritical drying sequence is elucidated within the literature [38].

3.1.2. Freeze Casting

The freeze-casting process is also frequently used in fabricating porous materials, including aerogels. The technique is thoroughly described in the literature by Li et al. and García-González et al. [39,40]. The aqueous gel is frozen slowly in a segmented pattern before drying. In the process, ice crystals of different sizes are formed in the segmented temperature zones, leading to a patterned meso/macro porosity of the aerogel (cryogel) monoliths after drying, as presented by Tetik and coworkers [41]. This method was used to prepare silk fibroin–silica aerogels by Maleki and coworkers [42] and crosslinked cellulose [13] aerogels for bone substitution by Osorio et al.

3.1.3. Supercritical Foaming

In some instances, supercritical carbon dioxide can also generate gelation and macropore formation. The generally used supercritical foaming technique is reviewed in the literature [43,44] and has been successfully applied for the preparation of aerogel–polymer composite scaffolds made from starch and polycaprolactone by Goimil et al. [45] or from silk fibroin/polycaprolactone by Goimil and coworkers [46].

3.1.4. Stereolithography, 3D Printing

Bio-ink technology and 3D printing stand as firmly established and widely embraced methodologies in the biomedical sphere, particularly in scaffold formulation and the provision of intricate structural arrangements. The outcome of a specific tissue replacement or tissue-mimicking application is contingent upon the materials' intrinsic nature, the 3D architecture of the scaffold, the involved cell types, and the presence or absence of stimulating factors. Computer-aided design tools rapidly generate the blueprint for a 3D framework, subsequently realized using specialized extrusion or syringe-type printers. These printers can utilize a singular bio-ink component capable of light-induced crosslinking, a pliable yet self-supporting paste that undergoes post-printing crosslinking, or a printer with a two-component coaxial head that triggers chemical reactions upon contact, or even a blend thereof. The fabrication of scaffolds for artificial bone or cartilage substitutes presents challenges due to the intricacies of identifying a suitable 3D-printable material. In biomedical practice, nanofibrous bioactive substances are frequently 3D printed and subsequently subjected to freeze drying, transforming them into aerogels or aerogel-like forms to maintain their structural integrity and functionality. An evaluation of the technique's merits and limitations has been comprehensively compiled by Badhe and colleagues [47].

Iglesias-Mejuto and García-González prepared an alginate–hydroxyapatite 3D-printed aerogel scaffold [10] as well as sterile dual crosslinked alginate–hydroxyapatite 3D-printed aerogel scaffolds with carbon dioxide gelling and glutaraldehyde crosslinking technology for bone tissue engineering. The as-prepared scaffolds showed enhanced fibroblast migration and good bioactivity; the latter correlated with the hydroxyapatite content [48].

Ng and coworkers developed a technique in which simultaneous 365 nm photo-crosslinking and microextrusion 3D printing of the mixture of methacrylated silk fibroin and methacrylated hollow silica nanoparticles provided a mechanically stable scaffold compared to the simple silk fibroin networks. Unidirectional freeze casting provided even more interconnection of the pores, after which the aerogel was made by freeze drying. The as-prepared material is expected to be osteoconductive and osteoinductive bone substitute material that can be loaded with ciprofloxacin or other drugs to treat bone-related diseases [49].

3.2. Drying Methods

Regardless of the specific synthetic methods employed, wet gels undergo diverse drying techniques to transform into aerogel-based materials. Among these approaches, freeze drying and supercritical carbon dioxide drying emerge as the most prevalent. To a somewhat lesser extent, alternative strategies such as subcritical drying, spring-back drying [50], and ambient pressure drying [51] have also been explored and subjected to systematic investigation.

3.2.1. Freeze Drying

Aqueous gels have been effectively transformed into aerogels through freeze drying, often referred to as cryogels. This method can be directly applied to aqueous gels, eliminating the need for the solvent exchange steps necessary in supercritical drying. An inherent benefit is that even highly heat-sensitive materials can be dried without undergoing decomposition. Thus, the freeze-drying technique has been harnessed to craft aerogels with successful outcomes.

Examples include the development of aerogels from nanocellulose-PEGDA by Tang et al. [52], from rGO-collagen by Bahrami et al. [53], from rGO network by Asha et al. [54], from PEGDA-CNF by Sun et al. [55], from nanocellulose–bioglass by Ferreira et al. [56], from CA and PCL nanofiber-reinforced chitosan by Zhang et al. [57], from crosslinked cellulose by Osorio et al. [13], and from silk fibroin–cellulose developed by Chen and coworkers [58]. In a number of cases, freeze drying was combined with a freeze-casting/cryotemplating technique.

3.2.2. Subcritical Drying

Subcritical drying of solvogels or aquagels is a recognized technique, albeit one employed with varying interpretations. It can be conveniently executed using cost-effective equipment at or near atmospheric pressure, or with pressures and temperatures slightly below the critical point. A shared characteristic across all variations is the wet gels' aging, followed by a solvent exchange step. The drying process takes several hours to a day or two, making subcritical drying comparable in time requirement to freeze drying. When ambient pressure drying is conducted, the resulting solid material can manifest as either an aerogel or a xerogel, contingent upon the solvent and the gel material's polarity. As an instance, sol–gel-synthesized silica monoliths with chemically modified hydrophobic surfaces can be subjected to the spring-back effect to yield aerogels [50].

The range of conditions and the quality of the dried material depends on the nature of the solvent that fills the pores, as was studied by Kirkbir and coworkers in making aerogels from atmospheric to supercritical conditions [51]. Shrinkage can be minimized to a few percentages under high-pressure conditions. Lower pressures result in higher shrinkage, which can be extensive at around the atmospheric pressure, as found by Singh and coworkers in making microsphere-based scaffolds for cartilage tissue regeneration [59]. In that situation, the dried product can be considered more a xerogel than an aerogel, but fairly frequently, it is also called an aerogel. The porosity of the low-pressure dried materials is well under or near 90%, compared to the 95–99% porosity of the supercritically or higher-pressure subcritically dried materials. The shrinkage itself is not necessarily a disadvantage. In some instances, it is a desirable feature to increase the stiffness. Subcritical CO_2 drying was applied to make polymeric microparticles for cartilage engineering by Bhamidipati and coworkers, for example [60].

Subcritical drying was applied by Vazhayal et al. during the synthesis of hierarchically porous aluminosiloxane particles in a sol–gel emulsion process, which was tested as a drug carrier and as an osteoconductive support matrix material for bone tissue engineering. The particles were dried from isopropanol at 50 °C under ambient pressure [61].

3.2.3. Supercritical Carbon Dioxide Drying

Supercritical carbon dioxide drying is one of the most widely used techniques to make aerogel materials. It was used in many cases; thus, only a few examples are listed here. The temperature range is approximately 40 to 80–90 °C, and the pressure range is 75–250 bar. This technique was used for the preparation of a wide range of aerogel materials including chitosan-GPTMS by Reyes-Peces et al. [62], collagen–alginate by Muñoz-Ruíz et al. [63], alginate–lignin by Quaraishi et al. [64], alginate by Martins et al. [65], starch and polycaprolactone by Goimil et al. [45], and silica-TCP-HAp by Lázár et al. [38].

3.3. *Post-Drying Workup and Shaping*

After the drying process is finished, aerogel materials frequently require further workup, i.e., cutting, mechanical shaping, or thermal treatment to meet the application-specific requirements. The most commonly used techniques are graphically summarized in Figure 2.

Due to the sensitivity of the fine aerogel structure to any kind of wetting liquids, solid-phase post-drying procedures can be used in most cases. Solution-phase soaking, leaching, wet grinding, and melting techniques cannot be applied when the original structure is to be maintained.

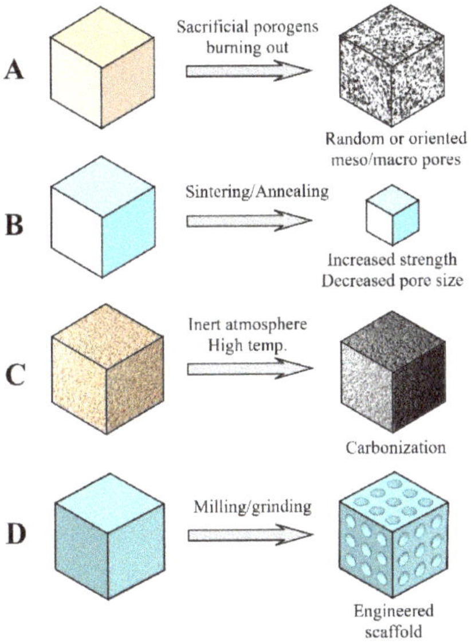

Figure 2. Major types of post-drying thermal or mechanical treatments of aerogels to generate application-specific properties. (**A**) Burning-out sacrificial porogen materials to provide macropores, (**B**) annealing of inorganic aerogels to provide increased mechanical strength, (**C**) high-temperature inert atmosphere carbonization of organic materials to change surface properties, (**D**) mechanical shaping (turning, drilling, milling) to make customized scaffolds.

Heat treatment is a simple and convenient way to change the porosity, mechanical strength, dissolvation, and degradation properties of inorganic aerogel-based materials. The process is viable only for thermally stable materials like silica, alumina, TCP, and HAp. Meso- and macropores are generated randomly or pre-arranged by burning out sacrificial porogen template materials at a temperature of a few hundred degrees Celsius (Figure 2A). A further increase in the temperature results in some degree of shrinking of the materials. That may increase the compressive strength and hardness to a high level, decrease the pore diameters, and reduce specific surface areas. Silica aerogel-based TCP composites containing the sacrificial porogen material microcrystalline cellulose or ashless filter paper or highly purified cotton fabric heated in the range of 500–1000 °C preserved their mesoporosity (average pore size: 26–46 nm) along with a decrease in specific surface area (from 400 to 184 m^2/g) and a significant increase in compressive strength (from 0.47 to 16 MPa). Biological activities of the heat-treated materials showed the maximum at 800 °C in cell studies and rat critical size calvaria defect model experiments. The preparation and biological activities of the heat-treated materials were presented by Szabó et al., Hegedűs et al., Kuttor et al., Lázár et al., and Hegedűs and coworkers (Figure 2B) [66–70].

Thermosetting polymer-based aerogel materials (resorcinol–formaldehyde, and polybenzoxazine) alone or in composites with other thermally stable material (i.e., TCP) may undergo an inert atmosphere thermal decomposition and carbonization process at a temperature near 1000 °C. The resulting carbon aerogel materials proved to be biocompatible and supported the growth of human osteoblast cells, as pointed out by Dong et al. and Rubinstein and coworkers [19,71] (Figure 2C).

High-mechanical-strength materials like heat-treated silica-TCP aerogel composites, successfully used in artificial bone substitution in vivo in animal models, can be implanted in load-bearing positions. A high number of applications, however, would require customized mechanical shaping of the specimens to fit in the shape of the defect. Machining, milling, and drilling can be performed with mechanically sufficiently strong materials to provide custom-shaped scaffolds. Silica-HAp or TCP composites, for example, can be drilled and shaped to provide 200–500 micron highly oriented channels that are expected to support bone ingrowth and vascularization in cortical bones [70] (Figure 2D).

4. Biomechanical, In Vitro and In Vivo Properties, Toxicity and Biocompatibility

The concept of "biocompatibility", outlined in 1986 as the "ability of a material to function with a suitable response within a given application", has remained unaltered and was reaffirmed during the 2018 Consensus Conference in Chengdu, organized under the auspices of International Union of Societies for Biomaterials Science & Engineering [72].

The assessment of biocompatibility encompasses two fundamental criteria: the absence of toxicity and the seamless integration of the material into the biological system. The latter implies that the material should not hinder cellular function and should possess mechanical, chemical, and physical attributes compatible with the facilitation of cell-specific functions [73]. The evolution of artificial bone substitute materials has followed a comprehensive path, traversing through various material generations, each with its distinct attributes and complexities, as comprehensively reviewed by Bongia et al. within the existing scholarly discourse [74].

Conversely, the history of aerogel-based materials is comparatively succinct. Nonetheless, their introduction to the domain introduces an array of distinctive benefits, primarily arising from their intricately porous architectures, which evoke specific tissue responses. The methodologies and protocols governing the scrutiny of biocompatibility and bioactivity commence with meticulously tailored inorganic solutions and culminate in intricate investigations involving living animal models [74].

4.1. Biomechanical Properties

The biomechanical properties of different bones and cartilage are well known for quite a long time [75]. Standardized experimental protocols and a wide range of instrumental techniques are used for their characterization. When aerogels are tested, some of the methods have to be significantly modified due to the much lower strength of aerogels. The aerogel-based materials and their scaffolds should match the mechanical properties of the connecting tissues to provide a cooperating and supportive medium for tissue ingrowth, provided the material is implanted in load-bearing positions. That task can be achieved with annealed aerogel-based bioceramic materials [67–70,76], while other aerogels should be placed in non-load-bearing positions. Most aerogel-based scaffolds contain one or more natural or synthetic polymeric components with or without inorganic counterparts like silica, graphene, carbon nanotube, calcium phosphates, etc. Although the physical properties are important in bone-related research, only a part of the papers contain relevant data [24,42,58,77–87]. Most recently, a critical review paper has been published describing and summarizing the syntheses, biomechanical properties, and their connection with the porosities of bone substitute aerogel materials by Souto-Lopez and coworkers [88].

The most frequently determined mechanical properties of aerogels are the following: compressive strength, Young's modulus, tensile strength, elastic modulus, stiffness, and shape recovery rate. Although the bone hardness scales (i.e., Vickers hardness, and Shore D) are essential indicators of bone quality [89], they are less frequently used for aerogels. Specialized measurement techniques for this family of materials are described in several papers. Many of them are traditionally related to silica hybrids and composites [90,91]; others deal with elastic organic aerogels [92,93].

4.2. In Vitro Testing Methods

The evaluation of artificial bone substitute materials through in vitro testing primarily encompasses distinct categories of assessments. Estimating the toxicity is always a vital step in determining the basic potential of a new preparation. Using diverse cell lines and cell types in controlled cultures is a standard way to assess various parameters, including viability, toxicity, potential immune reactions, adhesion, proliferation, and other pertinent characteristics. Additionally, an important part of the in vitro testing of artificial bone substitute materials revolves around the observation and characterization of the surface hydroxyapatite layer formation in diverse solutions termed as simulated body fluids (SBFs).

4.2.1. Biocompatibility, Cell Viability

Introducing an exogenous aerogel material into a living organism may induce more or less severe immune and inflammatory responses controlled by cytokines. While an initial inflammation of the damaged bone tissue is necessary for collecting osteoprogenitor cells, extended inflammation has adverse effects. Measuring the concentration of the cytokines may provide crucial information on the bone tissue compatibility of the materials [94].

Several methods based on the use of different living cells have been developed to test artificial bone substitute materials. Under controlled conditions, the tested materials are incubated with the selected cells, and from the immune response to the osteogenic differentiation, the studies follow several activities to determine the tested materials' toxicity, biocompatibility, and bioactivity. Przekora summarizes such examinations in a recent review paper [95].

Cell viability can be determined in the simplest cases by spectrophotometry absorbance measurements or by calculating the ratio of live and dead cells after specific staining and simple optical microscopy plus cell counting. Fluorescent dyes combined with fluorescence spectroscopy or computerized image analysis software may provide information on the number of live or dead cells and cellular activities [96,97]. Cell viability studies are so common that approximately half of the aerogel-related papers are involved; therefore, they are not listed here individually.

4.2.2. Antimicrobial Activity

In general, antimicrobial activity is not an expectation for artificial bone replacement materials. Still, its presence can be beneficial from the point of view of the use of the product. Only chitosan has inherent antimicrobial activity among the components of aerogels prepared for this purpose [98]. Other biopolymers, such as cellulose, gelatin, dextran, pectin, etc., are inactive but suitable for carrying selected antibiotics or gold, silver, platinum, TiO_2, or ZnO nanoparticles, all possessing antimicrobial activity [98–102].

4.2.3. Simulated Body Fluids

The concept of simulated body fluids finds widespread utilization in the exploration of bioactivity during bone mineralization processes. This testing methodology involves the immersion of samples in clear solutions containing the principal inorganic constituents found in human blood plasma over a span of several days. The inception of this technique traces back to the work of Kokubo and colleagues, who introduced the initial simulated body fluid (SBF) for such investigations [103]. Subsequently, a second publication accentuated the value of these tests [104] in approximating the in vivo bioactivity of distinct categories of bone substitute materials. Several works have effectively extended the application of Kokubo's test to aerogel-based bone substitute materials [62,66,77,105]. Expanding on the original formulation, researchers have sought modifications to more accurately mirror the comprehensive chemical composition of human blood plasma. Müller et al. ventured to vary the concentration of bicarbonate ions [106], whereas Győri et al. introduced amino acids and serum albumin to generate modified SBFs, thus rendering them more representative of in vivo conditions [107]. Practical considerations in the preparation and application of SBFs encompass crucial steps to prevent precipitation through proper component dissolution sequencing, an approach detailed by Kokubo et al. [104]. Further variations include adopting saturated stock solutions as recommended by Müller et al. [106], or devising a dual-component set of solutions as elucidated by Győri et al. and Vallés Lluch et al. [107,108], thereby simplifying the making of the SBF solution. Achieving a final pH of 7.4 is imperative, while the temperature must remain constant within the 36–37 °C range during the entire course of treatment. To mitigate the risk of bacterial contamination stemming from the presence of glucose, amino acids, or albumin, it is a prudent practice to supplement the SBFs with sodium azide or antibiotics like gentamycin or kanamycin [106,107]. Given the dynamic nature of the bicarbonate/carbon dioxide equilibrium dominant in unsealed containers, the composition and pH of SBFs might fluctuate over time. As a result, experimentation should be confined to sealed vessels, and periodic replenishment of the SBF solution with fresh aliquots becomes imperative. While SBF testing offers a straightforward avenue for estimating the potential for bone formation in artificial bone substitute materials, cautious interpretation is warranted due to the aforementioned intricacies. It is advisable not to exclusively rely on the outcomes of these tests in categorizing or sorting out materials [107].

4.2.4. In Vitro Cell Studies

In the realm of in vitro examinations, a significant portion of research entails the application of diverse cell lines as a fundamental approach. These cell-based investigations offer crucial insights into a range of parameters including cytotoxicity, inflammatory response, cellular metabolism, adhesion, proliferation, and an array of pertinent characteristics of the materials under scrutiny. However, it is important to note that individual cell culture tests may provide insights into only specific aspects of material behavior, with the broader context of intricate tissue reactions or the foreign body response remaining beyond their scope. Despite this limitation, such assays present a convenient and economical means of analysis, exempt from the complexities of permissions and ethical considerations, and are extensively reviewed in the literature [73,109,110]. The cornerstone of these investigations resides in human cell lines, which serve as a fundamental test for gauging and predicting the interactions of the materials in question. However, it is pertinent to acknowledge that access to primary human osteoblast cells remains limited. Consequently, alternative animal cell models have been adopted in the studies. This review does not aim to cover all the tests and protocols; a detailed account of non-aerogel-related areas is summarized in the literature [111–115].

In laboratory studies, a variety of bone tissue cells from both humans and animals are employed to assess cell adhesion, viability, and growth. Key human cell lines include primary osteosarcoma cell lines, such as SaOs-2 and MG-63, which are immortalized and malignant cells. Among non-human cell lines, there are immortalized osteoblast precursor MC3T3-E1 cells from mouse calvaria, primary osteoblast cells from animals like rats, mice, bovines, and rabbits [111], and induced osteoblasts from stem cells of different animal species or humans [115,116]. During these experiments, the focus is on examining cell attachment to surfaces, their viability and proliferation, as well as the potential of stem cells to transform into osteoblasts and the detection of specific indicators of bone metabolism [117].

Tang et al., for example, undertook an investigation involving a 3D-printed nanocellulose/PEGDA aerogel scaffold in conjunction with mouse bone marrow mesenchymal stem cells. Their study revealed that the scaffold exhibited supportive attributes for cell growth, stem cell proliferation, and chondrogenic induction, with outcomes influenced by the Poisson's ratio sign [52]. In a similar vein, Ge and colleagues crafted a silica aerogel-PCL composite, subjecting it to assessment with MC3T3 and primary mouse osteoblast cells. Their findings demonstrated that the silica aerogel contributed to heightened cell survival, attachment, and growth, while concurrently mitigating the cytotoxicity of the PCL film during prolonged contact [118]. Moreover, Bahrami et al. synthesized collagen aerogel scaffolds coated with reduced graphene oxide (rGO) using a combination of 3D printing and chemical crosslinking, followed by freeze drying. The scaffolds underwent evaluation for bioactivity and bone regeneration potential in both in vitro and in vivo settings. The incorporation of the rGO layer increased mechanical strength by a factor of 2.8 and did not lead to augmented cytotoxicity. Human mesenchymal stem cells displayed heightened viability and proliferation on the scaffold surface. When implanted into cranial bone defects in rabbits, the scaffolds exhibited enhanced bone formation after a 12-week observation period [53].

4.3. In Vivo Animal Testing Methods

4.3.1. General Considerations

Broadly, in vitro testing methods involve the utilization of diverse species possessing varying degrees of bone regeneration potential. These assessments encompass scenarios where bone substitute materials are either subjected to soft tissues without direct bone contact (heterotopic testing) or placed in direct proximity to bone tissue (orthotopic testing). These investigations, conducted across different mammalian species, yield insights into immunological responses, histochemical attributes, and cell regulatory mechanisms. By placing materials within artificial defects, the progression of bone remodeling and regeneration is monitored, spanning several months and occasionally extending beyond a year. The bone healing trajectory traverses three principal phases: the sterile inflammatory phase, the repair phase, and the remodeling phase.

4.3.2. The Role of Porosity

The porosity of materials emerges as a pivotal determinant in shaping the in vivo behavior of bioactive substances. Open-pore architectures stand as a remarkably effective conduit for facilitating the transport of materials to and from living tissues, ushering in vital elements like nutrients, oxygen, and signaling molecules. The significance of macroporosity has been expounded upon in the preceding section, elucidating its role in facilitating optimal bone ingrowth. A recent observation by Ratner underscores the role of material porosity in the early stages of regeneration. In instances where identical artificial materials are employed, densely compacted solid structures tend to trigger an inflammatory response in surrounding tissues, characterized as a foreign body reaction. In contrast, porous architectures tend to mitigate the occurrence of such an inflammatory phase [119]. Further insights, such as those offered by Matsiko et al., underscore that scaffold pore size significantly influences the differentiation process of stem cells [9].

The role of porosity of aerogel-based scaffolds has yet to be systematically studied. The need for large pores is well-known in bone tissue ingrowth [120]. However, besides providing a penetrable material-transport channel, the biological role of the much finer aerogel mesopores has yet to be discovered [88]. Comparative cellular studies with chemically identical aerogel samples exhibiting different narrow pore size distribution peaks walking through the entire mesopore and lower macropore region would be desirable to answer the questions.

Foreseen as instigators of minimal foreign body reactions upon implantation, the ab ovo porous aerogel-based materials may hold substantial promise in this context. The intricate structure of these scaffolds can potentially augment this advantage. Collectively, these observations underscore the pivotal role and potential of bioactive aerogel-based materials in the domain of artificial bone substitution.

4.3.3. Selection of the Animal Species

Animal models represent a cornerstone in the exploration of biocompatibility and regenerative potential, both for established commercial and novel experimental artificial bone substitute materials. This paradigm has also been embraced in the assessment of aerogel-based materials. Among the diverse array of animal species, including mouse, rat, rabbit, sheep, dog, goat, and pig, that have been employed in these inquiries, a comprehensive review of their usage, contexts, considerations, and outcomes is available within the literature [121–123]. In these experimental investigations, small laboratory animals, mostly rodents, are frequently used due to their accessibility within orthopedic surgical research facilities. However, it is worth noting that their inherent regenerative capabilities may significantly diverge from those of large animals. Although mature large-bodied animals exhibit bone structures akin to humans, their practical availability, expenses, and material demands introduce limiting constraints. The careful selection of the appropriate animal species for experimentation becomes crucial and hinges upon the specific objectives of the research endeavor [124].

4.3.4. Critical and Non-Critical Size Models

The dimensions and location of the bone defect assume a pivotal importance when gauging biological activity. Divergent compositions, structures, and qualities of bones across distinct animal species necessitate the careful selection of the right animal model for evaluating regenerative potentials, as comprehensively covered in the literature [123–127]. The size of the defect bears paramount significance in appraising regenerative capabilities. When working with bones, a defect that lacks spontaneous self-healing throughout the anticipated lifespan of the animal is classified as a critical size defect [69,127]. Contrarily, subcritical size defects may exhibit spontaneous healing. The interposition of the regenerative process with artificial bone substitute materials in critical-size models effectively demonstrates the regenerative potential inherent in the experimental materials. Typically, a single material is evaluated per animal, involving one or two defects. However, there are instances where multiple materials are concurrently tested within the same experimental animal [121].

The materials can undergo testing in load-bearing and non-load-bearing positions. A notably prevalent model in studies involving small animals encompasses the critical-size calvarial defect model. This model expedites material testing in an easily attainable and reproducible manner, obviating the need for precise positioning of experimental materials. In this approach, a disc-shaped sample is nestled within a circular opening atop the cranial bone (typically 6 or 8 mm in diameter), establishing contact with the native bone tissue. An instance of this technique involved the application of a calcium phosphate–silica aerogel composite in rats [67]. While this model is convenient, it does not furnish insights into the functional behavior of the materials, such as their mechanical properties. For investigations of such nature, a load-bearing defect position, such as within the femur, is selected to study the healing and remodeling dynamics of an aerogel material [66].

5. Aerogel-Based Materials and Structures for Bone Tissue Engineering

By the traditional IUPAC Gold Book definition, aerogel is a "gel comprised of a microporous solid in which the dispersed phase is a gas". A problem with the definition is that it does not follow the IUPAC definition of micropores. Aerogels are mostly mesoporous materials containing macropores in some cases. A large portion of aerogels does not have micropores at all. Besides the definition by Hüsing and Schubert, as mentioned in Section 1 [30], the aerogel definition needed fine-tuning. Following the most recent trends supported by several publications, an even broader definition of aerogels, which includes, i.e., the nanofibrous materials as they are appearing in the literature, is applied in this paper. According to the recommendations of Vareda et al. and García-Gonzalez et al., here we use the definition of aerogels as "solid, lightweight and coherent open porous networks of loosely packed, bonded particles or nanoscale fibers, obtained from a gel following the removal of the pore fluid without significant structural modification" [128,129].

Considering the materials and techniques used in bone and cartilage tissue engineering, wet gels, and aerogels have common roots and significant overlapping in many aspects. In this review, we focus only on dry aerogel materials. Independently from their features, only the gels that were dried to aerogels or cryogels by any means will be referred here.

5.1. Building Materials of Aerogels and Their Scaffolds Used in Hard Tissue Engineering

The majority of aerogels assessed for their potential in bone regeneration have been constructed from the same building materials widely employed and exhaustively investigated in practical applications. A significant proportion of the tested substances hold approval from the FDA for human usage. The most important characteristics of the materials utilized in the context of aerogel-based bone substitute materials, accompanied by references from the existing literature, are summarized in Table 1.

Table 1. Building block materials used for aerogel-based bone and cartilage tissue engineering. Their bulk bioactivities were tested in vivo, and many of them are used in clinical practice. The references in the table are mainly review papers summarizing the properties, in vivo effects, and therapeutic results achieved in the non-aerogel era.

Name	Properties	References
Alginate	The β-D-mannuronic acid and α-L-guluronic acid-containing alginates can be formulated into gels, particulate solids, nanofibers, or ordered microstructures. They are frequently combined with other biomolecules or chemically modified. Alginates exhibit excellent biocompatibility, biodegradability, and tunable cell-binding affinity, making them versatile materials in wound healing, drug delivery, cartilage, or bone tissue repair.	Sun and Tan; Martau et al. [130,131]
Aluminosilicate	Aluminosilicates show zeolite-like structures and link to the bone matrix. The coating on the alumina surface shows good biocompatibility with the osteoblasts that can sustain their bioactivity.	Oudadesse et al. [132]
Bioactive glass, Bioglass	Bioactive glasses exhibit excellent tissue binding and good bone regeneration properties. Their chemical composition is described with different SiO_2, Na_2O, CaO and P_2O_5 ratios. Depending on the composition, they may also bind to soft tissues. In combination with other bioactive materials, they are frequently used in bone scaffolds. Silicate ions liberated in the degradation process promote the formation of Type I collagen. Bioactive glasses are FDA-approved bone graft materials.	Bellucci et al.; Gerhardt and Boccaccini [114,133]
Carbon (amorphous, graphitized)	Carbon forms are insoluble and non-resorbable (thus permanent) bioinert materials made by high-temperature carbonization of resorcinol–formaldehyde or polybenzoxazine resins. Due to their electric conductance, they may find future applications as building materials in communicating fourth generation devices.	Dubey et al. [134]

Table 1. *Cont.*

Name	Properties	References
Cellulose acetate (CA)	Cellulose acetate is a hydrophilic and thermoplastic biodegradable cellulose derivative. It can be conveniently formulated into sheets, nanofibers, etc. CA scaffolds combined with other bioactive molecules, biopolymers, drugs, etc., support endothelial cell migration and adhesion, and do not promote platelet activation. Chemically modified CA mats bolster osteoconduction and osteoinduction and may help bone regeneration.	Laboy-López and Frenández; Shaban et al.; Rubenstein et al. [135–137]
Cellulose, bacterial cellulose nanofibrils (CNF)	Cellulose nanofibers (from plant or bacterial sources) are nontoxic, biocompatible, and biodegradable materials that can be produced in large quantities at low cost. Pristine and chemically modified or crosslinked CNFs have applications in controlled drug delivery, antibacterial wound dressing, and skin and bone tissue engineering.	Pandey; Torres et al.; Helenius et al. [138–140]
Chitosan	Chitosan is an amino group-containing polysaccharide derived from the natural chitin sources by deacetylation. It contains randomly ordered D-glucosamine and N-acetyl-D-glucosamine units. Chitosan is a highly biocompatible and biodegradable material that can be digested by either lysozyme or chitinase enzymes in the body. It is frequently used for drug delivery, antibacterial wound dressing, tissue engineering, and bone substitution purposes, in combination with other biopolymers like PEGDA, PLA, gelatin, and alginate. The higher degree of deacetylation increases the strength of cell membrane interactions and cellular uptake.	Rodrigues et al.; Venkatesan and Kim; Bojar et al. [141–143]
Collagen, Type-I and II	Collagen is a natural fibrous protein with excellent biocompatibility, biodegradability and bioactivity. Type I collagen is the major component of the extracellular matrix and the bones, while Type II collagen can be found in the cartilage tissues. Due to their excellent cellular interactions, both types were applied in bone scaffolds and cartilage repair preparations.	Ferreira et al.; Rezvani Ghomi et al.; Kilmer et al. [2,144,145]

Table 1. *Cont.*

Name	Properties	References
Gelatin	Gelatin is a partly hydrolyzed form of collagen containing interconnecting protein chains. It is isolated from animal skin, bone, or connecting tissues. The amino acid composition and sequence is changing with the origin of the tissue. Gelatin is mostly used with other bioactive polymers, i.e., alginate, chitosan, PLLA, and PCL. In scaffolds, it improves cell adhesion, proliferation, and infiltration.	Su and Wang; Peter et al. [146,147]
Glycosaminoglycan (GAG)	Glycosaminoglycans are long-chained polysaccharides built from repeating disaccharide units. They are present on cell surfaces and in the extracellular matrix. Due to their role in regulating the growth factor signaling, interaction with cytokines, and cell surface receptors, GAGs affect, for instance, the inflammation and cell growth processes. They are used in hydrogels, antibacterial surface layers, and porous scaffolds in tissue engineering.	Köwitsch et al. [148]
Graphene	Graphene nanosheets are made from graphite and consist of only a single layer of carbon atoms. Graphene is biocompatible, although it is not biodegradable. Graphene promotes stem cell growth and proliferation, as well as osteogenic differentiation. High concentrations of pristine graphene may decrease cell viability, but PEGylation may reduce that effect. Due to its electrical conductance, it might find application in the fourth generation of bioactive materials.	Dubey et al. [134]
Graphene oxide (GO)	GO is prepared from graphite or graphene by strong chemical oxidation. Epoxides, hydroxyl, and carboxylic groups are generated on the surface, providing connecting points to anchorage-dependent cells to adhere, spread and function.	Berrio et al.; Dubey et al. [8,134]
Graphene oxide, reduced (rGO)	rGO is made from GO by thermal decomposition or chemical reduction. Epoxide rings are removed, but carboxylic and phenolic groups remain on the perimeter. When combined with collagen type-I, the material becomes mechanically more robust and activates the differentiation of human osteoblast stem cells. Scaffolds made with them could be used in bone substitution.	Bahrami et al.; Norahan et al. [53,149]

Table 1. *Cont.*

Name	Properties	References
Pectin, Methoxyl pectin	Pectin is a highly hydrophilic, biocompatible, and biodegradable natural polysaccharide rich in carboxylic group-containing galacturonic acid. When more than half of the carboxylate groups are in the methyl ester form, the material is called high methoxyl pectin; otherwise, we talk about low methoxyl pectin. High methoxyl pectin can form hydrogels under mildly acidic conditions. Low methoxyl pectins can be crosslinked with calcium ions to make them less polar drug carriers. Pectins are used alone or in combination with other natural polymers in the 3D printing of scaffolds.	Martau et al.; Li et al.; Tortorella et al. [131,150,151]
Poly(lactic-co-glycolic acid) (PLGA)	PLGA is a highly biocompatible and biodegradable material approved by the FDA for drug delivery, gene engineering, and biomedical uses. Pristine polyglycolic acid would hydrolyze readily. Thus, it is blended with PLA or other polymers to improve hydrolytic and degradation properties. PLGA is combined with different bioactive materials (TCP, HA, gelatin, etc.) or bone morphogenetic proteins (BMPs) and is extensively used in artificial bone substitution applications to facilitate cell adhesion and proliferation. PLGA can easily be formulated into various matrices, from solid scaffolds to nanofiber mats.	Makadia and Siegel; Zhao et al.; Elmowafy et al.; Gentile et al.; Jin et al. [152–156]
Poly(lactic acid and poly(L-lactic acid) (PLA and PLLA)	PLA is a highly biocompatible and biodegradable thermoplastic polymeric material approved by the FDA for biomedical, drug delivery, and tissue engineering applications. Due to the less polar nature of PLA, it is frequently used in co-polymers with hydrophilic polyglycolic acid to improve hydrolytic behavior. When pristine PLA is used alone in the body, it often induces foreign body reactions. Electrospun PLA-copolymers and their microspheres and nanoparticles provide bioactive materials for drug delivery, wound healing, or bone substitution. PLA is widely used in 3D printing. In the human body, PLA implants degrade significantly slower than polyglycolic acid.	Makadia and Siegel; Zhao et al.; Elmowafy et al.; Gentile et al.; DaSilva et al.; Tyler et al.; Böstman and Pihlajamaki [152–155,157–159]

Table 1. *Cont.*

Name	Properties	References
Poly(methyl methacrylate) (PMMA)	PMMA is a bioinert polymeric material, the main component of acrylic bone cement. The mechanical properties can be improved by blending, i.e., with polystyrene. PMMA-based bone cement can be injected into the position and cured at room temperature. It can be mixed with antibiotics. PMMA is not biodegradable; it usually works as a spacer in joining implants. Fixation properties can be improved by chemical modification of the PMMA structure and by loading with TCP or other bioactive and degradable materials. PMMA cements are FDA-approved bone graft materials.	Arora; Magnan et al. [160,161]
Poly(ε-caprolactone) (PCL)	PCL is an FDA-approved biocompatible and biodegradable synthetic material for human drug delivery, suture, and adhesion barrier applications. The biodegradation is the slowest among the ester-type bone substitute materials. Thus, PCL is used in long-term implants. Orthopedics frequently combines it with bioactive components like silk fibroin, bioactive glasses, or TCP to improve cell adhesion. It can be formulated by molding, pressing, 3D printing, solution or melt electrospinning.	Janmohammadi and Nourbakhsh; Dwivedi et al. [162,163]
Polybenzoxazine (PBO)	The name polybenzoxazine covers a wide range of polymers in which the benzoxazine/polybenzoxazine moiety is the standard building block. PBO resins are prepared by thermal or catalytic ring opening and polymerization of substituted benzoxazine structures derived from synthetic or natural precursors, i.e., cellulose or chitosan. In thin films, PBOs show good antibacterial and antifungal activity. Carbonization at high temperatures results in carbon foams that offer good biocompatibility.	Ghosh et al.; Periyasamy et al.; Thirukumaran et al.; Lorjai et al. [164–167]
Poly(ethylene glycol diacrylate) (PEGDA)	Ethylene glycol diacrylate alone or combined with other acrylates can be easily polymerized or photopolymerized to PEGDA and copolymers. Crosslinking may increase the mechanical strength. PEGDA is a hydrophilic and low-immunogenic compound suitable for scaffolds and hydrogels. It is a good drug depot, and the drug release profile can be finely tuned. It can be used in bio-inks for 3D printing to provide biocompatible flow-through devices. It forms hydrogels that are used in cartilage tissue regeneration.	Rekowska et al.; Warr et al.; Qin et al.; Musumeci et al. [168–171]

Table 1. *Cont.*

Name	Properties	References
Silica	Silica is a biocompatible, biodegradable, and osteoconductive material. Silica enhances the osteogenic differentiation of stem cells and bone regeneration by promoting Type I collagen formation, stabilization, and matrix mineralization. Porous silica can be combined with various polymers, biomaterials, proteins, enzymes, drugs, and hormones. The surface can be covalently functionalized with bioactive agents. Higher concentrations of nano-silica particles may lead to bioaccumulation and cellular damage.	Zhou et al.; Jurkic et al.; Shadjou et al.; Vareda et al. [172–175]
Silk fibroin	Silk fibroin is a natural protein produced by insects. It is a lightweight but mechanically strong material and can be found, i.e., in spider webs and prepared from the cocoon of the domestic silkworm. Scaffolds made of it are biodegradable, can be functionalized, and support the attachment and growth of cells. In the form of fibers, nanofibers, mats, films, and porous structures, silk fibroin has many applications in cell cultures, tissue engineering, and cartilage tissue regeneration.	Nguyen et al.; Wang et al.; Wang et al.; Farokhi et al. [176–179]
Starch	Starch is a natural polysaccharide consisting of d-glucose units. It is produced mainly from potatoes, manioc, or seeds like rice, wheat, and corn. Starch is an edible, biocompatible, and readily biodegradable material. It supports cell growth on the surface. It can be formulated in different shapes and porosities with biodegradable polymeric materials. By 3D prototyping, custom-shaped bioactive scaffolds are created.	Martins et al.; Salgado et al. [180,181]
Strontium ranelate (SR)	SR is a medical drug to treat osteoporosis in men and women, regardless of age. It is capable of reducing the risk of fracture. Strontium ranelate promotes the osteoblastic differentiation of stem cells, inhibits osteoclasts, and improves the structure of bones.	Pilmane et al.; Kaufman et al.; Cianferotti et al. [182–184]

Table 1. *Cont.*

Name	Properties	References
Tricalcium phosphate (βTCP, TCP)	Beta tricalcium phosphate is the "gold standard" of bone grafts approved by the FDA. It is osteoinductive, biodegradable, and one of the most extensively used bone substitute materials in clinical practice. The physical appearance of TCP covers a wide range, from low-strength porous bodies to hard grafts. TCP shows no adverse effects and maintains normal calcium and phosphate ions level in the blood. The apparent in vivo behavior is affected to some extent by the purity and the way TCP was produced. TCP is insoluble under physiological conditions at pH 7.4 and is dissolved and resorbed by cell-mediated processes. The resorption time is in the 6–24 month range.	Lu et al.; Bohner et al.; Tanaka et al.; Gilmann and Jayasuriya [185–188]
Xanthan gum	Xanthan gum is a biodegradable branched polysaccharide produced in large quantities by industrial fermentation with the bacteria Xanthomonas campestris. The backbone is cellobiose, and the branches contain D-mannoses and D-glucuronic acid. The structure of the chain in solutions can be tuned from coiled to helical by increasing the temperature and the ionic strength. High-molecular-weight xanthan gums, frequently in combination with other biopolymers, have found application in the biomedical field, from drug delivery to bone substitute scaffolds.	Petri [189]

5.2. Aerogel-Based Materials for Bone Substitution

Given the wide range and diverse compositions of aerogel-based materials utilized in hard-tissue engineering, a systematic classification based on shared properties becomes necessary. The approach adopted here involves categorizing all aerogel-containing structures, except for single-phase aerogels, as composite materials, characterized by distinct physical phase boundaries. This classification proves especially relevant when natural or synthetic polymeric materials, and complex or layered structures are present. While the chemical composition remains the primary determinant, other factors, such as biocompatibility, bioactivity, cellular responses, and tissue reactions, and other parameters are deterministic and discussed in the previous sections. Pore structures, their multi-dimensional orientation, and the arrangement of different scaffold layers also exert significant influence. Table 1 presents the wide array of chemical components utilized in the field of aerogel-based tissue engineering. Their combination can yield numerous materials, the management of which is not always straightforward. Figure 3 provides an illustrative representation of potential classes and their interconnections, delineating increasing complexities.

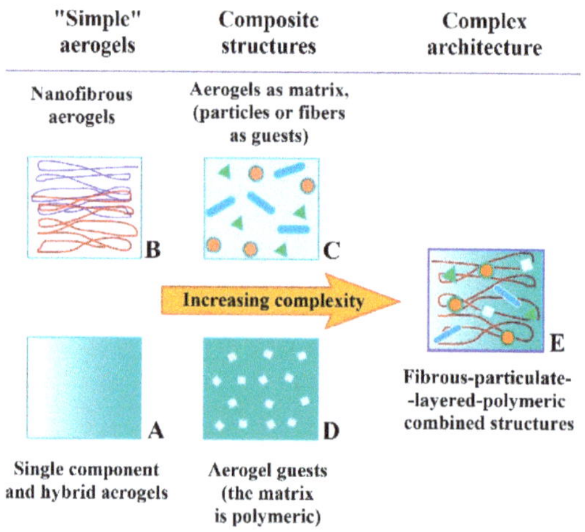

Figure 3. Schematic representation of the major types of chemical composition and structure of aerogel-based artificial bone substitute materials. (**A**) Homogeneous aerogels structures made of single- or multi-component material [15,17,42,61,62,64,65,71,77]. (**B**) Nanofibrous materials dried to an aerogel structure [52,55–57,79,190]. (**C**) Aerogel matrix material containing guest particles and/or fibers [14,18,19,53,63,66–70,80,191]). (**D**) Polymeric matrices containing guest aerogel particles [45,46,118,192]. (**E**) Highly complex structures made of aerogels, particles and nanofibers [13,24,27,41,54]. (The different symbols in the figure represent guest particles without further specification.)

A single-phase homogeneous material may be a chemically one-component pristine aerogel or a multi-component hybrid aerogel in which the components are mixed at the molecular level. In that meaning, there is no difference between aerogels of organic or inorganic origin (Figure 3A). Nanofibrous materials from mostly polymeric materials may also be distributed evenly in space by different treatments, forming a gel from which homogeneous aerogels are made by different drying techniques (Figure 3B). Such a homogeneous aerogel phase may serve as the matrix material in which guest particles are distributed (Figure 3C). Polymeric materials may also be combined or fortified with aerogel particles as guests to improve properties (Figure 3D). And finally, all the structures may be evolved into a very complex unit where the matrix and guest functions are combined, and new properties may appear due to the synergistic interaction of materials in the living environment.

The way aerogels are made for hard-tissue-engineering purposes depends on the material and the properties of the aerogel phase, as well as the complexity of the structure. However, independently from the nature of the materials and the final complexity of the structures, the common point is that all "pre-aerogels" go through a wet gel state, from which the final aerogel is prepared by a suitable drying technique. The technical implementation of wet gel-making procedures is summarized and shown in Figure 4. The simplest and most traditional way, as mentioned in Section 2, is the sol–gel technique (Figure 1).

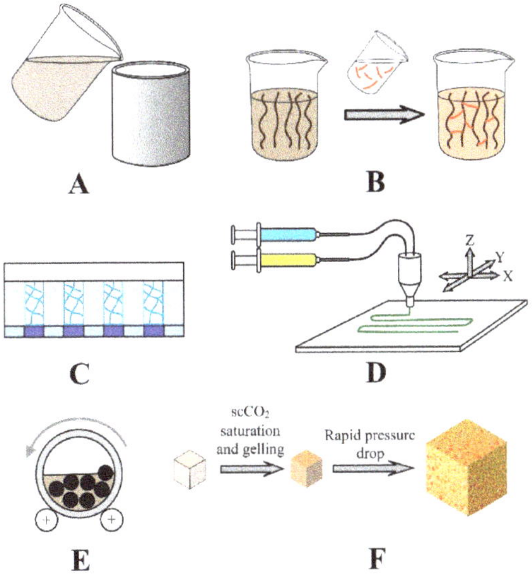

Figure 4. Major types of gel-making techniques leading to simple- or complex-shaped materials prior to drying. (**A**) Sol–gel process and gel casting, (**B**) gelling and crosslinking, (**C**) cryotemplating and freeze casting, (**D**) stereolithography and 3D printing, (**E**) ball milling, (**F**) supercritical CO_2 gelation and foaming. (The gel is made with $scCO_2$ then expanded by a rapid pressure drop).

In the gel-casting process (Figure 4A), the reaction mixture is poured into a mold and allowed to set there. The casting process may be combined with the addition of guest particles, fibers, or nanofibers, followed by crosslinking chemical reactions (Figure 4B). Freeze casting is the way to make controlled bimodal pore size distribution by programmed zone freezing of the solvent in the gelation phase (Figure 4C). Stereolithography processes use chemical crosslinking or photochemical polymerization in special 3D-printing techniques to provide custom shape and geometry of scaffolds with controlled macroporosity (Figure 4D). Nanofibrous gels are made from natural nanofibers or electrospun mats by ball milling in an adequately selected solvent (Figure 4E). A rarely used technique is the supercritical gelation and foaming initiated by a rapid pressure drop of gas-saturated polymeric materials combined with other gel-making steps (Figure 4F).

5.2.1. Single-Component and Hybrid Aerogels

Creating biocompatible aerogels for hard-tissue replacement can be achieved through a straightforward approach. One option involves using a single biocompatible or bioactive component, or alternatively, combining multiple such ingredients to form a hybrid structure without macroscopic or micron-level internal phase boundaries. Subsequently, these gels can be dried to aerogels without encountering any constraints in the drying process. This custom formulation allows for the development of the essential macroporous structure crucial for facilitating optimal bone tissue ingrowth.

Silica–chitosan hybrid aerogels were synthesized by Perez-Moreno and coworkers in a sol–gel process from TEOS and chitosan with the help of high-power ultrasound. Chitosan improved the mechanical properties of the gels, which were dried with supercritical CO_2 to monoliths with very high specific surface area (786–1072 m^2/g), and a 0.13–0.20 g/cm^3 density range. The aerogels were tested in simulated body fluid, and found that the surface silanol groups promoted the nucleation and formation of hydroxyapatite crystals on the surface, which is an indication of bioactivity. Human osteoblast cells were cultured on the aerogel surface and immunolabeled to monitor cytoskeletal changes and focal adhesion. The aerogels proved to be osteoconductive and osteoinductive in the cell studies [77].

Maleki et al. synthesized a silica–silk fibroin hybrid aerogel scaffold with honeycomb micromorphology and multiscale porosity manufactured from TEOS and silk fibroin in the presence of hexadecyltrimethylammonium bromide in a one-pot acetic acid-catalyzed sol–gel reaction and unidirectional freeze casting, which controlled the size of the macropores in the ten-micron range. The reason for the combination of silica and silk fibroin was to increase the pore size regime and the mechanical strength synergistically. Mechanical strength increased to a 4–7 MPa Young's modulus. The as-prepared aerogel proved to be cytocompatible and nonhemolytic, showed no toxicity, and triggered MG63 osteoblast cell attachment and proliferation in 14 days. Implantation of the material in rat femur bone defects resulted in bone formation in 25 days [42].

Polybenzoxazine (PBO) aerogel and its hybrid with resorcinol–formaldehyde (PBO-RF) were prepared and then carbonized at high temperature by Rubenstein and coworkers. Human calvarial osteoblasts were used in the biological studies. Results showed that PBO aerogel and its combination with RF and the carbonized aerogels are compatible with the osteoblasts. However, PBO-RF aerogel resulted in a low growth rate of cells. Carbonized PBO aerogel had better mechanical properties and high porosity. It proved to be the most advantageous for osteoblast growing, which makes the material a promising candidate for tissue-engineering applications [71].

Horvat and coworkers prepared a methoxyl pectin–xanthan aerogel layer on the surface of medical grade stainless steel from the aqueous solution of high-methoxyl pectin and xanthan in an optimized 1:1 ratio by an absolute ethanol-induced gelation process, after which the gel layer was dried with a continuous flow of supercritical CO_2. Non-steroidal anti-inflammatory drugs diclofenac sodium and indomethacin were loaded in the aerogel either from the saccharide solution directly or from an ethanol solution in the soaking phase. After drying, their release profiles were determined. The aerogel layer protected the steel surface from corrosion, and the loaded drugs were released in one day. The biocompatibility of the layer material was tested after dissolving the aerogels in a buffer with a human bone-derived osteoblast hFOB cell line. The results showed higher viability and better proliferation of the cells in the aerogel solutions than in the control samples [15].

Quraishi and coworkers prepared meso–macro porous alginate–lignin hybrid aerogels from a basic solution of alginate and lignin, containing calcium carbonate particles gelled under a CO_2 atmosphere (45 bar for 24 h), then foamed by a controlled release of pressure. The as-prepared gels were subjected to solvent exchange and then CO_2 supercritical drying. The biocompatibility of the materials was tested using a mouse fibroblast-like cell line L929. The aerogels proved to be non-cytotoxic in cell studies compared to tissue culture polystyrene reference. The cell viability was similar to that of the control, and the materials showed good cell adhesion and indicated no negative effect of the lignin component. The alginate–lignin aerogels are good candidates as scaffold materials for further in vivo tissue-engineering studies [64].

Calcium– alginate also served as one of the major components in new alginate–starch aerogels prepared by Martins et al. The wet gels were made from an aqueous solution of sodium alginate and starch in the presence of calcium carbonate particles. The gelation occurred under the acidification effect of high-pressure carbon dioxide. A rapid release of carbon dioxide produced a foamed material that was then dehydrated with anhydrous ethanol and dried to aerogel with supercritical carbon dioxide. The macropore formation sharply depended on the rate of depressurization. In simulated body fluid, the material developed surface hydroxyapatite crystals indicating bioactivity potential, which was attributed to the presence of calcium ions. Cell studies with fibroblast-like cell line L929 showed no cytotoxic effect, and the cells colonized the surface. Thus, the alginate–starch hybrid material may be applied in biomedical research and bone repair [65].

Vazhayal and coworkers synthesized mesochanneled and tunable bimodal pore size distribution aluminosiloxane microspheres from acidic pre-hydrolyzed aluminum iso-propoxide sol stabilized with PVA and aminopropyl trimethoxysilane solution injected in ammoniac paraffin oil that initiated self-assembly and solidified the droplets. After fortification of the structure by soaking in TEOS, the microspheres were washed, solvent exchanged, and dried to aerogel under subcritical conditions at 50 °C and ambient pressure. Finally, the microparticles were calcined at 600 °C to provide a pH-responsive, controlled-release drug carrier material. NSAIDs were adsorbed in the aerogel from hexane solutions, and the preparations were tested for release in simulated gastric and intestinal fluids. The biocompatibility and cytotoxicity of the aerogels were tested in vitro on normal H9c2 cells, while gastric ulceration was tested in vivo on albino male rats. Although it was not tested directly, the authors envisaged utilizing these aerogel microspheres in potential bone tissue engineering [61].

To defeat the mechanical limitations of the traditional hydroxyapatite scaffolds, a new highly porous and elastic single-phase aerogel material made from hydrothermally synthesized and freeze-dried ultra-long hydroxyapatite nanowires was prepared by Huang and coworkers. The biological activity was tested with rat bone marrow mesenchymal stem cells. The results showed that the material promotes cell adhesion, proliferation, and migration of the cells and elevate the expression of osteogenesis- and angiogenesis-related genes. The nanowire aerogel scaffold can promote the ingrowth of the new bone and neovascularization in the bone defect region, thus making this a promising material for bone tissue engineering [17].

Osorio and coworkers made sulfate or phosphate half-ester-functionalized cellulose nanocrystals and crosslinked them through the carboxylate derivative with adipic acid dihydrazide. The as-prepared materials were transformed into cryogel by freezing in molds at −5 °C, and then ice crystals were removed by soaking in absolute ethanol. Finally, the materials were dried with supercritical CO_2 to aerogels. The bioactivity was tested on SaOS-2 cells, and the materials showed an increase in cell metabolism for seven days. A simulated body fluid test showed hydroxyapatite layer formation after the materials were pre-treated with calcium chloride solution. The sulfated aerogel proved to be more advantageous regarding mechanical strength and stability under an aqueous environment. In vivo implantation in the calvaria of male rats showed a significant increase in bioactivity in 12 weeks, proving that the new and flexible materials can facilitate bone ingrowth [13].

Reyes-Peces et al. combined chitosan with hydrolyzing 3-glycidoxypropyl-trimethoxysilane (GPTMS) in an acid-catalyzed sol–gel process at 50 °C followed by supercritical CO_2 drying resulting in a mechanically exceptionally strong aerogel material. The crosslinking with GPTMS connects the amino and hydroxyl groups of the polysaccharide chains into a hybrid interconnected silica plus carbohydrate network. In vitro, biocompatibilities were proved by the hydroxyapatite layer formation in simulated body fluid. The in vivo bioactivities were tested on human osteoblast cells. No cytotoxicity was observed; the material induced cell adhesion and the cells showed cytoskeletal rearrangements and elongation with stress fibers [62].

5.2.2. Nanofiber Aerogels

Electrospun PLGA-collagen-gelatin nanofibers combined with Sr-Cu co-doped bioglass fibers and bone morphogenetic protein 2 (BMP-2) were combined in a 3D hybrid nanofiber aerogel network by Weng et al. The new material was tested for cranial bone healing using the critical-size rat calvaria model. The sustained slow-release of BMP-2 proteins from the degradable aerogel increased the rate of bone healing significantly and improved the vascularization. Histopathology data showed a near-complete degradation of the aerogel material in the regenerated tissue [79].

Xu and coworkers transformed electrospun polycaprolactone nanofibers into soft, elastic, and very porous aerogel scaffolds by freeze grinding the nanofibers and then by thermally inducing the self-agglomeration, and the as-prepared gels were freeze-dried. In vitro studies with mouse bone marrow mesenchymal stem cells showed high cell viability. Depending on their elasticities, the materials favored osteogenic or chondrogenic differentiation of the stem cells. In vivo experiments indicated that the highly porous and elastic scaffold can act as a favorable synthetic extracellular matrix for bone and cartilage regeneration [193].

Rong and coworkers prepared silk fibroin (SF)–chitosan (CS) aerogel scaffolds reinforced with different amounts of SF nanofibers (SF-CS/NF1%, SF-CS/NF2% and SF-CS/NF3%) for bone regeneration. In vitro cytotoxicity test against MC3T3-E1 cells confirmed that all samples were biocompatible while further experiments confirmed that by rougher surface, enhanced mechanical strength and well-regulated pores, this biocompatible scaffold significantly facilitated osteogenic differentiation [194].

5.2.3. Aerogels as Matrix Materials

Silica aerogel–tricalcium phosphate and hydroxyapatite composites were synthesized, and their potential in artificial bone substitution was systematically studied by Szabó et al., Győri et al., Hegedüs et al., Kuttor et al., and Lázár et al. The silica matrix was synthesized in a sol–gel process from TMOS under basic conditions. Microcrystalline or nanocrystalline TCP and/or HAp, which acted as bioactive components, in addition to microcrystalline cellulose, were all dispersed in the reaction mixture in the gelation phase. Large monoliths, small cylinders, spheres, and irregularly shaped particles were prepared and dried with supercritical CO_2 at 80 °C. Cellulose was a sacrificial porogen material and burned out at 500 °C. High-temperature annealing (in the range of 500–1000 °C) of the samples resulted in a change in their dissolution profile and mechanical strengths, but the mesoporous structure and high specific surface area were preserved at all temperatures. The highest temperature provided the highest rate of shrinkage and also the highest compressive strength (up to 102 MPa). The 900 ° and 1000 °C materials were strong enough to be tested in load-bearing positions. The in vitro SBF examination resulted in microcrystalline HAp layer formation on the surface. The cellular metabolism and proliferation were studied with MG-63 cells, while gene expression studies were also performed on SaOS-2 cells. In vivo small animal studies used 1.5 mm diameter cylinders in rat femurs and 8 mm discs in rat calvaria defect models. Both series of animal experiments proved the bioactivity and bone regeneration potential of the silica aerogel-TCP composites in a few months. The highest bone regeneration potential was observed with the 800 °C temperature sample versions [66–70,107].

Tevlek and coworkers synthesized electrically conductive carbon aerogels decorated with tricalcium phosphate nanocrystallites. The decorated aerogel was made from cellulose fibers, while TCP was also added. Freeze drying produced the pristine aerogels that were heated at 850 °C or 1100 °C under argon atmosphere. The new aerogels were not cytotoxic when tested on P9 L929 mouse fibroblast cells. Proliferation and attachment were tested using disk-shaped specimens with MC3T3-E1 mouse pre-osteoblast cells, providing, thus, a future possibility of applying electric stimuli that might have a significant effect on the cellular behavior [18].

Muñoz-Ruíz and coworkers synthesized a highly porous collagen–alginate aerogel-based scaffold with and without graphene oxide mixed in. The buffered solution of collagen and alginic acid (and graphene oxide) was crosslinked and gelled with calcium chloride, solvent-exchanged with ethanol, and dried under supercritical CO_2 conditions to form the aerogel. Osteoblast cells seeded on the surface of collagen–alginate aerogel showed adhesion, proliferation, and some degree of extracellular matrix formation after 48 h of incubation. In contrast, the graphene oxide-containing aerogel did not support the cellular growth and activity [63].

Chitosan (CH) matrix was combined with an electrospun nanomaterial of cellulose acetate (CA) and poly(ε-caprolactone) (PCL) by Zhang et al. in a ball-milling process and then freeze-dried to the aerogel CA/PCL/CH. The material showed increased mechanical strength and was bioactive in studies with the MC3T3-E1 cell line. It promoted cell adhesion, infiltration, and osteogenic differentiation [57].

Dong and coworkers prepared beta-tricalcium phosphate-based specimens with printing or compression and soaked them in the premix of a resorcinol–formaldehyde (RF) wet gel. After setting, the samples were dried under ambient conditions and carbonized at high temperature, resulting in the carbon aerogel-coated β-TCP scaffold, which was then used in photothermal therapy. The material was not only effective in ablation of osteosarcoma tumors but promoted osteogenesis as well [19].

A new composite aerogel composed of nano-hydroxyapatite (n-HAp), silk fibroin, and cotton cellulose, crosslinked with epichlorohydrin, and freeze-dried from tert-butanol was developed by Chen et al. to overcome the mechanical problems of the previously synthesized n-HAp biopolymeric composites. Uniaxial compressing of the aerogel showed increased mechanical strength and toughness, making the values similar to that of the cancellous bones. HEK-293T cell studies of the material showed a high ability of cell adhesion, proliferation, and differentiation [58].

Tetik et al. prepared bioinspired aerogels resembling the pore structure of the bones using the unidirectional freeze-casting process followed by freeze drying, resulting in layered mesoporous and macroporous regions. Colloidal silica and graphene oxide were used as base materials. The study focused on the technical aspects of the process. The as-prepared structured aerogels were not tested for bone substitution potential [41].

Graphene oxide (GO) (in the 0–0.2% range) and Type I collagen-containing composite aerogels with enhanced stiffness were prepared by Liu et al. to improve the bone repairing potential of large monolithic aerogel pieces. The aerogel materials were tested in the rat cranial defect model, which proved its biocompatibility and osteogenic activity. The graphene oxide content positively affected the mechanical properties, and the 0.1% GO content produced the highest biological activity [80].

Wu and coworkers synthesized an alginate aerogel combined with in situ-prepared octahedral metallic copper nanocrystals stabilized with carbon dots and loaded with the antibiotic tigecycline. The aerogel proved to be an efficient slow-release antibacterial agent in which the antibiotics and the copper ions acted synergistically. The as-prepared aerogel material showed low cytotoxicity and may be important in preventing bone infections leading to osteomyelitis [14].

Nanoparticles consisting of the miR-26a and a cationic polymeric gene delivery vector (HA–SS–PGEA) were embedded by Li et al. in an electrospun 3D matrix made of poly(lactic-co-glycolic acid) (PLGA)–collagen–gelatin (PCG) and bioactive glass (BG). The scaffold proved to be a promising bone graft candidate in the rat cranial defect model. The molecular mechanism of the mesenchymal stem cells is governed by the microRNAs. In the osteoblastogenesis process, miRNA-26a acts as the promoter of the osteogenic differentiation of bone marrow derived from mesenchymal stem cells [27].

Scaffolds made from type I collagen aerogel (Col) and reduced graphene-oxide–collagen aerogel (Col-rGO) were synthesized by Bahrami et al. in a two-step crosslinking and freeze-drying process. The addition of rGO improved the mechanical strength, and the aerogel showed no cytotoxicity and increased the viability and proliferation of human bone marrow mesenchymal stem cells. The rabbit cranial defect model showed an increased rate of bone formation [53].

5.2.4. Aerogels as Guest Particles

The matrix material polyethylene glycol diacrylate was combined with hydrophilic and highly biocompatible cellulose nanofibrils (PEGDA/CNF) by Sun et al. in different compositions and printed out in a self-built stereolithographic method using a hexagonal mask pattern irradiated with white light, followed by freeze drying to dry aerogel scaffolds. Soaking the aerogels in water resulted in significant water uptake, leading to aerogel–wet gel combo materials. Mechanical properties and the biocompatibility of the as-prepared wet materials were tested. The non-toxic aerogel–wet gel scaffolds were of a porous nature that proved to be advantageous for the adhesion of bone mesenchymal stem cells [55].

PMMA-based bone cements are widely used in the medical practice in filling bone cavities or fixing metallic implants in position. Although such bone cements are bioinert materials, ossification is not induced on their surface. Lázár and coworkers embedded functionalized silica aerogels as guest particles in in situ polymerized PMMA matrix and tested them in simulated body fluids for bioactivity. Results showed that the compressive strengths were increased compared to the neat PMMA. SBF solution resulted in a dissolution of the hydrophilic silica aerogel from the polymeric matrix, leaving a highly porous surface behind. In contrast to the smooth surfaces of the PMMA bone cements, the newly developed porous surface may be advantageous, providing a better bone tissue adherence [192].

Goimil and coworkers embedded starch aerogel microspheres in a supercritically foamed poly(ε-caprolactone) (PCL) highly porous scaffold, increasing the interconnectivity of the pores and the specific surface area. The composite was loaded with ketoprofen under supercritical CO_2 conditions and showed a sustained ketoprofen release at pH 7.4. Starch aerogel microspheres mildly decreased the mechanical strength and increased the drug release rate compared to the pristine PCL matrix [45].

Silica aerogel was embedded in poly(ε-caprolactone) (PCL) by Ge et al., and its presence prevented any cytotoxic effect of PCL in a long period of time in contact with tissue cultures. It improved the survival and growth of 3T3 cells and primary mouse osteoblast cells. Silica aerogel helped maintain the pH and prevented the acidification of the connecting tissues for four weeks [118].

Silk fibroin aerogel is embedded in supercritically foamed poly(ε-caprolactone) (PCL) scaffolds loaded with dexamethasone under $scCO_2$ conditions by Goimil et al. Silk fibroin is a cell-adhesion promoter, while dexamethasone is an osteogenic differentiation agent. The aerogels improved the pore structure and, thus, the biological fluid transport and facilitated the cell infiltration. The in vivo calvarial test showed the importance of the form of dexamethasone, which promoted bone tissue regeneration [46].

Finely ground heat-treated silica aerogel-TCP composites were embedded in PVA/chitosan electrospun nanofiber (147+/−50 nm diameter) meshes and crosslinked with citric acid by Boda and coworkers. Dental pulp stem cells were seeded onto the surfaces and proved the bioactivity of the materials. Rat critical-size calvarial models were used to test the role of the meshes on bone regeneration. After six months, significant new bone formation was observed, proving that the hybrid nanospun scaffolds containing bioactive aerogel guest particles may be used as new experimental bone substitute bioactive materials [195].

5.2.5. Complex Aerogel Structures

Incorporating complex structures defies straightforward categorization. Often, classification appears arbitrary due to varying perspectives. This challenge is particularly evident with composites or scaffolds, where determining whether the aerogel phase serves as the host or the guest becomes intricate, especially within multicomponent systems.

Zhang and coworkers prepared 3D fibrous composite aerogels in a three-layer gradient structure from poly(L-lactide)/gelatin composite fibers, glycosaminoglycan in the top layer, and apatite in the middle and lower layers. The properties of the materials are described in detail in Section 5 [24].

Li et al. prepared strontium ranelate (SR) and incorporated it in mushroom tyrosinase enzyme-induced crosslinked gelatine nanoparticles/silk fibroin gel that was freeze-dried to aerogel. Rapid deposition of HAp on the surface of the scaffold took place, but initial burst release of strontium did not occur. Instead, increased osteogenic differentiation of osteoblasts and inhibiting the activity of osteoclasts was observed in ovariectomized rats using the calvaria defect model [196].

Asha and coworkers made reduced graphene oxide (A-rGO) aerogel from rGO with citric acid at 90 °C in aqueous solution. After gelation, the first aerogel was made by freeze drying. After that, it was functionalized with chitosan by soaking A-rGO in chitosan solution, then freeze-dried again. HAp particle decoration was made by soaking in SBF. MG63 cell studies indicated that the chitosan interfacial layer improves biocompatibility, and the mineralized chitosan layer increased the cell viability and proliferation [54].

Ferreira et al. combined a colloidal aqueous suspension of cellulose nanofibrils (20%) and bioglass particles (80%) in an interconnected 3D network, freeze-dried to a porous cryogel structure. A hydroxyapatite layer was formed on the surface in SBF, shown by the red stain (Alizarin Red) and the IR spectroscopy, indicating good in vitro biocompatibility of the material. The bioglass content provided the necessary ions to facilitate BMP-2 production in cells. The combined material in the in vivo experiments showed no liver or kidney toxicity. Rat calvarial defect experiments proved that the composite material induced new bone formation in 57 days [56].

6. Aerogel-Based Materials for Cartilage Tissue Engineering

In comparison to bone substitution, only a limited number of publications address the utilization of aerogel-based materials for cartilage tissue regeneration. This endeavor encounters notable mechanical, chemical, and biological complexities. Articular cartilage comprises four key constituents: type II collagen, proteoglycans (glycosaminoglycans bound to proteins), water, and chondrocyte cells embedded within the extracellular matrix. Unlike bone, articular cartilage lacks intrinsic self-repair capabilities. Consequently, all essential elements must be supplied externally, with material transport occurring gradually through the synovial fluid. Notably, chondrocyte cell density is much higher on or near the gliding surface compared to the base layer, with cellular morphology and orientation varying by depth. Artificial scaffold materials necessitate seeding with chondrogenic cells sourced from the patient or, more recently, utilizing undifferentiated stem cells. Furthermore, maintaining a continuous mechanical stimulus in vitro is crucial to foster the development of a compression and impact-resistant surface characterized by aligned chondrocytes and collagen fibers parallel to the surface [197,198].

Chen et al. made 3D aerogel-like scaffolds from electrospun nanofibers containing gelatin–polylactic acid and gelatine–polylactic acid–hyaluronic acid and studied their bioactivity. In vitro examinations proved the adhesion, growth, and proliferation of chondrocyte cells. The materials were elastic and showed a sort of shape memory effect. The rabbit articular cartilage injury model indicated that gelatin-PLA had only a limited cartilage-repair effect. However, the hyaluronic acid-modified gelatin-PLA scaffold proved to be more active in cartilage regeneration [190].

Scaffold materials with tunable mechanical properties were synthesized by Tang and coworkers from nanocellulose fibers and polyethylene glycol diacrylate (PEGDA) by stereolithography (SLA), and the wet gels were freeze-dried to an aerogel. The macropore sizes were basically determined by the parameters of the SLA process. Mouse bone marrow mesenchymal stem cells showed proliferation and induction, making the material a promising candidate for the cartilage repair [52].

Zhang and coworkers prepared aerogel-based gradient scaffolds to provide an artificial bioactive interface between the bone and the cartilage tissue. Three layers of 3D fibrous aerogel structure were constructed in a gradient arrangement from electrospun poly(L-lactide), gelatin, glycosaminoglycan, and hydroxyapatite. The aerogel layers were prepared separately from electrospun mats by homogenization, freeze drying, and crosslinking with heat, then mineralized and glued together with gelatine. The hierarchical aerogel scaffold induced the bone mesenchymal stromal cells, which differentiated into chondrogenic and osteogenic phenotypes specific to the zone they were in contact with. Cell affinity peptide E7, intended to enhance cell migration, was grafted in the gradient structure by soaking the pre-treated aerogel in its aqueous solution. Aerogel scaffolds without composition gradient and scaffolds with gradient aerogels were implanted in rabbit knees and monitored for 12 weeks for tissue regeneration. The results showed that the gradient aerogels could reconstruct an osteochondral interface, and the E7 peptide-containing aerogel scaffolds are promising candidates in tissue engineering [24].

Three-dimensional porous nanocomposite scaffolds based on cellulose nanofibers for cartilage tissue engineering were prepared by Naseri and coworkers containing freeze-dried cellulose nanofibers as the major component in a gelatin and chitosan matrix crosslinked with genipin. The scaffold showed a macroporous structure of interconnected pores. The dry material's mechanical strength (compression modulus) was higher than that of the natural cartilage tissue and lowered in phosphate-buffered saline solution. The high porosity and compatibility with the chondrocytes made the material interesting for cell attachment and extracellular matrix production [198].

7. Challenges, Opportunities and Future Trends

Bone and cartilage tissue regeneration materials have evolved through three distinct developmental stages. Initially, they served as simple bioinert tissue support (first generation). Subsequently, advancements led to the emergence of nano-engineered resorbable composite materials containing bioactive molecules and growth factors (third generation), as elucidated by Hench and Polak [199]. Now, we look towards the future of this biomedical sector of aerogel research, seeking answers to several key questions. How can the tissue regeneration potential be further enhanced in the fourth generation of materials? How can aerogel-based materials, structures, and devices take advantage on these developments? And how can the new technical extensions be effectively integrated or combined with aerogels?

Aerogel-based bone and cartilage substitution has encountered various challenges from its inception. The use of substances already approved for clinical applications proves invaluable in selecting aerogel building materials with the desired bioactivity. One of the remarkable advantages of aerogels lies in their high porosity and interconnected open-pore structure, which offers additional benefits. The porosity plays a crucial role in enhancing their bioactivity and tissue reactions, facilitating the efficient transport of dissolved oxygen and nutrients to surrounding tissues, and potentially reducing the occurrence of foreign body reactions. Another noteworthy feature is the high specific surface area of aerogels, allowing them to be loaded with bioactive small molecules during the gelation or nanofiber-making stages, or even after drying. Supercritical adsorption is a convenient method for loading aerogels while preserving their original structure. Additionally, during the gelation phase, protein-like molecules, macromolecules, growth factors, and even living cells may be embedded within the aerogel, which can later be freeze-dried.

Improving the benefits of aerogel-based materials could be further refined by crafting oriented, multi-component, and function-specific layered structures or scaffolds, featuring concentration gradients, precisely tailored macropores, and channels that mirror the intended tissue's architecture for regeneration. The inclusion of oriented macro-channels could facilitate tissue ingrowth and promote vascularization. Various techniques, such as successive casting, electrospinning, stereolithography, 3D printing, selective leaching, cryotemplating, and employing sacrificial porogens, along with post-drying manufacturing, can be employed to achieve these specialized materials. Through the integration of materials within a carefully planned hierarchical structure and the utilization of additive manufacturing techniques, the bioactivity can be heightened. Looking ahead, the potential for improved efficacy in bioactive aerogels may also stem from the discovery of new building materials and unexplored synergistic interactions. Utilizing aerogels made from materials that have already gained approval and clinical licensure may offer the advantage of expediting the approval process for animal experiments. However, the continuous quest for novel materials remains paramount in uncovering novel interactions with living tissues, underscoring the persistent need for innovation in this area.

In addition to their advantages, aerogel-based materials may have some drawbacks when compared to traditional tissue-engineering materials. One significant concern is their mechanical properties, which may not be suitable for load-bearing positions, except for aerogel-based bioceramics. Moreover, the variable sensitivity of aerogel materials to wetting liquids like water, body fluids, or blood can also pose challenges for certain applications. In terms of manufacturing, subcritical and freeze-drying techniques have shown potential for upscaling to economically feasible high volume levels due to their relative simplicity and lower costs. However, supercritical drying, while not impossible, is more difficult and costly to be upscaled to produce large quantities of aerogel-based materials. Despite these limitations, ongoing research and advancements in aerogel technology continue to address these challenges and open up new possibilities for their use in tissue-engineering applications. As the field progresses, we can expect to see further developments to overcome these drawbacks and fully harness the potential of aerogel-based materials in regenerative medicine.

In the foreseeable future, the field of aerogel-based bone and cartilage tissue engineering is poised for ongoing development. The research will continue to create new materials, combine existing ones, and explore synergistic interactions to enhance outcomes. Additionally, there will be a concerted effort to fabricate intricate 3D scaffold structures that closely emulate the composition, hierarchy, and functions of the target tissues.

In the more distant future, the trajectory of aerogel-based tissue-engineering materials appears to be closely linked with the advancement of the fourth generation of bone and cartilage tissue-engineering materials, as described or anticipated by Ning et al. [200]. A potential outcome of future investigations could involve the integration of next-generation aerogel-based scaffolds with implantable and biodegradable power sources, along with microelectronic circuits capable of continuously monitoring the progress of the healing process. While the fundamental components of such electronic devices and power sources have been developed, their incorporation into implants remains an ongoing endeavor. In relation to aerogels, a few promising examples exist that could potentially open new avenues for innovative solutions in the years ahead.

Hong and colleagues have already developed graphene aerogels incorporating electrically conducting polydimethylsiloxane sheets. These materials bear structural and functional resemblance to cartilage tissue found in articular joints. These innovative constructs have found application in sensor technology, capable of transmitting signals concerning mechanical force intensity during joint loading [33].

Another avenue for exploration is the potential to enhance or stimulate bone healing through external stimuli. As demonstrated by Caliogna et al., pulsed electromagnetic fields can initiate or bolster the healing process [201]. Although dedicated aerogel-based composite devices designed to generate or support external stimuli are not yet available, a noteworthy advancement is exemplified by the electrically conductive carbon aerogel adorned with ceramic tricalcium phosphate nanocrystallites, as developed by Tevlek and colleagues. This work could potentially pave the way for upcoming advancements [18].

The existing devices and therapies have already demonstrated certain aspects of the concept. For instance, microwave devices are employed for the sensing the bone-healing process [202], and low-dose microwaves have also been tested to promote bone healing [203,204]. Electrical stimulation has been extensively studied in bone therapy [205], while infrared laser has been found to aid in bone healing when combined with bone morphogenetic protein [206]. Even red visible light has shown potential in promoting bone regeneration [207]. These examples highlight the diverse array of approaches being explored to enhance bone healing and tissue regeneration.

8. Conclusions

Aerogel-based materials continue to play a significant and evolving role in orthopedic and dental research. The strategic combination of bioactive inorganic, organic, natural, and synthetic polymeric materials within aerogel matrices has notably enhanced the biocompatibility and bioactivity of engineered bone and cartilage substitutes. Leveraging their exceptional porosity and customizable surface properties, these grafts and scaffolds create a conducive milieu for stem cell growth, proliferation, and differentiation, fostering osteogenic development. In vivo animal studies have underscored that aerogel-based materials exhibit not only biocompatibility but also osteoinductive properties and active bioresorption, leading to the regeneration of deficient bone tissues.

By incorporating aerogels with established bioactive materials, the adverse effects linked to the degradation of polymeric materials have been mitigated. Recent works have yielded highly oriented and layered aerogel architectures that closely emulate the intricacies of living tissue environments. These materials have already demonstrated their potential in healing and regenerating bone and cartilage tissue defects.

In the near future, the refinement of scaffold designs tailored to specific application sites, coupled with novel material combinations, is poised to amplify their therapeutic efficacy and biomedical utility. Beyond advancements in chemical composition and structural intricacy, the next developments of aerogel bone substitute materials may involve external interactions after implantation, to both bolster and monitor the healing process.

Author Contributions: I.L., conceptualization; I.L. and M.M., writing—original draft preparation; I.L., M.M. and L.Č., writing—review and editing. All authors have read and agreed to the published version of the manuscript.

Funding: This research received no external funding.

Institutional Review Board Statement: Not applicable.

Informed Consent Statement: Not applicable.

Data Availability Statement: No new data were created or analyzed in this study. Data sharing is not applicable to this article.

Acknowledgments: The work was carried out in the frame of the COST-Action "Advanced Engineering of aeroGels for Environment and Life Sciences (AERoGELS, ref. CA18125) funded by the European Commission".

Conflicts of Interest: The authors declare no conflict of interest.

References

1. Jayakumar, P.; Di Silvio, L. Osteoblasts in Bone Tissue Engineering. *Proc. Inst. Mech. Eng.* **2010**, *224*, 1415–1440. [CrossRef]
2. Ferreira, A.M.; Gentile, P.; Chiono, V.; Ciardelli, G. Collagen for Bone Tissue Regeneration. *Acta Biomater.* **2012**, *8*, 3191–3200. [CrossRef]
3. Amini, A.R.; Laurencin, C.T.; Nukavarapu, S.P. Bone Tissue Engineering: Recent Advances and Challenges. *Crit. Rev. Biomed. Eng.* **2012**, *40*, 363–408. [CrossRef]
4. Kazimierczak, P.; Przekora, A. Osteoconductive and Osteoinductive Surface Modifications of Biomaterials for Bone Regeneration: A Concise Review. *Coatings* **2020**, *10*, 971. [CrossRef]
5. Lopes, D.; Martins-Cruz, C.; Oliveira, M.B.; Mano, J.F. Bone Physiology as Inspiration for Tissue Regenerative Therapies. *Biomaterials* **2018**, *185*, 240–275. [CrossRef]
6. Albrektsson, T.; Johansson, C. Osteoinduction, Osteoconduction and Osseointegration. *Eur. Spine J.* **2001**, *10*, S96–S101. [CrossRef]
7. Ansari, M. Bone Tissue Regeneration: Biology, Strategies and Interface Studies. *Prog. Biomater.* **2019**, *8*, 223–237. [CrossRef]
8. Berrio, M.E.; Oñate, A.; Salas, A.; Fernández, K.; Meléndrez, M.F. Synthesis and Applications of Graphene Oxide Aerogels in Bone Tissue Regeneration: A Review. *Mater. Today Chem.* **2021**, *20*, 100422. [CrossRef]
9. Matsiko, A.; Gleeson, J.P.; O'Brien, F.J. Scaffold Mean Pore Size Influences Mesenchymal Stem Cell Chondrogenic Differentiation and Matrix Deposition. *Tissue Eng. Part A* **2015**, *21*, 486–497. [CrossRef]
10. Iglesias-Mejuto, A.; García-González, C.A. 3D-Printed Alginate-Hydroxyapatite Aerogel Scaffolds for Bone Tissue Engineering. *Mater. Sci. Eng. C* **2021**, *131*, 112525. [CrossRef]
11. Włodarczyk-Biegun, M.K.; del Campo, A. 3D Bioprinting of Structural Proteins. *Biomaterials* **2017**, *134*, 180–201. [CrossRef] [PubMed]
12. Puppi, D.; Chiellini, F.; Piras, A.M.; Chiellini, E. Polymeric Materials for Bone and Cartilage Repair. *Prog. Polym. Sci.* **2010**, *35*, 403–440. [CrossRef]
13. Osorio, D.A.; Lee, B.E.J.; Kwiecien, J.M.; Wang, X.; Shahid, I.; Hurley, A.L.; Cranston, E.D.; Grandfield, K. Cross-Linked Cellulose Nanocrystal Aerogels as Viable Bone Tissue Scaffolds. *Acta Biomater.* **2019**, *87*, 152–165. [CrossRef] [PubMed]
14. Wu, X.-X.; Zhang, Y.; Hu, T.; Li, W.-X.; Li, Z.-L.; Hu, H.-J.; Zhu, S.-R.; Chen, W.-Z.; Zhou, C.-S.; Jiang, G.-B. Long-Term Antibacterial Composite via Alginate Aerogel Sustained Release of Antibiotics and Cu Used for Bone Tissue Bacteria Infection. *Int. J. Biol. Macromol.* **2021**, *167*, 1211–1220. [CrossRef] [PubMed]
15. Horvat, G.; Xhanari, K.; Finšgar, M.; Gradišnik, L.; Maver, U.; Knez, Ž.; Novak, Z. Novel Ethanol-Induced Pectin–Xanthan Aerogel Coatings for Orthopedic Applications. *Carbohydr. Polym.* **2017**, *166*, 365–376. [CrossRef]
16. Maleki, H.; Durães, L.; García-González, C.A.; del Gaudio, P.; Portugal, A.; Mahmoudi, M. Synthesis and Biomedical Applications of Aerogels: Possibilities and Challenges. *Adv. Colloid Interface Sci.* **2016**, *236*, 1–27. [CrossRef]
17. Huang, G.-J.; Yu, H.-P.; Wang, X.-L.; Ning, B.-B.; Gao, J.; Shi, Y.-Q.; Zhu, Y.-J.; Duan, J.-L. Correction: Highly Porous and Elastic Aerogel Based on Ultralong Hydroxyapatite Nanowires for High-Performance Bone Regeneration and Neovascularization. *J. Mater. Chem. B* **2021**, *9*, 1277–1287. [CrossRef]
18. Tevlek, A.; Atya, A.M.N.; Almemar, M.; Duman, M.; Gokcen, D.; Ganin, A.Y.; Yiu, H.H.P.; Aydin, H.M. Synthesis of Conductive Carbon Aerogels Decorated with β-Tricalcium Phosphate Nanocrystallites. *Sci. Rep.* **2020**, *10*, 5758. [CrossRef]

19. Dong, S.; Zhang, Y.; Wan, J.; Cui, R.; Yu, X.; Zhao, G.; Lin, K. A Novel Multifunctional Carbon Aerogel-Coated Platform for Osteosarcoma Therapy and Enhanced Bone Regeneration. *J. Mater. Chem. B* **2020**, *8*, 368–379. [CrossRef]
20. Wan, Y.; Chang, P.; Yang, Z.; Xiong, G.; Liu, P.; Luo, H. Constructing a Novel Three-Dimensional Scaffold with Mesoporous TiO_2 Nanotubes for Potential Bone Tissue Engineering. *J. Mater. Chem. B* **2015**, *3*, 5595–5602. [CrossRef]
21. Richardson, J.B.; Caterson, B.; Evans, E.H.; Ashton, B.A.; Roberts, S. Repair of Human Articular Cartilage after Implantation of Autologous Chondrocytes. *J. Bone Jt. Surg. Br.* **1999**, *81*, 1064–1068. [CrossRef]
22. Feng, B.; Ji, T.; Wang, X.; Fu, W.; Ye, L.; Zhang, H.; Li, F. Engineering Cartilage Tissue Based on Cartilage-Derived Extracellular Matrix cECM/PCL Hybrid Nanofibrous Scaffold. *Mater. Des.* **2020**, *193*, 108773. [CrossRef]
23. Chen, Y.; Shafiq, M.; Liu, M.; Morsi, Y.; Mo, X. Advanced Fabrication for Electrospun Three-Dimensional Nanofiber Aerogels and Scaffolds. *Bioact. Mater.* **2020**, *5*, 963–979. [CrossRef] [PubMed]
24. Zhang, L.; Fang, J.; Fu, L.; Chen, L.; Dai, W.; Huang, H.; Wang, J.; Zhang, X.; Cai, Q.; Yang, X. Gradient Fibrous Aerogel Conjugated with Chemokine Peptide for Regulating Cell Differentiation and Facilitating Osteochondral Regeneration. *Chem. Eng. J.* **2021**, *422*, 130428. [CrossRef]
25. Chainani, A.; Hippensteel, K.J.; Kishan, A.; Garrigues, N.W.; Ruch, D.S.; Guilak, F.; Little, D. Multilayered Electrospun Scaffolds for Tendon Tissue Engineering. *Tissue Eng. Part A* **2013**, *19*, 2594–2604. [CrossRef]
26. Shim, I.K.; Suh, W.H.; Lee, S.Y.; Lee, S.H.; Heo, S.J.; Lee, M.C.; Lee, S.J. Chitosan Nano-/Microfibrous Double-Layered Membrane with Rolled-up Three-Dimensional Structures for Chondrocyte Cultivation. *J. Biomed. Mater. Res. A* **2009**, *90*, 595–602. [CrossRef]
27. Li, R.; Wang, H.; John, J.V.; Song, H.; Teusink, M.J.; Xie, J. 3D Hybrid Nanofiber Aerogels Combining with Nanoparticles Made of a Biocleavable and Targeting Polycation and MiR-26a for Bone Repair. *Adv. Funct. Mater.* **2020**, *30*, 2005531. [CrossRef]
28. Wang, L.; Qiu, Y.; Lv, H.; Si, Y.; Liu, L.; Zhang, Q.; Cao, J.; Yu, J.; Li, X.; Ding, B. 3D Superelastic Scaffolds Constructed from Flexible Inorganic Nanofibers with Self-Fitting Capability and Tailorable Gradient for Bone Regeneration. *Adv. Funct. Mater.* **2019**, *29*, 1901407. [CrossRef]
29. Kistler, S.S. Coherent Expanded Aerogels and Jellies. *Nature* **1931**, *127*, 741. [CrossRef]
30. Hüsing, N.; Schubert, U. Aerogels—Airy Materials: Chemistry, Structure, and Properties. *Angew. Chem. Int. Ed.* **1998**, *37*, 22–45. [CrossRef]
31. Karamikamkar, S.; Naguib, H.E.; Park, C.B. Advances in Precursor System for Silica-Based Aerogel Production toward Improved Mechanical Properties, Customized Morphology, and Multifunctionality: A Review. *Adv. Colloid Interface Sci.* **2020**, *276*, 102101. [CrossRef]
32. Ganesamoorthy, R.; Vadivel, V.K.; Kumar, R.; Kushwaha, O.S.; Mamane, H. Aerogels for Water Treatment: A Review. *J. Clean. Prod.* **2021**, *329*, 129713. [CrossRef]
33. Hong, J.-Y.; Yun, S.; Wie, J.J.; Zhang, X.; Dresselhaus, M.S.; Kong, J.; Park, H.S. Cartilage-Inspired Superelastic Ultradurable Graphene Aerogels Prepared by the Selective Gluing of Intersheet Joints. *Nanoscale* **2016**, *8*, 12900–12909. [CrossRef]
34. Boday, D.J.; Keng, P.Y.; Muriithi, B.; Pyun, J.; Loy, D.A. Mechanically Reinforced Silica Aerogel Nanocomposites via Surface Initiated Atom Transfer Radical Polymerizations. *J. Mater. Chem.* **2010**, *20*, 6863. [CrossRef]
35. Leventis, N. Three-Dimensional Core-Shell Superstructures: Mechanically Strong Aerogels. *Acc. Chem. Res.* **2007**, *40*, 874–884. [CrossRef]
36. Randall, J.P.; Meador, M.A.B.; Jana, S.C. Tailoring Mechanical Properties of Aerogels for Aerospace Applications. *ACS Appl. Mater. Interfaces* **2011**, *3*, 613–626. [CrossRef] [PubMed]
37. Leventis, N.; Sotiriou-Leventis, C.; Zhang, G.; Rawashdeh, A.-M.M. Nanoengineering Strong Silica Aerogels. *Nano Lett.* **2002**, *2*, 957–960. [CrossRef]
38. Lázár, I.; Fábián, I. A Continuous Extraction and Pumpless Supercritical CO_2 Drying System for Laboratory-Scale Aerogel Production. *Gels* **2016**, *2*, 26. [CrossRef] [PubMed]
39. Scotti, K.L.; Dunand, D.C. Freeze Casting—A Review of Processing, Microstructure and Properties via the Open Data Repository, FreezeCasting.Net. *Prog. Mater. Sci.* **2018**, *94*, 243–305. [CrossRef]
40. Li, W.L.; Lu, K.; Walz, J.Y. Freeze Casting of Porous Materials: Review of Critical Factors in Microstructure Evolution. *Int. Mater. Rev.* **2012**, *57*, 37–60. [CrossRef]
41. Tetik, H.; Feng, D.; Oxandale, S.W.; Yang, G.; Zhao, K.; Feist, K.; Shah, N.; Liao, Y.; Leseman, Z.C.; Lin, D. Bioinspired Manufacturing of Aerogels with Precisely Manipulated Surface Microstructure through Controlled Local Temperature Gradients. *ACS Appl. Mater. Interfaces* **2021**, *13*, 924–931. [CrossRef] [PubMed]
42. Maleki, H.; Shahbazi, M.-A.; Montes, S.; Hosseini, S.H.; Eskandari, M.R.; Zaunschirm, S.; Verwanger, T.; Mathur, S.; Milow, B.; Krammer, B.; et al. Mechanically Strong Silica-Silk Fibroin Bioaerogel: A Hybrid Scaffold with Ordered Honeycomb Micromorphology and Multiscale Porosity for Bone Regeneration. *ACS Appl. Mater. Interfaces* **2019**, *11*, 17256–17269. [CrossRef] [PubMed]
43. Zhou, Y.; Tian, Y.; Peng, X. Applications and Challenges of Supercritical Foaming Technology. *Polymers* **2023**, *15*, 402. [CrossRef] [PubMed]
44. García-González, C.A.; Concheiro, A.; Alvarez-Lorenzo, C. Processing of Materials for Regenerative Medicine Using Supercritical Fluid Technology. *Bioconjug. Chem.* **2015**, *26*, 1159–1171. [CrossRef] [PubMed]

45. Goimil, L.; Braga, M.E.M.; Dias, A.M.A.; Gómez-Amoza, J.L.; Concheiro, A.; Alvarez-Lorenzo, C.; de Sousa, H.C.; García-González, C.A. Supercritical Processing of Starch Aerogels and Aerogel-Loaded Poly(ε-Caprolactone) Scaffolds for Sustained Release of Ketoprofen for Bone Regeneration. *J. CO2 Util.* **2017**, *18*, 237–249. [CrossRef]
46. Goimil, L.; Santos-Rosales, V.; Delgado, A.; Évora, C.; Reyes, R.; Lozano-Pérez, A.A.; Aznar-Cervantes, S.D.; Cenis, J.L.; Gómez-Amoza, J.L.; Concheiro, A.; et al. scCO_2-Foamed Silk Fibroin Aerogel/Poly(ε-Caprolactone) Scaffolds Containing Dexamethasone for Bone Regeneration. *J. CO2 Util.* **2019**, *31*, 51–64. [CrossRef]
47. Badhe, R.V.; Chatterjee, A.; Bijukumar, D.; Mathew, M.T. Current Advancements in Bio-Ink Technology for Cartilage and Bone Tissue Engineering. *Bone* **2023**, *171*, 116746. [CrossRef]
48. Iglesias-Mejuto, A.; García-González, C.A. 3D-Printed, Dual Crosslinked and Sterile Aerogel Scaffolds for Bone Tissue Engineering. *Polymers* **2022**, *14*, 1211. [CrossRef]
49. Ng, P.; Pinho, A.R.; Gomes, M.C.; Demidov, Y.; Krakor, E.; Grume, D.; Herb, M.; Lê, K.; Mano, J.; Mathur, S.; et al. Fabrication of Antibacterial, Osteo-Inductor 3D Printed Aerogel-Based Scaffolds by Incorporation of Drug Laden Hollow Mesoporous Silica Microparticles into the Self-Assembled Silk Fibroin Biopolymer. *Macromol. Biosci.* **2022**, *22*, 2100442. [CrossRef]
50. Bhagat, S.D.; Oh, C.-S.; Kim, Y.-H.; Ahn, Y.-S.; Yeo, J.-G. Methyltrimethoxysilane Based Monolithic Silica Aerogels via Ambient Pressure Drying. *Microporous Mesoporous Mater.* **2007**, *100*, 350–355. [CrossRef]
51. Kirkbir, F.; Murata, H.; Meyers, D.; Chaudhuri, S.R. Drying of Aerogels in Different Solvents between Atmospheric and Supercritical Pressures. *J. Non-Cryst. Solids* **1998**, *225*, 14–18. [CrossRef]
52. Tang, A.; Ji, J.; Li, J.; Liu, W.; Wang, J.; Sun, Q.; Li, Q. Nanocellulose/PEGDA Aerogels with Tunable Poisson's Ratio Fabricated by Stereolithography for Mouse Bone Marrow Mesenchymal Stem Cell Culture. *Nanomaterials* **2021**, *11*, 603. [CrossRef] [PubMed]
53. Bahrami, S.; Baheiraei, N.; Shahrezaee, M. Biomimetic Reduced Graphene Oxide Coated Collagen Scaffold for in Situ Bone Regeneration. *Sci. Rep.* **2021**, *11*, 16783. [CrossRef] [PubMed]
54. Asha, S.; Ananth, A.N.; Jose, S.P.; Rajan, M.A.J. Reduced Graphene Oxide Aerogel Networks with Soft Interfacial Template for Applications in Bone Tissue Regeneration. *Appl. Nanosci.* **2018**, *8*, 395–405. [CrossRef]
55. Sun, D.; Liu, W.; Tang, A.; Guo, F.; Xie, W. A New PEGDA/CNF Aerogel-Wet Hydrogel Scaffold Fabricated by a Two-Step Method. *Soft Matter* **2019**, *15*, 8092–8101. [CrossRef]
56. Ferreira, F.V.; Souza, L.P.; Martins, T.M.M.; Lopes, J.H.; Mattos, B.D.; Mariano, M.; Pinheiro, I.F.; Valverde, T.M.; Livi, S.; Camilli, J.A.; et al. Nanocellulose/Bioactive Glass Cryogels as Scaffolds for Bone Regeneration. *Nanoscale* **2019**, *11*, 19842–19849. [CrossRef]
57. Zhang, Y.; Yin, C.; Cheng, Y.; Huang, X.; Liu, K.; Cheng, G.; Li, Z. Electrospinning Nanofiber-Reinforced Aerogels for the Treatment of Bone Defects. *Adv. Wound Care* **2020**, *9*, 441–452. [CrossRef]
58. Chen, Z.-J.; Shi, H.-H.; Zheng, L.; Zhang, H.; Cha, Y.-Y.; Ruan, H.-X.; Zhang, Y.; Zhang, X.-C. A New Cancellous Bone Material of Silk Fibroin/Cellulose Dual Network Composite Aerogel Reinforced by Nano-Hydroxyapatite Filler. *Int. J. Biol. Macromol.* **2021**, *182*, 286–297. [CrossRef]
59. Singh, M.; Sandhu, B.; Scurto, A.; Berkland, C.; Detamore, M.S. Microsphere-Based Scaffolds for Cartilage Tissue Engineering: Using Subcritical CO_2 as a Sintering Agent. *Acta Biomater.* **2010**, *6*, 137–143. [CrossRef]
60. Bhamidipati, M.; Sridharan, B.; Scurto, A.M.; Detamore, M.S. Subcritical CO2 Sintering of Microspheres of Different Polymeric Materials to Fabricate Scaffolds for Tissue Engineering. *Mater. Sci. Eng. C* **2013**, *33*, 4892–4899. [CrossRef]
61. Vazhayal, L.; Talasila, S.; Abdul Azeez, P.M.; Solaiappan, A. Mesochanneled Hierarchically Porous Aluminosiloxane Aerogel Microspheres as a Stable Support for pH-Responsive Controlled Drug Release. *ACS Appl. Mater. Interfaces* **2014**, *6*, 15564–15574. [CrossRef] [PubMed]
62. Reyes-Peces, M.V.; Pérez-Moreno, A.; de-los-Santos, D.M.; Mesa-Díaz, M.d.M.; Pinaglia-Tobaruela, G.; Vilches-Pérez, J.I.; Fernández-Montesinos, R.; Salido, M.; de la Rosa-Fox, N.; Piñero, M. Chitosan-GPTMS-Silica Hybrid Mesoporous Aerogels for Bone Tissue Engineering. *Polymers* **2020**, *12*, 2723. [CrossRef] [PubMed]
63. Muñoz-Ruíz, A.; Escobar-García, D.M.; Quintana, M.; Pozos-Guillén, A.; Flores, H. Synthesis and Characterization of a New Collagen-Alginate Aerogel for Tissue Engineering. *J. Nanomater.* **2019**, *2019*, 2875375. [CrossRef]
64. Quraishi, S.; Martins, M.; Barros, A.A.; Gurikov, P.; Raman, S.P.; Smirnova, I.; Duarte, A.R.C.; Reis, R.L. Novel Non-Cytotoxic Alginate–Lignin Hybrid Aerogels as Scaffolds for Tissue Engineering. *J. Supercrit. Fluids* **2015**, *105*, 1–8. [CrossRef]
65. Martins, M.; Barros, A.A.; Quraishi, S.; Gurikov, P.; Raman, S.P.; Smirnova, I.; Duarte, A.R.C.; Reis, R.L. Preparation of Macroporous Alginate-Based Aerogels for Biomedical Applications. *J. Supercrit. Fluids* **2015**, *106*, 152–159. [CrossRef]
66. Szabó, B.A.; Kiss, L.; Manó, S.; Jónás, Z.; Lázár, I.; Fábián, I.; Dezső, B.; Csernátony, Z. The Examination of Aerogel Composite Artificial Bone Substitutes in Animal Models. *Biomech. Hung.* **2013**, *6*, 52. [CrossRef]
67. Hegedűs, C.; Czibulya, Z.; Tóth, F.; Dezső, B.; Hegedűs, V.; Boda, R.; Horváth, D.; Csík, A.; Fábián, I.; Tóth-Győri, E.; et al. The Effect of Heat Treatment of β-Tricalcium Phosphate-Containing Silica-Based Bioactive Aerogels on the Cellular Metabolism and Proliferation of MG63 Cells. *Biomedicines* **2022**, *10*, 662. [CrossRef]
68. Kuttor, A.; Szalóki, M.; Rente, T.; Kerényi, F.; Bakó, J.; Fábián, I.; Lázár, I.; Jenei, A.; Hegedüs, C. Preparation and Application of Highly Porous Aerogel-Based Bioactive Materials in Dentistry. *Front. Mater. Sci.* **2014**, *8*, 46–52. [CrossRef]
69. Lázár, I.; Manó, S.; Jónás, Z.; Kiss, L.; Fábián, I.; Csernátony, Z. Mesoporous Silica-Calcium Phosphate Composites for Experimental Bone Substitution. *Biomech. Hung.* **2010**, *3*, 151. [CrossRef]
70. Hegedűs, V.; Kerényi, F.; Boda, R.; Horváth, D.; Lázár, I.; Tóth-Győri, E.; Dezső, B.; Hegedus, C. β-Tricalcium Phosphate-Silica Aerogel as an Alternative Bioactive Ceramic for the Potential Use in Dentistry. *Adv. Appl. Ceram.* **2020**, *119*, 364–371. [CrossRef]

71. Rubenstein, D.A.; Lu, H.; Mahadik, S.S.; Leventis, N.; Yin, W. Characterization of the Physical Properties and Biocompatibility of Polybenzoxazine-Based Aerogels for Use as a Novel Hard-Tissue Scaffold. *J. Biomater. Sci. Polym. Ed.* **2012**, *23*, 1171–1184. [CrossRef] [PubMed]
72. Williams, D.; Zhang, X. (Eds.) *Definitions of Biomaterials for the Twenty-First Century: Proceedings of a Consensus Conference Held in Chengdu, People's Republic of China, June 11th and 12th 2018, Organized under the Auspices of the International Union of Societies for Biomaterials Science & Engineering; Hosted and Supported by Sichuan University, Chengdu and the Chinese Society for Biomaterials, China;* Materials Today; Elsevier: Amsterdam, The Netherlands, 2019; ISBN 978-0-12-818291-8.
73. Peters, K.; Unger, R.E.; Kirkpatrick, C.J. Biocompatibility Testing. In *Biomedical Materials*; Narayan, R., Ed.; Springer International Publishing: Cham, 2021; pp. 423–453, ISBN 978-3-030-49205-2.
74. Bongio, M.; van den Beucken, J.J.J.P.; Leeuwenburgh, S.C.G.; Jansen, J.A. Development of Bone Substitute Materials: From 'Biocompatible' to 'Instructive'. *J. Mater. Chem.* **2010**, *20*, 8747. [CrossRef]
75. Cullinane, D.M.; Einhorn, T.A. Biomechanics of Bone. In *Principles of Bone Biology*; Elsevier: Amsterdam, The Netherlands, 2002; pp. 17–32, ISBN 978-0-12-098652-1.
76. Manó, S.; Ferencz, G.; Lázár, I.; Fábián, I.; Csernátony, Z. Determination of the Application Characteristics of the Slooff-Technique with Nano-Composite Bone Substitution Material by Biomechanical Tests. *Biomech. Hung.* **2013**, *6*, 64. [CrossRef]
77. Perez-Moreno, A.; Reyes-Peces, M.d.l.V.; de los Santos, D.M.; Pinaglia-Tobaruela, G.; de la Orden, E.; Vilches-Pérez, J.I.; Salido, M.; Piñero, M.; de la Rosa-Fox, N. Hydroxyl Groups Induce Bioactivity in Silica/Chitosan Aerogels Designed for Bone Tissue Engineering. In Vitro Model for the Assessment of Osteoblasts Behavior. *Polymers* **2020**, *12*, 2802. [CrossRef]
78. Karamat-Ullah, N.; Demidov, Y.; Schramm, M.; Grumme, D.; Auer, J.; Bohr, C.; Brachvogel, B.; Maleki, H. 3D Printing of Antibacterial, Biocompatible, and Biomimetic Hybrid Aerogel-Based Scaffolds with Hierarchical Porosities via Integrating Antibacterial Peptide-Modified Silk Fibroin with Silica Nanostructure. *ACS Biomater. Sci. Eng.* **2021**, *7*, 4545–4556. [CrossRef]
79. Weng, L.; Boda, S.K.; Wang, H.; Teusink, M.J.; Shuler, F.D.; Xie, J. Novel 3D Hybrid Nanofiber Aerogels Coupled with BMP-2 Peptides for Cranial Bone Regeneration. *Adv. Healthc. Mater.* **2018**, *7*, 1701415. [CrossRef]
80. Liu, S.; Zhou, C.; Mou, S.; Li, J.; Zhou, M.; Zeng, Y.; Luo, C.; Sun, J.; Wang, Z.; Xu, W. Biocompatible Graphene Oxide–Collagen Composite Aerogel for Enhanced Stiffness and in Situ Bone Regeneration. *Mater. Sci. Eng. C* **2019**, *105*, 110137. [CrossRef]
81. Liu, M.; Shafiq, M.; Sun, B.; Wu, J.; Wang, W.; EL-Newehy, M.; EL-Hamshary, H.; Morsi, Y.; Ali, O.; Khan, A.U.R.; et al. Composite Superelastic Aerogel Scaffolds Containing Flexible SiO_2 Nanofibers Promote Bone Regeneration. *Adv. Healthc. Mater.* **2022**, *11*, 2200499. [CrossRef]
82. Souto-Lopes, M.; Grenho, L.; Manrique, Y.A.; Dias, M.M.; Fernandes, M.H.; Monteiro, F.J.; Salgado, C.L. Full Physicochemical and Biocompatibility Characterization of a Supercritical CO2 Sterilized Nano-Hydroxyapatite/Chitosan Biodegradable Scaffold for Periodontal Bone Regeneration. *Biomater. Adv.* **2023**, *146*, 213280. [CrossRef]
83. Ruphuy, G.; Souto-Lopes, M.; Paiva, D.; Costa, P.; Rodrigues, A.E.; Monteiro, F.J.; Salgado, C.L.; Fernandes, M.H.; Lopes, J.C.; Dias, M.M.; et al. Supercritical CO_2 Assisted Process for the Production of High-purity and Sterile Nano-hydroxyapatite/Chitosan Hybrid Scaffolds. *J. Biomed. Mater. Res. B Appl. Biomater.* **2018**, *106*, 965–975. [CrossRef]
84. Reyes-Peces, M.V.; Fernández-Montesinos, R.; Mesa-Díaz, M.D.M.; Vilches-Pérez, J.I.; Cárdenas-Leal, J.L.; De La Rosa-Fox, N.; Salido, M.; Piñero, M. Structure-Related Mechanical Properties and Bioactivity of Silica–Gelatin Hybrid Aerogels for Bone Regeneration. *Gels* **2023**, *9*, 67. [CrossRef] [PubMed]
85. Pérez-Moreno, A.; Piñero, M.; Fernández-Montesinos, R.; Pinaglia-Tobaruela, G.; Reyes-Peces, M.V.; Mesa-Díaz, M.D.M.; Vilches-Pérez, J.I.; Esquivias, L.; De La Rosa Fox, N.; Salido, M. Chitosan Silica Hybrid Biomaterials for Bone Tissue Engineering: A Comparative Study of Xerogels and Aerogels. *Gels* **2023**, *9*, 383. [CrossRef] [PubMed]
86. Chen, Z.-J.; Zhang, Y.; Zheng, L.; Zhang, H.; Shi, H.-H.; Zhang, X.-C.; Liu, B. Mineralized Self-Assembled Silk Fibroin/Cellulose Interpenetrating Network Aerogel for Bone Tissue Engineering. *Biomater. Adv.* **2022**, *134*, 112549. [CrossRef] [PubMed]
87. Liu, S.; Li, D.; Chen, X.; Jiang, L. Biomimetic Cuttlebone Polyvinyl Alcohol/Carbon Nanotubes/Hydroxyapatite Aerogel Scaffolds Enhanced Bone Regeneration. *Colloids Surf. B Biointerfaces* **2022**, *210*, 112221. [CrossRef] [PubMed]
88. Souto-Lopes, M.; Fernandes, M.H.; Monteiro, F.J.; Salgado, C.L. Bioengineering Composite Aerogel-Based Scaffolds That Influence Porous Microstructure, Mechanical Properties and In Vivo Regeneration for Bone Tissue Application. *Materials* **2023**, *16*, 4483. [CrossRef]
89. Ibrahim, A.; Magliulo, N.; Groben, J.; Padilla, A.; Akbik, F.; Abdel Hamid, Z. Hardness, an Important Indicator of Bone Quality, and the Role of Collagen in Bone Hardness. *J. Funct. Biomater.* **2020**, *11*, 85. [CrossRef]
90. Moner-Girona, M.; Roig, A.; Molins, E.; Martínez, E.; Esteve, J. Micromechanical Properties of Silica Aerogels. *Appl. Phys. Lett.* **1999**, *75*, 653–655. [CrossRef]
91. Woignier, T.; Primera, J.; Alaoui, A.; Despetis, F.; Calas-Etienne, S.; Faivre, A.; Duffours, L.; Levelut, C.; Etienne, P. Techniques for Characterizing the Mechanical Properties of Aerogels. *J. Sol-Gel Sci. Technol.* **2020**, *93*, 6–27. [CrossRef]
92. Revin, V.V.; Pestov, N.A.; Shchankin, M.V.; Mishkin, V.P.; Platonov, V.I.; Uglanov, D.A. A Study of the Physical and Mechanical Properties of Aerogels Obtained from Bacterial Cellulose. *Biomacromolecules* **2019**, *20*, 1401–1411. [CrossRef] [PubMed]
93. Xia, Y.; Gao, C.; Gao, W. A Review on Elastic Graphene Aerogels: Design, Preparation, and Applications. *J. Polym. Sci.* **2022**, *60*, 2239–2261. [CrossRef]
94. Liu, C.; Chu, D.; Kalantar-Zadeh, K.; George, J.; Young, H.A.; Liu, G. Cytokines: From Clinical Significance to Quantification. *Adv. Sci.* **2021**, *8*, 2004433. [CrossRef] [PubMed]

95. Przekora, A. The Summary of the Most Important Cell-Biomaterial Interactions That Need to Be Considered during in Vitro Biocompatibility Testing of Bone Scaffolds for Tissue Engineering Applications. *Mater. Sci. Eng. C* **2019**, *97*, 1036–1051. [CrossRef] [PubMed]
96. Kamiloglu, S.; Sari, G.; Ozdal, T.; Capanoglu, E. Guidelines for Cell Viability Assays. *Food Front.* **2020**, *1*, 332–349. [CrossRef]
97. Tian, M.; Ma, Y.; Lin, W. Fluorescent Probes for the Visualization of Cell Viability. *Acc. Chem. Res.* **2019**, *52*, 2147–2157. [CrossRef]
98. Confederat, L.G.; Tuchilus, C.G.; Dragan, M.; Sha'at, M.; Dragostin, O.M. Preparation and Antimicrobial Activity of Chitosan and Its Derivatives: A Concise Review. *Molecules* **2021**, *26*, 3694. [CrossRef]
99. Yahya, E.B.; Jummaat, F.; Amirul, A.A.; Adnan, A.S.; Olaiya, N.G.; Abdullah, C.K.; Rizal, S.; Mohamad Haafiz, M.K.; Khalil, H.P.S.A. A Review on Revolutionary Natural Biopolymer-Based Aerogels for Antibacterial Delivery. *Antibiotics* **2020**, *9*, 648. [CrossRef]
100. Abdul Khalil, H.P.S.; Adnan, A.S.; Yahya, E.B.; Olaiya, N.G.; Safrida, S.; Hossain, M.d.S.; Balakrishnan, V.; Gopakumar, D.A.; Abdullah, C.K.; Oyekanmi, A.A.; et al. A Review on Plant Cellulose Nanofibre-Based Aerogels for Biomedical Applications. *Polymers* **2020**, *12*, 1759. [CrossRef]
101. García-González, C.A.; Barros, J.; Rey-Rico, A.; Redondo, P.; Gómez-Amoza, J.L.; Concheiro, A.; Alvarez-Lorenzo, C.; Monteiro, F.J. Antimicrobial Properties and Osteogenicity of Vancomycin-Loaded Synthetic Scaffolds Obtained by Supercritical Foaming. *ACS Appl. Mater. Interfaces* **2018**, *10*, 3349–3360. [CrossRef]
102. Zahran, M.; Marei, A.H. Innovative Natural Polymer Metal Nanocomposites and Their Antimicrobial Activity. *Int. J. Biol. Macromol.* **2019**, *136*, 586–596. [CrossRef]
103. Kokubo, T.; Miyaji, F.; Kim, H.-M.; Nakamura, T. Spontaneous Formation of Bonelike Apatite Layer on Chemically Treated Titanium Metals. *J. Am. Ceram. Soc.* **1996**, *79*, 1127–1129. [CrossRef]
104. Kokubo, T.; Takadama, H. How Useful Is SBF in Predicting in Vivo Bone Bioactivity? *Biomaterials* **2006**, *27*, 2907–2915. [CrossRef] [PubMed]
105. Reséndiz-Hernández, P.J.; Cortés-Hernández, D.A.; Méndez Nonell, J.; Escobedo-Bocardo, J.C. Bioactive and Biocompatible Silica/Pseudowollastonite Aerogels. *Adv. Sci. Technol.* **2014**, *96*, 21–26. [CrossRef]
106. Müller, L.; Müller, F.A. Preparation of SBF with Different HCO3- Content and Its Influence on the Composition of Biomimetic Apatites. *Acta Biomater.* **2006**, *2*, 181–189. [CrossRef] [PubMed]
107. Győri, E.; Fábián, I.; Lázár, I. Effect of the Chemical Composition of Simulated Body Fluids on Aerogel-Based Bioactive Composites. *J. Compos. Sci.* **2017**, *1*, 15. [CrossRef]
108. Vallés Lluch, A.; Gallego Ferrer, G.; Monleón Pradas, M. Biomimetic Apatite Coating on P(EMA-Co-HEA)/SiO2 Hybrid Nanocomposites. *Polymer* **2009**, *50*, 2874–2884. [CrossRef]
109. Pizzoferrato, A.; Ciapetti, G.; Stea, S.; Cenni, E.; Arciola, C.R.; Granchi, D. Lucia Cell Culture Methods for Testing Biocompatibility. *Clin. Mater.* **1994**, *15*, 173–190. [CrossRef]
110. Kirkpatrick, C.J.; Bittinger, F.; Wagner, M.; Köhler, H.; van Kooten, T.G.; Klein, C.L.; Otto, M. Current Trends in Biocompatibility Testing. *Proc. Inst. Mech. Eng.* **1998**, *212*, 75–84. [CrossRef]
111. Regenerative Medicine Institute, National Centre for Biomedical Engineering Science, National University of Ireland, Galway; Czekanska, E.; Stoddart, M.; Richards, R.; Hayes, J. In Search of an Osteoblast Cell Model for In Vitro Research. *Eur. Cell Mater.* **2012**, *24*, 1–17. [CrossRef]
112. Lindner, C.; Pröhl, A.; Abels, M.; Löffler, T.; Batinic, M.; Jung, O.; Barbeck, M. Specialized Histological and Histomorphometrical Analytical Methods for Biocompatibility Testing of Biomaterials for Maxillofacial Surgery in (Pre-) Clinical Studies. *Vivo* **2020**, *34*, 3137–3152. [CrossRef]
113. Wittkowske, C.; Reilly, G.C.; Lacroix, D.; Perrault, C.M. In Vitro Bone Cell Models: Impact of Fluid Shear Stress on Bone Formation. *Front. Bioeng. Biotechnol.* **2016**, *4*, 87. [CrossRef]
114. Bellucci, D.; Veronesi, E.; Dominici, M.; Cannillo, V. On the in Vitro Biocompatibility Testing of Bioactive Glasses. *Materials* **2020**, *13*, 1816. [CrossRef] [PubMed]
115. Kargozar, S.; Mozafari, M.; Hamzehlou, S.; Brouki Milan, P.; Kim, H.-W.; Baino, F. Bone Tissue Engineering Using Human Cells: A Comprehensive Review on Recent Trends, Current Prospects, and Recommendations. *Appl. Sci.* **2019**, *9*, 174. [CrossRef]
116. Colnot, C. Cell Sources for Bone Tissue Engineering: Insights from Basic Science. *Tissue Eng. Part B Rev.* **2011**, *17*, 449–457. [CrossRef] [PubMed]
117. Christenson, R.H. Biochemical Markers of Bone Metabolism: An Overview. *Clin. Biochem.* **1997**, *30*, 573–593. [CrossRef]
118. Ge, J.; Li, M.; Zhang, Q.; Yang, C.Z.; Wooley, P.H.; Chen, X.; Yang, S.-Y. Silica Aerogel Improves the Biocompatibility in a Poly-ε-Caprolactone Composite Used as a Tissue Engineering Scaffold. *Int. J. Polym. Sci.* **2013**, *2013*, 402859. [CrossRef]
119. Ratner, B.D. A Pore Way to Heal and Regenerate: 21st Century Thinking on Biocompatibility. *Regen. Biomater.* **2016**, *3*, 107–110. [CrossRef] [PubMed]
120. Karageorgiou, V.; Kaplan, D. Porosity of 3D Biomaterial Scaffolds and Osteogenesis. *Biomaterials* **2005**, *26*, 5474–5491. [CrossRef]
121. Nuss, K.M.; Auer, J.A.; Boos, A.; Rechenberg, B. von An Animal Model in Sheep for Biocompatibility Testing of Biomaterials in Cancellous Bones. *BMC Musculoskelet. Disord.* **2006**, *7*, 67. [CrossRef]
122. Mendes, S.C.; Reis, R.L.; Bovell, Y.P.; Cunha, A.M.; van Blitterswijk, C.A.; de Bruijn, J.D. Biocompatibility Testing of Novel Starch-Based Materials with Potential Application in Orthopaedic Surgery: A Preliminary Study. *Biomaterials* **2001**, *22*, 2057–2064. [CrossRef]

123. Peric, M.; Dumic-Cule, I.; Grcevic, D.; Matijasic, M.; Verbanac, D.; Paul, R.; Grgurevic, L.; Trkulja, V.; Bagi, C.M.; Vukicevic, S. The Rational Use of Animal Models in the Evaluation of Novel Bone Regenerative Therapies. *Bone* **2015**, *70*, 73–86. [CrossRef]
124. Gomes, P.S.; Fernandes, M.H. Rodent Models in Bone-Related Research: The Relevance of Calvarial Defects in the Assessment of Bone Regeneration Strategies. *Lab. Anim.* **2011**, *45*, 14–24. [CrossRef] [PubMed]
125. AO Research Institute, AO Foundation, Clavadelerstrasse 8, Davos, Switzerland; Pearce, A.; Richards, R.; Milz, S.; Schneider, E.; Pearce, S. Animal Models for Implant Biomaterial Research in Bone: A Review. *Eur. Cells Mater.* **2007**, *13*, 1–10. [CrossRef] [PubMed]
126. Li, Y.; Chen, S.-K.; Li, L.; Qin, L.; Wang, X.-L.; Lai, Y.-X. Bone Defect Animal Models for Testing Efficacy of Bone Substitute Biomaterials. *J. Orthop. Transl.* **2015**, *3*, 95–104. [CrossRef] [PubMed]
127. Spicer, P.P.; Kretlow, J.D.; Young, S.; Jansen, J.A.; Kasper, F.K.; Mikos, A.G. Evaluation of Bone Regeneration Using the Rat Critical Size Calvarial Defect. *Nat. Protoc.* **2012**, *7*, 1918–1929. [CrossRef]
128. Vareda, J.P.; Lamy-Mendes, A.; Durães, L. A Reconsideration on the Definition of the Term Aerogel Based on Current Drying Trends. *Microporous Mesoporous Mater.* **2018**, *258*, 211–216. [CrossRef]
129. García-González, C.A.; Budtova, T.; Durães, L.; Erkey, C.; Del Gaudio, P.; Gurikov, P.; Koebel, M.; Liebner, F.; Neagu, M.; Smirnova, I. An Opinion Paper on Aerogels for Biomedical and Environmental Applications. *Molecules* **2019**, *24*, 1815. [CrossRef]
130. Sun, J.; Tan, H. Alginate-Based Biomaterials for Regenerative Medicine Applications. *Materials* **2013**, *6*, 1285–1309. [CrossRef]
131. Martău, G.A.; Mihai, M.; Vodnar, D.C. The Use of Chitosan, Alginate, and Pectin in the Biomedical and Food Sector—Biocompatibility, Bioadhesiveness, and Biodegradability. *Polymers* **2019**, *11*, 1837. [CrossRef]
132. Oudadesse, H.; Derrien, A.C.; Martin, S.; Chaair, H.; Cathelineau, G. Surface and Interface Investigation of Aluminosilicate Biomaterial by the "in Vivo" Experiments. *Appl. Surf. Sci.* **2008**, *255*, 593–596. [CrossRef]
133. Gerhardt, L.-C.; Boccaccini, A.R. Bioactive Glass and Glass-Ceramic Scaffolds for Bone Tissue Engineering. *Materials* **2010**, *3*, 3867–3910. [CrossRef]
134. Dubey, N.; Bentini, R.; Islam, I.; Cao, T.; Castro Neto, A.H.; Rosa, V. Graphene: A Versatile Carbon-Based Material for Bone Tissue Engineering. *Stem Cells Int.* **2015**, *2015*, 804213. [CrossRef] [PubMed]
135. Laboy-López, S.; Méndez Fernández, P.O.; Padilla-Zayas, J.G.; Nicolau, E. Bioactive Cellulose Acetate Electrospun Mats as Scaffolds for Bone Tissue Regeneration. *Int. J. Biomater.* **2022**, *2022*, 3255039. [CrossRef]
136. Shaban, N.Z.; Kenawy, M.Y.; Taha, N.A.; Abd El-Latif, M.M.; Ghareeb, D.A. Cellulose Acetate Nanofibers: Incorporating Hydroxyapatite (HA), HA/Berberine or HA/Moghat Composites, as Scaffolds to Enhance In Vitro Osteoporotic Bone Regeneration. *Polymers* **2021**, *13*, 4140. [CrossRef]
137. Rubenstein, D.A.; Venkitachalam, S.M.; Zamfir, D.; Wang, F.; Lu, H.; Frame, M.D.; Yin, W. In Vitro Biocompatibility of Sheath–Core Cellulose-Acetate-Based Electrospun Scaffolds Towards Endothelial Cells and Platelets. *J. Biomater. Sci. Polym. Ed.* **2010**, *21*, 1713–1736. [CrossRef] [PubMed]
138. Pandey, A. Pharmaceutical and Biomedical Applications of Cellulose Nanofibers: A Review. *Environ. Chem. Lett.* **2021**, *19*, 2043–2055. [CrossRef]
139. Torres, F.; Commeaux, S.; Troncoso, O. Biocompatibility of Bacterial Cellulose Based Biomaterials. *J. Funct. Biomater.* **2012**, *3*, 864–878. [CrossRef] [PubMed]
140. Helenius, G.; Bäckdahl, H.; Bodin, A.; Nannmark, U.; Gatenholm, P.; Risberg, B. In Vivo Biocompatibility of Bacterial Cellulose. *J. Biomed. Mater. Res. A* **2006**, *76A*, 431–438. [CrossRef]
141. Rodrigues, S.; Dionísio, M.; López, C.R.; Grenha, A. Biocompatibility of Chitosan Carriers with Application in Drug Delivery. *J. Funct. Biomater.* **2012**, *3*, 615–641. [CrossRef]
142. Venkatesan, J.; Kim, S.-K. Chitosan Composites for Bone Tissue Engineering—An Overview. *Mar. Drugs* **2010**, *8*, 2252–2266. [CrossRef]
143. Bojar, W.; Ciach, T.; Flis, S.; Szawiski, M.; Jagielak, M. Novel Chitosan-Based Bone Substitute. A Summary of in Vitro and in Vivo Evaluation. *Dent. Res. Oral Health* **2021**, *4*, 12–24. [CrossRef]
144. Rezvani Ghomi, E.; Nourbakhsh, N.; Akbari Kenari, M.; Zare, M.; Ramakrishna, S. Collagen-based Biomaterials for Biomedical Applications. *J. Biomed. Mater. Res. B Appl. Biomater.* **2021**, *109*, 1986–1999. [CrossRef] [PubMed]
145. Kilmer, C.E.; Battistoni, C.M.; Cox, A.; Breur, G.J.; Panitch, A.; Liu, J.C. Collagen Type I and II Blend Hydrogel with Autologous Mesenchymal Stem Cells as a Scaffold for Articular Cartilage Defect Repair. *ACS Biomater. Sci. Eng.* **2020**, *6*, 3464–3476. [CrossRef]
146. Su, K.; Wang, C. Recent Advances in the Use of Gelatin in Biomedical Research. *Biotechnol. Lett.* **2015**, *37*, 2139–2145. [CrossRef] [PubMed]
147. Peter, M.; Binulal, N.S.; Nair, S.V.; Selvamurugan, N.; Tamura, H.; Jayakumar, R. Novel Biodegradable Chitosan–Gelatin/Nano-Bioactive Glass Ceramic Composite Scaffolds for Alveolar Bone Tissue Engineering. *Chem. Eng. J.* **2010**, *158*, 353–361. [CrossRef]
148. Köwitsch, A.; Zhou, G.; Groth, T. Medical Application of Glycosaminoglycans: A Review: Medical Application of Glycosaminoglycans. *J. Tissue Eng. Regen. Med.* **2018**, *12*, e23–e41. [CrossRef]
149. Norahan, M.H.; Amroon, M.; Ghahremanzadeh, R.; Rabiee, N.; Baheiraei, N. Reduced Graphene Oxide: Osteogenic Potential for Bone Tissue Engineering. *IET Nanobiotechnol.* **2019**, *13*, 720–725. [CrossRef]
150. Li, D.; Li, J.; Dong, H.; Li, X.; Zhang, J.; Ramaswamy, S.; Xu, F. Pectin in Biomedical and Drug Delivery Applications: A Review. *Int. J. Biol. Macromol.* **2021**, *185*, 49–65. [CrossRef]

151. Tortorella, S.; Inzalaco, G.; Dapporto, F.; Maturi, M.; Sambri, L.; Vetri Buratti, V.; Chiariello, M.; Comes Franchini, M.; Locatelli, E. Biocompatible Pectin-Based Hybrid Hydrogels for Tissue Engineering Applications. *New J. Chem.* **2021**, *45*, 22386–22395. [CrossRef]
152. Makadia, H.K.; Siegel, S.J. Poly Lactic-Co-Glycolic Acid (PLGA) as Biodegradable Controlled Drug Delivery Carrier. *Polymers* **2011**, *3*, 1377–1397. [CrossRef]
153. Zhao, D.; Zhu, T.; Li, J.; Cui, L.; Zhang, Z.; Zhuang, X.; Ding, J. Poly(Lactic-Co-Glycolic Acid)-Based Composite Bone-Substitute Materials. *Bioact. Mater.* **2021**, *6*, 346–360. [CrossRef]
154. Elmowafy, E.M.; Tiboni, M.; Soliman, M.E. Biocompatibility, Biodegradation and Biomedical Applications of Poly(Lactic Acid)/Poly(Lactic-Co-Glycolic Acid) Micro and Nanoparticles. *J. Pharm. Investig.* **2019**, *49*, 347–380. [CrossRef]
155. Gentile, P.; Chiono, V.; Carmagnola, I.; Hatton, P. An Overview of Poly(Lactic-Co-Glycolic) Acid (PLGA)-Based Biomaterials for Bone Tissue Engineering. *Int. J. Mol. Sci.* **2014**, *15*, 3640–3659. [CrossRef] [PubMed]
156. Jin, S.; Xia, X.; Huang, J.; Yuan, C.; Zuo, Y.; Li, Y.; Li, J. Recent Advances in PLGA-Based Biomaterials for Bone Tissue Regeneration. *Acta Biomater.* **2021**, *127*, 56–79. [CrossRef]
157. Da Silva, D.; Kaduri, M.; Poley, M.; Adir, O.; Krinsky, N.; Shainsky-Roitman, J.; Schroeder, A. Biocompatibility, Biodegradation and Excretion of Polylactic Acid (PLA) in Medical Implants and Theranostic Systems. *Chem. Eng. J.* **2018**, *340*, 9–14. [CrossRef]
158. Tyler, B.; Gullotti, D.; Mangraviti, A.; Utsuki, T.; Brem, H. Polylactic Acid (PLA) Controlled Delivery Carriers for Biomedical Applications. *Adv. Drug Deliv. Rev.* **2016**, *107*, 163–175. [CrossRef] [PubMed]
159. Böstman, O.; Pihlajamäki, H. Clinical Biocompatibility of Biodegradable Orthopaedic Implants for Internal Fixation: A Review. *Biomaterials* **2000**, *21*, 2615–2621. [CrossRef] [PubMed]
160. Arora, M. Polymethylmethacrylate Bone Cements and Additives: A Review of the Literature. *World J. Orthop.* **2013**, *4*, 67. [CrossRef]
161. Magnan, B.; Bondi, M.; Maluta, T.; Samaila, E.; Schirru, L.; Dall'Oca, C. Acrylic Bone Cement: Current Concept Review. *Musculoskelet. Surg.* **2013**, *97*, 93–100. [CrossRef]
162. Janmohammadi, M.; Nourbakhsh, M.S. Electrospun Polycaprolactone Scaffolds for Tissue Engineering: A Review. *Int. J. Polym. Mater. Polym. Biomater.* **2019**, *68*, 527–539. [CrossRef]
163. Dwivedi, R.; Kumar, S.; Pandey, R.; Mahajan, A.; Nandana, D.; Katti, D.S.; Mehrotra, D. Polycaprolactone as Biomaterial for Bone Scaffolds: Review of Literature. *J. Oral Biol. Craniofacial Res.* **2020**, *10*, 381–388. [CrossRef]
164. Ghosh, N.N.; Kiskan, B.; Yagci, Y. Polybenzoxazines—New High Performance Thermosetting Resins: Synthesis and Properties. *Prog. Polym. Sci.* **2007**, *32*, 1344–1391. [CrossRef]
165. Periyasamy, T.; Asrafali, S.; Shanmugam, M.; Kim, S.-C. Development of Sustainable and Antimicrobial Film Based on Polybenzoxazine and Cellulose. *Int. J. Biol. Macromol.* **2021**, *170*, 664–673. [CrossRef] [PubMed]
166. Thirukumaran, P.; Shakila Parveen, A.; Atchudan, R.; Kim, S.-C. Sustainability and Antimicrobial Assessments of Bio Based Polybenzoxazine Film. *Eur. Polym. J.* **2018**, *109*, 248–256. [CrossRef]
167. Lorjai, P.; Wongkasemjit, S.; Chaisuwan, T. Preparation of Polybenzoxazine Foam and Its Transformation to Carbon Foam. *Mater. Sci. Eng. A* **2009**, *527*, 77–84. [CrossRef]
168. Rekowska, N.; Teske, M.; Arbeiter, D.; Brietzke, A.; Konasch, J.; Riess, A.; Mau, R.; Eickner, T.; Seitz, H.; Grabow, N. Biocompatibility and Thermodynamic Properties of PEGDA and Two of Its Copolymer. In Proceedings of the 2019 41st Annual International Conference of the IEEE Engineering in Medicine and Biology Society (EMBC), Berlin, Germany, 23–27 July 2019; IEEE: Berlin, Germany, 2019; pp. 1093–1096.
169. Warr, C.; Valdoz, J.C.; Bickham, B.P.; Knight, C.J.; Franks, N.A.; Chartrand, N.; Van Ry, P.M.; Christensen, K.A.; Nordin, G.P.; Cook, A.D. Biocompatible PEGDA Resin for 3D Printing. *ACS Appl. Bio Mater.* **2020**, *3*, 2239–2244. [CrossRef] [PubMed]
170. Qin, X.; He, R.; Chen, H.; Fu, D.; Peng, Y.; Meng, S.; Chen, C.; Yang, L. Methacrylated Pullulan/Polyethylene (Glycol) Diacrylate Composite Hydrogel for Cartilage Tissue Engineering. *J. Biomater. Sci. Polym. Ed.* **2021**, *32*, 1057–1071. [CrossRef]
171. Musumeci, G.; Loreto, C.; Castorina, S.; Imbesi, R.; Leonardi, R. New Perspectives in the Treatment of Cartilage Damage: Poly(Ethylene Glycol) Diacrylate (PEGDA) Scaffold: A Review. *IJAE Ital. J. Anat. Embryol.* **2013**, *118*, 204–210. [CrossRef]
172. Zhou, X.; Zhang, N.; Mankoci, S.; Sahai, N. Silicates in Orthopedics and Bone Tissue Engineering Materials. *J. Biomed. Mater. Res. A* **2017**, *105*, 2090–2102. [CrossRef]
173. Jurkić, L.M.; Cepanec, I.; Pavelić, S.K.; Pavelić, K. Biological and Therapeutic Effects of Ortho-Silicic Acid and Some Ortho-Silicic Acid-Releasing Compounds: New Perspectives for Therapy. *Nutr. Metab.* **2013**, *10*, 2. [CrossRef]
174. Shadjou, N.; Hasanzadeh, M. Silica-Based Mesoporous Nanobiomaterials as Promoter of Bone Regeneration Process: Bone Regeneration Process Using Silica-Based Mesoporous Nanobiomaterials. *J. Biomed. Mater. Res. A* **2015**, *103*, 3703–3716. [CrossRef]
175. Vareda, J.P.; García-González, C.A.; Valente, A.J.M.; Simón-Vázquez, R.; Stipetic, M.; Durães, L. Insights on Toxicity, Safe Handling and Disposal of Silica Aerogels and Amorphous Nanoparticles. *Environ. Sci. Nano* **2021**, *8*, 1177–1195. [CrossRef]
176. Nguyen, T.P.; Nguyen, Q.V.; Nguyen, V.-H.; Le, T.-H.; Huynh, V.Q.N.; Vo, D.-V.N.; Trinh, Q.T.; Kim, S.Y.; Le, Q.V. Silk Fibroin-Based Biomaterials for Biomedical Applications: A Review. *Polymers* **2019**, *11*, 1933. [CrossRef]
177. Wang, Y.; Rudym, D.D.; Walsh, A.; Abrahamsen, L.; Kim, H.-J.; Kim, H.S.; Kirker-Head, C.; Kaplan, D.L. In Vivo Degradation of Three-Dimensional Silk Fibroin Scaffolds. *Biomaterials* **2008**, *29*, 3415–3428. [CrossRef] [PubMed]
178. Wang, Q.; Han, G.; Yan, S.; Zhang, Q. 3D Printing of Silk Fibroin for Biomedical Applications. *Materials* **2019**, *12*, 504. [CrossRef] [PubMed]

179. Farokhi, M.; Mottaghitalab, F.; Fatahi, Y.; Saeb, M.R.; Zarrintaj, P.; Kundu, S.C.; Khademhosseini, A. Silk Fibroin Scaffolds for Common Cartilage Injuries: Possibilities for Future Clinical Applications. *Eur. Polym. J.* **2019**, *115*, 251–267. [CrossRef]
180. Martins, A.; Chung, S.; Pedro, A.J.; Sousa, R.A.; Marques, A.P.; Reis, R.L.; Neves, N.M. Hierarchical Starch-Based Fibrous Scaffold for Bone Tissue Engineering Applications. *J. Tissue Eng. Regen. Med.* **2009**, *3*, 37–42. [CrossRef]
181. Salgado, A.J.; Coutinho, O.P.; Reis, R.L. Novel Starch-Based Scaffolds for Bone Tissue Engineering: Cytotoxicity, Cell Culture, and Protein Expression. *Tissue Eng.* **2004**, *10*, 465–474. [CrossRef] [PubMed]
182. Pilmane, M.; Salma-Ancane, K.; Loca, D.; Locs, J.; Berzina-Cimdina, L. Strontium and Strontium Ranelate: Historical Review of Some of Their Functions. *Mater. Sci. Eng. C* **2017**, *78*, 1222–1230. [CrossRef]
183. Kaufman, J.-M.; Audran, M.; Bianchi, G.; Braga, V.; Diaz-Curiel, M.; Francis, R.M.; Goemaere, S.; Josse, R.; Palacios, S.; Ringe, J.D.; et al. Efficacy and Safety of Strontium Ranelate in the Treatment of Osteoporosis in Men. *J. Clin. Endocrinol. Metab.* **2013**, *98*, 592–601. [CrossRef]
184. Cianferotti, L.; D'Asta, F.; Brandi, M.L. A Review on Strontium Ranelate Long-Term Antifracture Efficacy in the Treatment of Postmenopausal Osteoporosis. *Ther. Adv. Musculoskelet. Dis.* **2013**, *5*, 127–139. [CrossRef]
185. Lu, H.; Zhou, Y.; Ma, Y.; Xiao, L.; Ji, W.; Zhang, Y.; Wang, X. Current Application of Beta-Tricalcium Phosphate in Bone Repair and Its Mechanism to Regulate Osteogenesis. *Front. Mater.* **2021**, *8*, 698915. [CrossRef]
186. Bohner, M.; Santoni, B.L.G.; Döbelin, N. β-Tricalcium Phosphate for Bone Substitution: Synthesis and Properties. *Acta Biomater.* **2020**, *113*, 23–41. [CrossRef] [PubMed]
187. Tanaka, T.; Komaki, H.; Chazono, M.; Kitasato, S.; Kakuta, A.; Akiyama, S.; Marumo, K. Basic Research and Clinical Application of Beta-Tricalcium Phosphate (β-TCP). *Morphologie* **2017**, *101*, 164–172. [CrossRef] [PubMed]
188. Gillman, C.E.; Jayasuriya, A.C. FDA-Approved Bone Grafts and Bone Graft Substitute Devices in Bone Regeneration. *Mater. Sci. Eng. C* **2021**, *130*, 112466. [CrossRef] [PubMed]
189. Petri, D.F.S. Xanthan Gum: A Versatile Biopolymer for Biomedical and Technological Applications. *J. Appl. Polym. Sci.* **2015**, *132*, 42035. [CrossRef]
190. Chen, W.; Chen, S.; Morsi, Y.; El-Hamshary, H.; El-Newhy, M.; Fan, C.; Mo, X. Superabsorbent 3D Scaffold Based on Electrospun Nanofibers for Cartilage Tissue Engineering. *ACS Appl. Mater. Interfaces* **2016**, *8*, 24415–24425. [CrossRef]
191. Li, D.; Chen, K.; Duan, L.; Fu, T.; Li, J.; Mu, Z.; Wang, S.; Zou, Q.; Chen, L.; Feng, Y.; et al. Strontium Ranelate Incorporated Enzyme-Cross-Linked Gelatin Nanoparticle/Silk Fibroin Aerogel for Osteogenesis in OVX-Induced Osteoporosis. *ACS Biomater. Sci. Eng.* **2019**, *5*, 1440–1451. [CrossRef]
192. Lázár, I.; Bereczki, H.F.; Manó, S.; Daróczi, L.; Deák, G.; Fábián, I.; Csernátony, Z. Synthesis and Study of New Functionalized Silica Aerogel Poly(Methyl Methacrylate) Composites for Biomedical Use. *Polym. Compos.* **2015**, *36*, 348–358. [CrossRef]
193. Xu, T.; Miszuk, J.M.; Zhao, Y.; Sun, H.; Fong, H. Electrospun Polycaprolactone 3D Nanofibrous Scaffold with Interconnected and Hierarchically Structured Pores for Bone Tissue Engineering. *Adv. Healthc. Mater.* **2015**, *4*, 2238–2246. [CrossRef]
194. Rong, R.; Li, H.; Dong, X.; Hu, L.; Shi, X.; Du, Y.; Deng, H.; Sa, Y. Silk Fibroin-Chitosan Aerogel Reinforced by Nanofibers for Enhanced Osteogenic Differentiation in MC3T3-E1 Cells. *Int. J. Biol. Macromol.* **2023**, *233*, 123501. [CrossRef]
195. Boda, R.; Lázár, I.; Keczánné-Üveges, A.; Bakó, J.; Tóth, F.; Trencsényi, G.; Kálmán-Szabó, I.; Béresová, M.; Sajtos, Z.; Tóth, E.D.; et al. β-Tricalcium Phosphate-Modified Aerogel Containing PVA/Chitosan Hybrid Nanospun Scaffolds for Bone Regeneration. *Int. J. Mol. Sci.* **2023**, *24*, 7562. [CrossRef]
196. Wasyłeczko, M.; Sikorska, W.; Chwojnowski, A. Review of Synthetic and Hybrid Scaffolds in Cartilage Tissue Engineering. *Membranes* **2020**, *10*, 348. [CrossRef] [PubMed]
197. Balko, S.; Weber, J.F.; Waldman, S.D. Mechanical Stimulation Methods for Cartilage Tissue Engineering. In *Orthopedic Biomaterials*; Li, B., Webster, T., Eds.; Springer International Publishing: Cham, Switzerland, 2018; pp. 123–147. ISBN 978-3-319-89541-3.
198. Naseri, N.; Poirier, J.-M.; Girandon, L.; Fröhlich, M.; Oksman, K.; Mathew, A.P. 3-Dimensional Porous Nanocomposite Scaffolds Based on Cellulose Nanofibers for Cartilage Tissue Engineering: Tailoring of Porosity and Mechanical Performance. *RSC Adv.* **2016**, *6*, 5999–6007. [CrossRef]
199. Hench, L.L.; Polak, J.M. Third-Generation Biomedical Materials. *Science* **2002**, *295*, 1014–1017. [CrossRef]
200. Ning, C.; Zhou, L.; Tan, G. Fourth-Generation Biomedical Materials. *Mater. Today* **2016**, *19*, 2–3. [CrossRef]
201. Caliogna, L.; Medetti, M.; Bina, V.; Brancato, A.M.; Castelli, A.; Jannelli, E.; Ivone, A.; Gastaldi, G.; Annunziata, S.; Mosconi, M.; et al. Pulsed Electromagnetic Fields in Bone Healing: Molecular Pathways and Clinical Applications. *Int. J. Mol. Sci.* **2021**, *22*, 7403. [CrossRef] [PubMed]
202. Akdoğan, V.; Özkaner, V.; Alkurt, F.Ö.; Karaaslan, M. Theoretical and Experimental Sensing of Bone Healing by Microwave Approach. *Int. J. Imaging Syst. Technol.* **2022**, *32*, 2255–2261. [CrossRef]
203. Ye, D.; Xu, Y.; Zhang, H.; Fu, T.; Jiang, L.; Bai, Y. Effects of Low-Dose Microwave on Healing of Fractures with Titanium Alloy Internal Fixation: An Experimental Study in a Rabbit Model. *PLoS ONE* **2013**, *8*, e75756. [CrossRef]
204. Leon, S.A.; Asbell, S.O.; Arastu, H.H.; Edelstein, G.; Packel, A.J.; Sheehan, S.; Daskai, I.; Guttmann, G.G.; Santos, I. Effects of Hyperthermia on Bone. II. Heating of Bone *in Vivo* and Stimulation of Bone Growth. *Int. J. Hyperth.* **1993**, *9*, 77–87. [CrossRef]
205. Khalifeh, J.M.; Zohny, Z.; MacEwan, M.; Stephen, M.; Johnston, W.; Gamble, P.; Zeng, Y.; Yan, Y.; Ray, W.Z. Electrical Stimulation and Bone Healing: A Review of Current Technology and Clinical Applications. *IEEE Rev. Biomed. Eng.* **2018**, *11*, 217–232. [CrossRef]

206. Pinheiro, A.L.B.; Gerbi, M.E.M.; Ponzi, E.A.C.; Ramalho, L.M.P.; Marques, A.M.C.; Carvalho, C.M.; Santos, R.D.C.; Oliveira, P.C.; Nóia, M. Infrared Laser Light Further Improves Bone Healing When Associated with Bone Morphogenetic Proteins and Guided Bone Regeneration: An In Vivo Study in a Rodent Model. *Photomed. Laser Surg.* **2008**, *26*, 167–174. [CrossRef]
207. Garavello-Freitas, I.; Baranauskas, V.; Joazeiro, P.P.; Padovani, C.R.; Dal Pai-Silva, M.; Da Cruz-Höfling, M.A. Low-Power Laser Irradiation Improves Histomorphometrical Parameters and Bone Matrix Organization during Tibia Wound Healing in Rats. *J. Photochem. Photobiol. B* **2003**, *70*, 81–89. [CrossRef] [PubMed]

gels

Article

Preparation of Vancomycin-Loaded Aerogels Implementing Inkjet Printing and Superhydrophobic Surfaces

Patricia Remuiñán-Pose [1], Clara López-Iglesias [1], Ana Iglesias-Mejuto [1], Joao F. Mano [2], Carlos A. García-González [1,*] and M. Isabel Rial-Hermida [1,*]

[1] I + D Farma Group (GI-1645), Departamento de Farmacoloxía, Farmacia e Tecnoloxía Farmacéutica, Faculty of Pharmacy, iMATUS and Health Research Institute of Santiago de Compostela (IDIS), Universidade de Santiago de Compostela, 15782 Santiago de Compostela, Spain; patricia.remuinan@rai.usc.es (P.R.-P.); clara.lopez.iglesias@rai.usc.es (C.L.-I.); ana.iglesias.mejuto@rai.usc.es (A.I.-M.)

[2] CICECO Aveiro Institute of Materials, Chemistry Department, University of Aveiro, 3810-193 Aveiro, Portugal; jmano@ua.pt

* Correspondence: carlos.garcia@usc.es (C.A.G.-G.); mariaisabel.rial@usc.es (M.I.R.-H.); Tel.: +34-881815252 (M.I.R.-H.)

Abstract: Chronic wounds are physical traumas that significantly impair the quality of life of over 40 million patients worldwide. Aerogels are nanostructured dry porous materials that can act as carriers for the local delivery of bioactive compounds at the wound site. However, aerogels are usually obtained with low drug loading yields and poor particle size reproducibility and urges the implementation of novel and high-performance processing strategies. In this work, alginate aerogel particles loaded with vancomycin, an antibiotic used for the treatment of *Staphylococcus aureus* infections, were obtained through aerogel technology combined with gel inkjet printing and water-repellent surfaces. Alginate aerogel particles showed high porosity, large surface area, a well-defined spherical shape and a reproducible size (609 ± 37 μm). Aerogel formulation with vancomycin loadings of up to 33.01 ± 0.47 μg drug/mg of particle were obtained with sustained-release profiles from alginate aerogels for more than 7 days (PBS pH 7.4 medium). Overall, this novel green aerogel processing strategy allowed us to obtain nanostructured drug delivery systems with improved drug loading yields that can enhance the current antibacterial treatments for chronic wounds.

Keywords: aerogels; chronic wounds; vancomycin; gel inkjet printing; superhydrophobic surfaces; 3D droplet printing; alginate; bioaerogels

check for updates

Citation: Remuiñán-Pose, P.; López-Iglesias, C.; Iglesias-Mejuto, A.; Mano, J.F.; García-González, C.A.; Rial-Hermida, M.I. Preparation of Vancomycin-Loaded Aerogels Implementing Inkjet Printing and Superhydrophobic Surfaces. *Gels* **2022**, *8*, 417. https://doi.org/10.3390/gels8070417

Academic Editors: István Lázár and Melita Menelaou

Received: 31 May 2022
Accepted: 29 June 2022
Published: 4 July 2022

1. Introduction

Wounds are physical traumas where the integrity of the skin or any other tissue is compromised. The normal wound healing process has different overlapping phases, namely, hemostasis/inflammatory, proliferative and remodelling phases. A chronic wound appears when the injury is not capable of healing during those phases in a determined period of time [1,2]. Common chronic wounds include ulcers, diabetic foot ulcers, pressure ulcers, surgical wounds or infectious wounds. These injuries represent a global health problem due to the reduction in quality of life of patients, with pain, stress and usually, several work leaves [3]. From a health-economics perspective, chronic wounds could require daily health workers' attention and possible surgical interventions. Economic forecasts expect expense growth of up to USD 27.8 billion by 2026 [4].

Chronic wounds can be prevented with suitable, effective and space- and time-accurate treatments [5]. Ideal drug release in the wound healing process must be a controlled and local release with two phases: (i) a burst release with immediate therapeutic effect at the infected site and (ii) a sustained release for a prolonged period of time [6]. Currently, there are several commercially available wound dressings, including foams, gauzes and hydrocolloids. However, they do not solve simultaneously chronic wound problems such as odour, pain, limited ability to absorb blood or exudates, control bacterial infections or promotion of cell migration and skin regeneration. Moreover, new technologies such as regenerative medicine and antimicrobial or bioactive materials are being used to develop next-generation dressings [7].

Aerogels are defined as solid, lightweight, open porous networks endowed with unique properties such as low bulk density (0.05–0.3 g/cm^3), high porosity (>95%), very high specific surface area (>200 m^2/g) and wide capacity of swelling (i.e., absorbent) [8]. Therefore, aerogels could be excellent materials in biomedicine because they allow fast initial biological fluid absorption and can also act as a carrier for bioactive compounds with a high loading capacity [7,9]. Aerogel-based formulations have been proposed as effective therapeutic solutions for wound treatment [7,10–15]. Compared with hydrogels (i.e., crosslinked hydrophilic polymeric networks where the internal phase is a hydrophilic solvent [16]), aerogels rely on an extraordinary degree of swelling of the dried network. They have a glove-like fitting capacity to the exudative wound morphology as well as a triggered release of their drug payload after contact with the wound fluid [13,17]. Furthermore, aerogels open the possibility of including hydrophilic and hydrophobic active agents within the matrix [18].

Biopolymer aerogels (bioaerogels) are usually preferred for these purposes because they can be obtained from sustainable sources, can have their own biological activity, and are well tolerated and non-toxic. Bioaerogels also usually have at their disposal a wide variety of available functional groups at the surface for the interaction with the biological environment and further chemical tailoring [5,19,20]. Particularly, alginate is a superabsorbent and haemostatic biopolymer, widely used in the development of drug delivery systems [16]. Moreover, alginate is able to stimulate fibroblasts, targeting growth factors and promoting granulated tissue formation [21,22]. Due to these characteristics, the use of alginate aerogels in wound healing could be suitable due to the capacity to maintain a balance between humidity, amount of exudate and gas permeation—important factors for good wound healing. Their high surface area is an advantage in absorbing a high amount of exudate and improving contact with the wound area. Furthermore, they have the simultaneous capacity of being flexible enough to fit the natural shape of wounds and of avoiding all potential sources of infection [8,19]. Particularly, aerogels in the form of small particles or beads are very interesting for reaching the application site in the case of deep chronic wounds [7].

Bioaerogels are usually prepared by extraction of the liquid phase of a hydrogel without alteration of the inner polymer structure. The most common preparation scheme involves the formation of the hydrogel precursor, followed by a solvent exchange and an extraction of the solvent. Conventional methods to obtain hydrogel particle precursors (electrostatic/vibrating nozzle, atomisation or jet cutting) are based on dripping a polymer solution on a bath containing the crosslinking agent, so that droplets are gelled in the form of spheres [23]. Despite the simplicity of these methods, droplets freely falling under gravitational forces may lead to polydisperse particles. To prepare drug-loaded hydrogel spheres, the drug can be loaded in the initial polymer solution; however, high volumes of gelation bath may lead to a prompt drug diffusion, resulting in decreased loading efficiency [24]. Finally, the use of a supercritical drying process is advantageous for the extraction step as it leads to inner mesoporous structures, differently from atmospheric and freeze-drying processes [7,23,25,26].

Gel inkjet 3D printing is an alternative method that allows the production of gel particles with narrow size distributions [27–29]. Gel inkjet printing is an accurate and flexible technique with a drop volume range in the micro- to picoliter range at high throughputs [27,30]. On the other hand, superhydrophobic surfaces bring new possibilities to enhance the drug entrapment, showing a virtual 100% yield since the encapsulation of the active occurs in the air–solid interface, reducing the use of solvents. These surfaces are able to repel polar dispersions of hydrophilic polymers into polar solvents. The contact angles formed between the drop and the surface are higher than 150°, so the obtained drops have a spherical shape. These surfaces have been used for the development of polymeric hydrogels with a perfectly spherical shape [31]. This technology is simple, reproducible and biocompatible and widely used in several drug delivery formulations and tissue engineering approaches [30–32].

Local administration of antibiotics is commonly preferred to promote wound healing and fight against possible infections, while avoiding systemic side effects and potential bacteria drug resistance [33]. There are several products in the market that have implemented a topical delivery, as they can be a good help to remove biofilm and avoid multidrug-bacteria resistance in the wound site [34]. Vancomycin is a common choice to treat *Staphylococcus aureus* infections, the most frequent Gram-positive bacteria in chronic wound diseases [10]. Vancomycin topical administration allows therapeutic levels (minimum inhibitory concentration (MIC) for *Staphylococcus aureus* of 2 μg/mL) without systemic side effects [35]. Conventional dropping methods to obtain drug-loaded aerogels (including vancomycin) demonstrated a trend of low loading efficiency, with a load percentage of vancomycin and other actives of around 7–12% [5,10,11,23].

An innovative technological combination of 3D-printing and water-repellent surfaces is herein proposed as a proof-of-concept to obtain drug-loaded aerogel microspheres at uniform particle size, shape reproducibility and expected high drug loading efficiency. To the best of our knowledge, there is not another aerogel system developed implementing the combination of both technologies. Alginate aerogels loaded with vancomycin were obtained through gel inkjet printing of an aqueous alginate solution (0.1–1.0 wt.%) into water-repellent surfaces and followed by supercritical drying. Several drug loading strategies (in the ink, bath or ink + bath) were tested. The obtained aerogels were characterised in terms of particle size by optical and scanning electron microscopies, textural properties by N_2 adsorption–desorption analysis and drug loading yield, as well as their release profile in a simulated body fluid medium (PBS pH 7.4).

2. Results and Discussion

2.1. *Definition of Operating Window for Alginate Aerogel Preparation*

The classical ionic gelation mechanism of alginate has been integrated into the technological development with inkjet printing herein explored to obtain hydrogel particles of high sphericity and reproducible size. Alginate is able to form links between adjacent chains in an *egg-box* conformation [36]. Generally, these bonds are established with divalent and trivalent cations (typically Ca^{2+}). These links allow us to encapsulate a broad type of bioactive agents, including, for example, monoclonal antibodies [16]

Injection pressure, nozzle-to-bath distance, output cycle time and printhead speed were crucial parameters for inkjet printing to produce spheres with uniform size on the superhydrophobic surface. High injection pressures and distances between the nozzle and the gelation bath resulted in flattened drops due to the heavy impact of solutions above the superhydrophobic surface. If the printhead speed was high and drop output cycle time (i.e., frequency of droplet ejection) was low, drops were printed within a short time of each other and agglomeration of several drops was observed [27,37,38].

The optimum concentration of alginate within the aqueous ink formulations was established by visual observation of 50 drops of each alginate concentration printed using the method described in Section 4.3.1. Inks with alginate concentrations of 0.1 and 0.25% (*w/v*) resulted in very diluted hydrogels, flattened and without enough consistency for handling. Hydrogel particles obtained from inks with 0.5% (*w/v*) concentration also resulted in deformed particles, although to a lesser degree. Instead, hydrogels obtained from inks with 0.75 and 1% (*w/v*) alginate were mostly spherical and with good consistency. However, hydrogels from inks of 0.75% (*w/v*) alginate concentration hydrogels were chosen as the optimum ones (Figure 1) since the pressure used (40 kPa) was half of the value needed to be able to print drops from 1% (*w/v*) alginate inks.

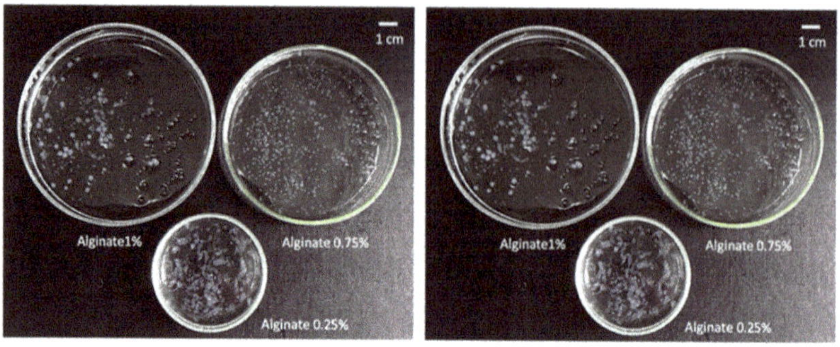

(**a**) (**b**)

Figure 1. Images of gels from the different alginate concentrations tested for the search of the optimum concentration (**a**) before (hydrogels) and (**b**) after (aerogels) the $scCO_2$ drying.

Drops of 0.75% (*w/v*) alginate solution were printed and gelified using 5 mL $CaCl_2$ aqueous bath solutions of different concentrations (0.2, 0.5, 0.8 and 1 M) over a superhydrophobic surface (cf. Section 4.3.2). The use of 0.2 M $CaCl_2$ solutions resulted in a slow gelation process. The use of 1 M $CaCl_2$ solutions resulted in a very fast gelation only able to gelify the outer shell of the sphere, resulting in hollow particles. Trade-off $CaCl_2$ concentrations were set to the 0.5–0.8 M range, with 0.8 M selected for ulterior tests because of faster gelation. At a 0.8 M $CaCl_2$ concentration, drops began to gelify with a transparent appearance as soon as they were in contact with the gelation bath. As gel particles aged, their appearance changed to a whitish colour that was maintained after the solvent exchange and supercritical drying steps reported in Section 4.3.3.

2.2. Production of Vancomycin-Loaded Aerogels Using Superhydrophobic Surfaces

Alginate aerogels were loaded with vancomycin in situ during the gelation process. Vancomycin is a very water-soluble drug and is non-soluble in ethanol. Therefore, vancomycin can be dissolved in high amounts in both the aqueous ink and in the gelation bath. Solvent exchange from the aqueous matrix of hydrogels to pure ethanol avoided additional losses during the aerogel obtaining process. The development of three types of vancomycin-loaded aerogels was intended to compare the capacities of those types to load increasing amounts of the bioactive agent.

After supercritical drying, vancomycin loading of aerogel formulations was 26.59 $\pm$ 0.17, 17.22 $\pm$ 0.35 and 33.01 $\pm$ 0.47 µg vancomycin/mg of type I, II and III particles, respectively (Table 1). These values are quite high if compared to vancomycin-loaded bioaerogels previously reported (ca. 8.5 $\pm$ 4.0, 12.9 $\pm$ 1.0 and 27.3 $\pm$ 2.8 µg vancomycin/mg of particle [10]). The results obtained correlate with the initial composition of the forming solutions. Type III aerogels, where vancomycin was initially present in the alginate ink and in the $CaCl_2$ bath, were the ones with the highest loading. Type I aerogels had a lower drug loading as with this strategy; the drug is only encapsulated by impregnation when the hydrogels adsorbed the vancomycin in the gelation bath. Finally, type II aerogels were the ones with the lowest loading due to the possible diffusion of vancomycin to the crosslinking bath during the process, as the vancomycin is water-soluble. Type II and III aerogels had an encapsulation yield of 15.63 ($\pm$0.10) and 19.41 ($\pm$0.28), respectively, which are coherent yields if conventional methods were used (12.0 $\pm$ 2.0%) [5,10,11,23]. Despite the loading yields being higher than the reported ones, we hypothesise that vancomycin diffusion took place when the gels were in contact with the crosslinking bath to a certain extent. Nevertheless, the volume of this bath was 8 mL, i.e., the minimum possible at the moment, so there is room for improvement of load yields in further studies.

Table 1. Vancomycin loading and entrapment yield of different aerogels using superhydrophobic surfaces.

	Type I	Type II	Type III
Vancomycin loading (μg/mg particles)	26.59 (±0.17)	17.22 (±0.35)	33.01 (±0.47)
Entrapment yield (%)	15.63 (±0.10)	10.13 (±0.20)	19.41 (±0.28)

2.3. Morphological and Physicochemical Characterisation of Alginate Gels and Aerogels

Aerogel and hydrogel Feret diameters were measured using an optical microscope to determine the particle size distribution and the volume shrinkage taking place upon aerogel processing. The average diameter of the hydrogels was 861 ± 32 μm, while the average diameter of the aerogels was 609 ± 37 μm (Figure 2). The standard deviation of the aerogels was very low; consequently, they can be considered uniform due to their narrow size distribution compared with traditional methods of gel preparation [23]. The degree of the volume shrinkage was 64.61%, a similar value compared with literature (57.0 ± 5.0%) [10]. Moreover, aerogel circularity was 80.2 ± 0.7%, while hydrogel circularity was 73.7 ± 0.6%.

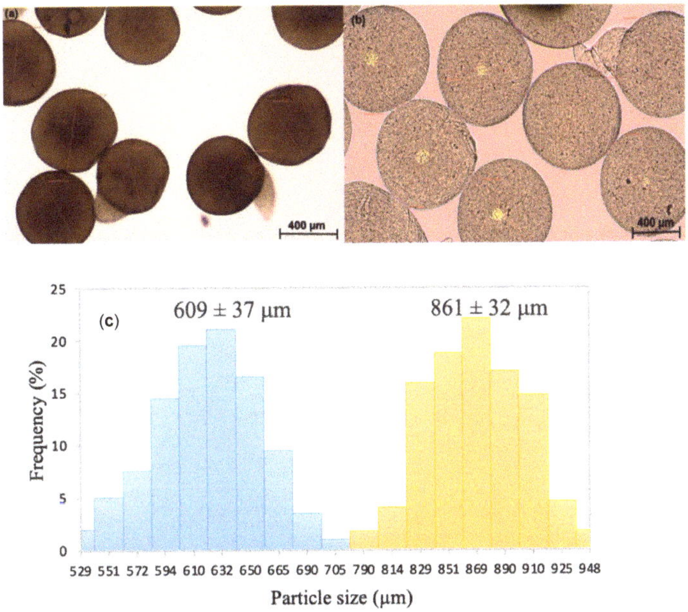

Figure 2. Optical images of 0.75% (*w/v*) alginate (crosslinked in 0.8 M $CaCl_2$ baths) (**a**) aerogels and (**b**) hydrogels obtained by inkjet printing, with (**c**) the particle size distribution (in number) of aerogels (blue) and hydrogels (yellow), respectively.

Aerogel surface morphology and inner structure were analysed by SEM for both blank and loaded ones (Figure 3). A well-defined spherical shape was shown in all formulations. Moreover, the characteristic porous structure with the strongly interconnected fibrous network of alginate aerogels was also observed. These fibres are mixed with each other, and the generated voids are responsible for the porosity both in the outer and inner aerogel structure. The presence of mesoporosity was confirmed by nitrogen adsorption-desorption analysis (Table 2, Figures S1 and S2), with typical parameters for these systems [39,40]. The amino groups of vancomycin confer the ability to bond via electrostatic interactions to carboxyl groups of alginate of the ink [41]. This could be an explanation for this effect of vancomycin in the textural properties, as the said drug may act as an extra crosslinking agent, which would stabilise the structure and better prevent shrinkage during the solvent exchange and supercritical drying processes.

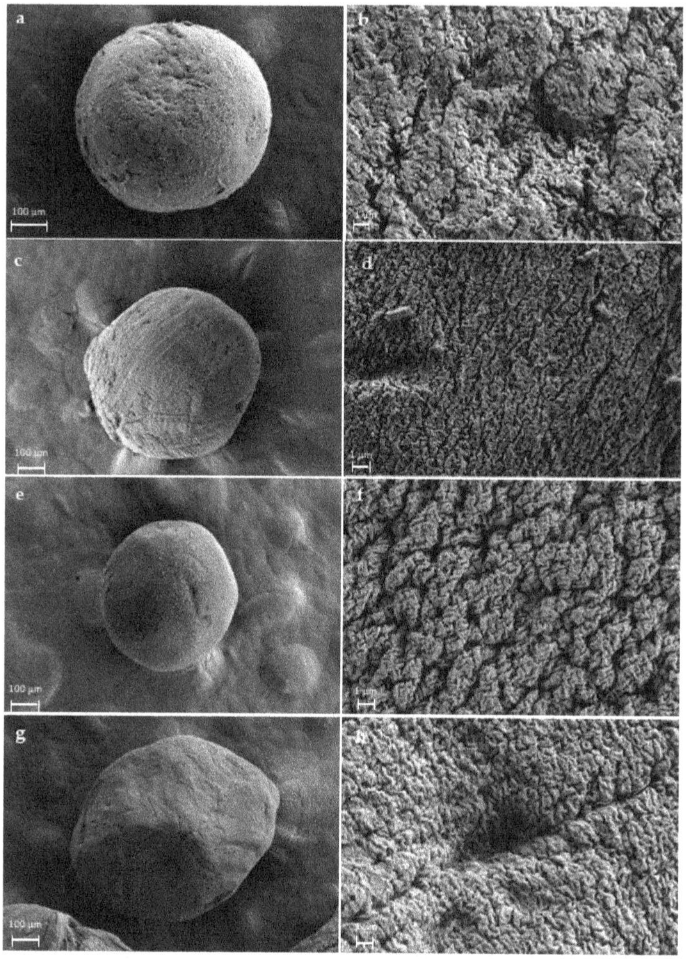

Figure 3. Blank 0.75% (*w/v*) alginate aerogels (**a**,**b**); type I aerogels (**c**,**d**); type II aerogels (**e**,**f**); type III aerogel (**g**,**h**) structures. In higher magnification images, porous structure is observed in all formulations.

Table 2. Textural properties of blank and vancomycin-loaded alginate aerogel microspheres evaluated by nitrogen adsorption–desorption tests. In parenthesis, standard deviation values.

Formulations	a_{BET} (m^2/g) *	$V_{P,BJH}$ (cm^3/g) **	$D_{P,BJH}$ (nm) ***
Blank aerogels	312 (16)	0.80 (0.04)	11.2 (0.6)
Type I	742 (37)	3.90 (0.20)	21.8 (1.1)
Type II	268 (13)	2.54 (0.13)	35.2 (1.8)
Type III	530 (26)	2.65 (0.13)	21.0 (1.0)

* Specific surface area by the BET method; ** Overall specific pore volume obtained by the BJH-method from the desorption curve; *** Mean pore diameter by the BJH-method from the desorption curve.

Aerogels evaluated by nitrogen adsorption–desorption showed type IV isotherms according to IUPAC recommendations (Figure S1). The isotherm morphology from nitrogen adsorption–desorption tests is typical of mesoporous materials and is characterised by a huge volume of nitrogen molecules adsorbed at high relative pressures [42]. Physisorption studies using non-reactive molecules, like nitrogen adsorption–desorption tests, are usually preferred for textural analysis over other chemisorption tests (such as ammonia or deuterated water) as they do not react with the functional groups of the polymer chains [43]. In the initial part of the isotherm, the adsorption of the monolayer was completed, and then multilayer adsorption started. An H1 hysteresis loop was observed, which is associated with capillary condensation in the mesopores. Aerogels also presented a log-normal, unimodal pore size distribution according to the BJH method. The surface properties of biopolymer aerogels have a great influence on biological processes such as cell attachment and proliferation, protection against bacteria and fluid sorption capacity [7].

FTIR results show the presence of the vancomycin within the structure of the aerogels compared with the blank ones (Figure 4). The absorption bands of the vancomycin at 3450, 1654, 1504 and 1230 cm^{-1} for hydroxyl stretching, C=O stretching, C=C and phenols were observed in the vancomycin spectrum [10]. However, due to overlapping with the alginate bands, very slight changes in the formulations were observed compared with the spectrum of the polysaccharide at ca. 1540 cm^{-1} (single asterisk, Figure 4) and 1230 cm^{-1} (double asterisk, Figure 4). However, in the physical mixture of alginate and vancomycin (Figure 4f), the peaks at these wavenumbers were clearer. Since the physical mixture was evaluated at the same ratio of alginate/vancomycin as the ink, this difference in band intensities could be explained by the loss of vancomycin in the aerogel processing during the crosslinking in the gelation bath by means of diffusion.

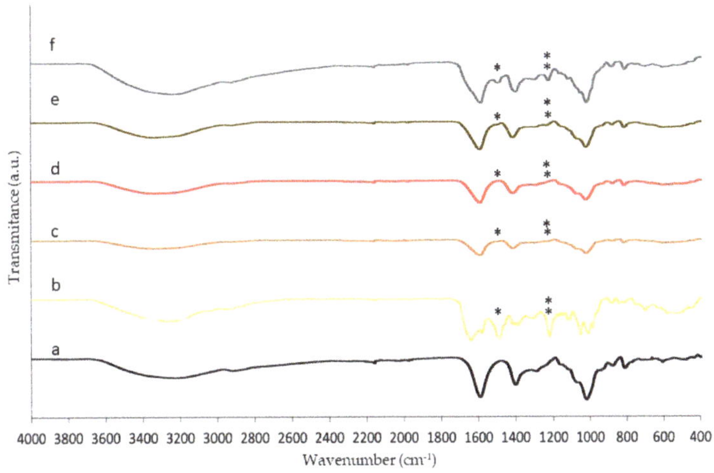

Figure 4. FTIR spectra of (**a**) alginate blank aerogels; (**b**) vancomycin; printed aerogels; (**c**) type I: 0.75% alginate in the ink and a saturated bath of vancomycin in $CaCl_2$ O.8 M; (**d**) type II: 0.75% alginate and 17% (*w/w* alginate) vancomycin forming part of the ink; (**e**) type III: drug both in the saturated vancomycin aqueous bath and forming part of the ink and (**f**) the physical mixture of alginate and vancomycin powders. Single asterisk and double asterisks indicate the presence in all formulations of typical bands of vancomycin at 1540 cm^{-1} and 1230 cm^{-1} respectively.

2.4. Vancomycin Release Studies from Alginate Aerogels

Drug release profile from the aerogel matrices will depend on several factors such as drug and aerogel material hydrophilicity, drug crystallinity, degree of gel crosslinking, drug mass transport mechanisms, specific interactions between the drug and the aerogel, pH and temperature, among others [9]. The three types of vancomycin-loaded aerogels processed using superhydrophobic surfaces had a similar release profile with three stages.

Type III aerogels provided a sustained release during a longer release period with respect to the other aerogel formulations (Figure 5).

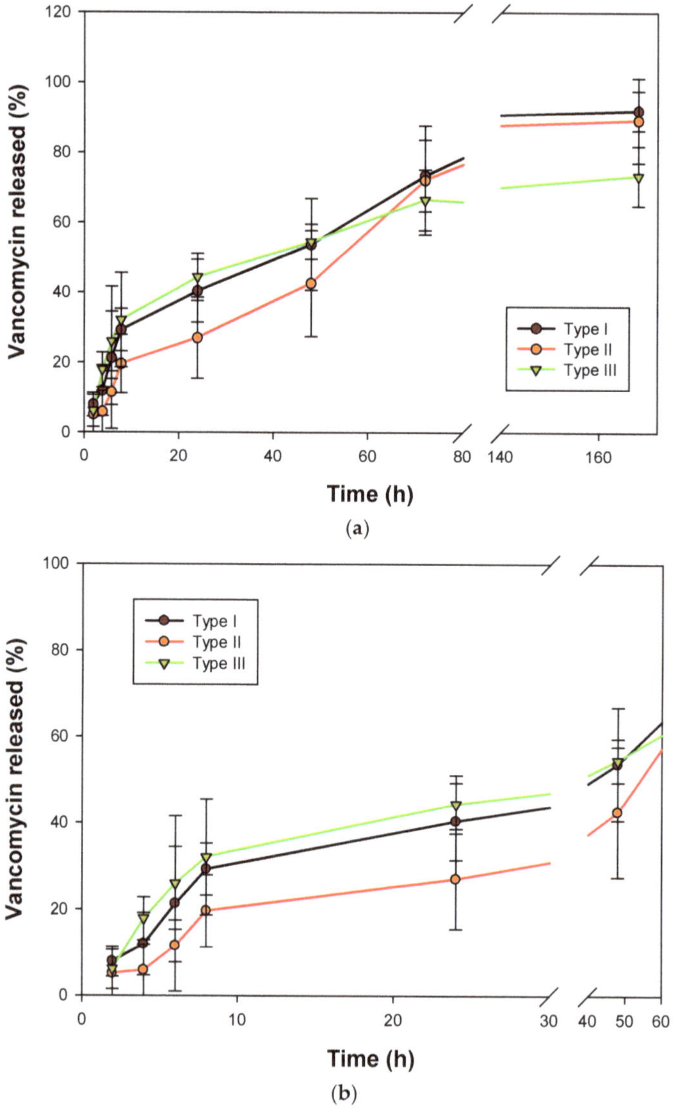

Figure 5. (**a**) Vancomycin released (%) from the 3 aerogel formulations in PBS pH 7.4 at 37 °C for 7 days and (**b**) magnification of the release profile in the first hours, showing the burst release and the initial phase of the plateau region.

Several release kinetics models intended for spherical drug delivery systems with swelling matrices were used to fit the vancomycin release profiles (Table 3). The Ritger–Peppas equation (Equation (1)) was the kinetic model that better characterised the release behaviour, revealing a classical kinetic release for swelling matrices:

$$Mt/M\infty = k \cdot t^{n} \quad (1)$$

where $Mt/M\infty$ is the fraction (in %) of drug released at a certain time t (in h.), k is the release kinetic coefficient (h^{-1}) and n is the diffusion coefficient [44].

Table 3. Kinetic fitting parameters of the vancomycin release from drug-loaded alginate aerogels in PBS solution (pH 7.4) to different kinetic models. Bold letters highlight the more adequate parameters.

Kinetic Model	Parameters	Type I	Type II	Type III
Ritger-Peppas	k (h^{-1})	8.24	13.88	3.35
	n	0.48	0.39	0.41
	R^2	0.96	0.94	0.96
Higuchi	k ($h^{-1/2}$)	0.38	11.18	8.85
	n	8.13	8.91	6.82
	R^2	0.95	0.93	0.97
Zero order	k (h^{-1})	13.42	6.30	22.93
	n	0.914	0.87	0.64
	R^2	0.87	0.96	0.92
First order	k (h^{-1})	87.35	100.48	79.04
	n	0.011	0.016	0.012
	R^2	0.93	0.93	0.97

This type of release kinetics has three typical stages; the first one is a burst release, in this case, during the first 8 h. This could be attributed to the fraction of drug in the external surface of the aerogels, which is weakly bound to the structure and also, it is easy to diffuse due to the absence of resistance of the polymeric matrix. In this release stage, there is a slower drug release from type II aerogels that can be related to the effect that the vancomycin loading has on the textural properties of the aerogels. As the surface area is higher in the case of type I and III aerogels, a faster initial release can be expected. In the case of the type II aerogels, the surface area, as well as the pore size, is lower, and thus a slower release was obtained. Then, there is a sustained release from 8 to 72 h because of a combination of Fickian diffusion and the slow erosion of the polymeric matrix. As time progressed, the matrix changed with large sizes of pores in the inner region due to the erosion of the polymeric matrix. Then, this erosion controls the release with a slow drug release ratio [45,46].

In the Ritger–Peppas model, the governing mass transport mechanism of the drug from the carrier system to the release medium depends on the geometry and structure of the said carrier (Figure S3). In spherical-shape systems, the value of the diffusion exponent (n) indicates the drug release mechanism. As the n-value of aerogels type I is in the range of 0.43 to 0.85, it could be classified as anomalous transport, thus non-Fickian diffusion. In the case of type II and III aerogels, the mechanism could be assumed as Fickian diffusion. [44]

The minimum inhibitory concentration (MIC) for sensitive bacterial strains *Staphylococcus aureus* (2 μg/mL) was outperformed in all drug release tests after the first 2 h so that the aerogels exhibited initially rapid and subsequently sustained antimicrobial activity over time [10]. The test was performed for 7 days due to the intended biomedical application.

3. Conclusions

For the first time, vancomycin-loaded alginate aerogels were successfully developed using a novel 3D printing method implementing superhydrophobic surfaces. The use of this class of surfaces is an efficient method to obtain spherical and uniform size aerogel particles with very low variability (aerogel circularity was 80.2 ± 0.7%). This could lead to the preparation of more efficient drug-loaded aerogels to deliver the agent due to the increase of the specific surface area and the mesoporosity achieved. Furthermore, the implementation of the combination of 3D printing and superhydrophobic surfaces is versatile in terms of the use of different methods for the preparation of drug-loaded aerogels, as is proved in this work with vancomycin loading. All the loading strategies used herein were able to deliver the bioactive agent in PBS pH 7.4 medium for 1 week. The quantity of drug released from the alginate aerogels is enough to treat the most common infection in chronic wounds, reaching therapeutic levels. The possibility of a theoretical 100% encapsulation yield in the hydrogel formation could lead to aerogels with an improvement in the drug payload and will be the subject of future studies. Further research will also focus on the evaluation of the antimicrobial efficiency of the obtained aerogels and the evaluation of the versatility of this technological combination for the incorporation of different wound healing promoting agents.

4. Materials and methods

4.1. Materials

Alginic acid sodium salt from brown algae (guluronic/mannuronic acid ratio of 70/30) was supplied by Sigma Life Science (Irvine, UK). Vancomycin hydrochloride ($C_{66}H_{75}Cl_2N_9O_{24}\cdot HCl$, 94.3% purity) was supplied by Guinama (Valencia, Spain). Calcium chloride anhydrous ($CaCl_2$, >99% purity) was supplied by Scharlab (Barcelona, Spain). Absolute ethanol (EtOH, >99.9% purity) and CO_2 (99.8% purity) were purchased from VWR Chemicals (Fontenay-sous-Bois, France) and Nippon Gases (Madrid, Spain), respectively. Tetraethylorthosilicate (TEOS, 98% purity), ammonium aqueous solution 30–33% and 1H,1H,2H,2H-perfluorodecyltriethoxisilane (PFDTS, 97% purity) were from Sigma–Aldrich (Darmstadt, Germany). Water was purified using reverse osmosis (resistivity > 18 MΩ·cm, Milli-Q, Millipore®, Madrid, Spain).

Ultrapure nitrogen (N_2 (g), >99% purity) supplied by Praxair (Madrid, Spain) was used for the adsorption-desorption textural analysis. Phosphate buffered saline (PBS) pH 7.4, and potassium dihydrogen phosphate (KH_2PO_4, 98.0–100.5% purity) were both supplied by ITW Reagents (Barcelona, Spain). Sodium hydroxide (NaOH, 99% purity) and acetonitrile (CH_3CN, ≥99.9% purity) were both purchased from VWR Chemicals (Barcelona, Spain).

4.2. Preparation of Superhydrophobic Surfaces

The production of polystyrene superhydrophobic surfaces was performed using a simple, economical and fast procedure based on the protocol previously described [47]. In brief, polystyrene Petri dish plates were spray-coated with UV-resistant FluoroThane-MW reagent (WX 2100™) that provides a contact angle of about 150°, as described by the manufacturer (Cytonix, Beltsville, MD, USA). The Petri dish surface was spray-coated and left to dry overnight in a chemical safety fume hood at room temperature. On the following day, the surface was washed with ethanol (99%) and oven dried at 37 °C for 5 days.

4.3. Preparation of Alginate Aerogel Microspheres

4.3.1. Determination of Optimal Concentration of The Alginate and $CaCl_2$ Solutions

Alginate hydrogel particles were prepared following the sol-gel method and consequently crosslinked with an ionic crosslinker ($CaCl_2$). Alginate solutions with concentrations of 0.1, 0.25, 0.5, 0.75 and 1.0% (*w/v*) were used as inks and loaded into the 6 mL cartridge of the inkjet printhead of BIO X™ printer (Cellink, Gothenburg, Sweden) to evaluate their printability. Drops were printed in a Petri dish at room temperature to analyse them visually. Printing parameters, such as printhead speed (8–20 mm/s), injection pressure (10–55 kPa), drop output cycle time (250–450 ms), microvalve opening time (5–15 ms) and pattern density (1.5–3.5%), were modified for each alginate concentration.

Drops with the optimum alginate concentration (0.75% (*w/v*)) were inkjet printed on a superhydrophobic surface and crosslinked in gelation baths of $CaCl_2$ at different concentrations (0.2, 0.5, 0.8 and 1 M). Gelation time of alginate drops was evaluated. Optimum concentrations of the alginate and $CaCl_2$ solutions were chosen, taking into account the best printing parameters obtained among the different concentrations of the alginate and $CaCl_2$ tested in our laboratory.

4.3.2. Preparation of Alginate Hydrogels by Gel Inkjet Printing Using Superhydrophobic Surfaces

Three millilitres of 0.75 % (*w/v*) alginate concentration were loaded into the printing cartridge. The superhydrophobic surface was placed at a slant, and 8 mL of 0.8 M $CaCl_2$ was added at the bottom of it. Drops were printed from the top of the surface and rolled along it (ca. 8 cm) until the gelation bath. Optimum printing parameters were 16 mm/s printhead speed, 40 kPa pressure, 300 ms drop output cycle time, 10 ms microvalve opening time and 3.5% pattern density. Grid patterns were 30 mm × 30 mm × 1 mm (ca. 65 drops). Alginate beads were left in the gelation bath for 24 h at room temperature and pressure. These parameters were experimentally determined. (Video S1).

Three different printing strategies were evaluated to produce vancomycin-loaded hydrogels. In the first type of formulation (Type I), vancomycin at a saturated concentration (25 mg/mL) was placed in the 0.8 M $CaCl_2$ bath. In type II, 17.0 % (*w/w*) vancomycin was included within the alginate ink. The third type (Type III) of sample had vancomycin in the ink and in the $CaCl_2$ bath at the same concentration as in formulations I and II (Figure 6).

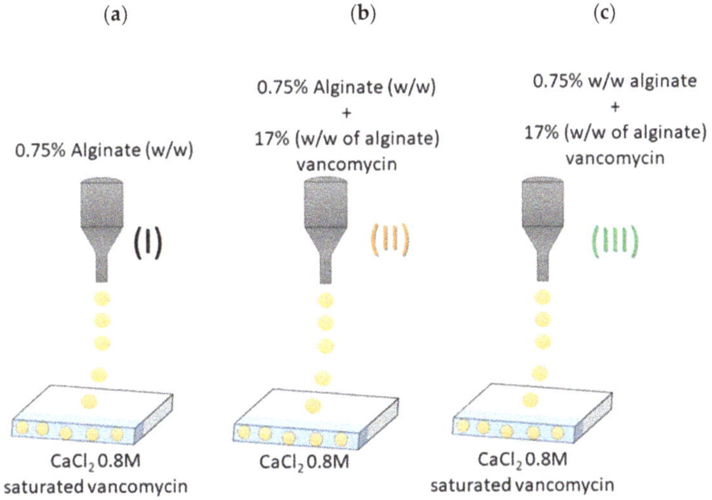

Figure 6. Different ink and gelation bath compositions for inkjet-printing of hydrogels; (**a**) Type I: 0.75% (*w/v*) alginate in the ink and a saturated bath of vancomycin in $CaCl_2$ O.8 M; (**b**) Type II: 0.75% (*w/v*) alginate and 17% (*w/w* alginate) vancomycin forming part of the ink; and (**c**) Type III: drug both in the saturated vancomycin bath and being part of the ink.

4.3.3. Solvent Exchange and Supercritical Drying of Alginate Gels

Ninety millilitres of each vancomycin-loaded gel type (Section 4.3.2.) were produced. Then, two sequential solvent exchanges of the alginate gels with absolute EtOH were carried out at a frequency of 24 h to eliminate water from the gel particles. Alcogel particles were introduced into paper cartridges and put into a 100 mL autoclave (TharSFC, Pittsburg, PA, USA). Twenty millilitres of EtOH were previously added to avoid the premature evaporation of the EtOH contained in alcogels. During the drying process (3.5 h), temperature and pressure were 40 °C and 120 bar, respectively, with a CO_2 flow of 5–7 g/min passing through the autoclave. Ethanol extracts were taken out and weighed at selected drying times to monitor the supercritical process. Finally, the equipment was depressurised, and the aerogels were collected from the autoclave for further characterisation [10].

4.4. *Alginate Gels Characterization*

4.4.1. Morphological and Physicochemical Properties of Alginate Beads

Morphological and physicochemical characteristics were analysed similarly to other aerogel formulations [14,48]. Firstly, hydrogel and aerogel particle diameters were determined by using a CKC53 optical microscope equipped with an EP50 camera and using EPview image analysis software v.1.3 (Olympus, Tokyo, Japan). Surface structure of aerogel microspheres was studied by scanning electron microscopy (SEM) using a FESEM Ultra-Plus microscope (Zeiss, Jena, Germany). Aerogels were previously sputter-coated with a 10 nm layer of iridium to improve the contrast (Q150 T S/E/ES equipment, Quorum Technologies, Lewes, UK). Then, textural properties of aerogel particles were characterised by nitrogen adsorption–desorption analysis (ASAP 2000, Micromeritics, Norcross, GA, USA). Samples were degassed under vacuum at 40 °C for 24 h. The Brunauer–Emmet–Teller (BET) and Barrett–Joyner–Halenda (BJH) methods were applied to determine the specific surface area (a_{BET}), pore size distribution, pore diameter (Dp) and pore volume (Vp). Lastly, attenuated total reflectance/Fourier-transform infrared spectroscopy (ATR/FT-IR) was performed with a Gladi-ATR accessory using a diamond crystal (Pike, Madison, WI, USA). Raw vancomycin, blank alginate aerogels, a physical mixture and the three formulations in the powdered form were characterised in the 400–4000 cm^{-1} IR-spectrum range using 32 scans at a resolution of 2 cm^{-1}.

4.4.2. Vancomycin Drug Content and Loading Efficiency

Vancomycin entrapment yield of the different aerogel formulations was evaluated with ca. 20 mg of each sample placed in Eppendorf tubes with 5 mL of PBS pH 7.4 buffer solution. Tests were carried out in triplicate. Samples were introduced in ultrasound equipment (Branson Ultrasonics, Danbury, CT, USA) for 30 min to completely dissolve the drug. Then, 1 mL of sample was filtered (PTFE hydrophilic, 13 mm, 0.22 μm) and introduced into HPLC glass vials. Jasco LC-4000 HPLC (Madrid, Spain) equipped with a C_{18} column (symmetry columns, 5 μm, 3.9 × 150 mm) was used to measure the drug content. HPLC method conditions were set at 25 °C, a mobile phase of phosphate buffer (30 mM, pH 2.2) and acetonitrile (86:14% *v/v*) operating at an isocratic flow of 0.72 mL/min for 7 min. Chromatograms were obtained at the wavelength of 205 nm [49]. Previously, a calibration curve of vancomycin was obtained and validated in the 0.5–50 μg/mL range ($R^2 > 0.999$) [50].

4.4.3. Vancomycin Release from Alginate Aerogels

Vertical Franz diffusion cells with a 6.2 mL volume in the receptor compartment fitted with cellulose nitrate filters (pore size 0.45 μm) as membranes were employed in the test. Tests were carried out in quadruplicate at 37 °C and 70 rpm of continuous stirring with Heidolph Incubator 1000 equipment (Schwabach, Germany). Franz diffusion cells were filled with PBS buffer pH 7.4 as a release medium, and ca. 20 mg of aerogels were placed into the cells. Then, 200 μL of PBS was added to the donor compartment containing the aerogels to mimic the wet wound environment (time 0). Aliquots of 0.9 mL of release medium were taken at pre-established times (2, 4, 6, 8, 24, 48, 72, 96 and 168 h) in the receptor chamber, and vancomycin content was monitored by HPLC. The extracted volume was immediately replaced with equal volumes of fresh PBS medium. Vancomycin content was measured using the same HPLC method reported in Section 4.4.2.

For the kinetic fitting, the following equations were applied [10,51]:

Zero order (Equation (2)):

$$Mt/M\infty = kt \tag{2}$$

where Mt/M∞ is the amount of vancomycin released (%) at time t, k the release kinetic coefficient and t the time in h.

First order (Equation (3)):

$$Mt/M\infty = 1 + e^{-k_1 t} \tag{3}$$

where $Mt/M\infty$ is the amount of drug released (%) at time t, k_1 is the release kinetic coefficient and t, the time in h.

Higuchi model (Equation (4)):

$$Mt/M\infty = k_2\, t^{\frac{1}{2}} \quad (4)$$

where $Mt/M\infty$ is the amount of drug released (%) at time t, k_2 is the release kinetic coefficient and t, the time in h.

Supplementary Materials: The following supporting information can be downloaded at: https://www.mdpi.com/article/10.3390/gels8070417/s1, Figure S1: Nitrogen adsorption–desorption isotherm of blank alginate aerogels. This isotherm is representative of all aerogel formulations; Figure S2: Pore size distribution of blank and drug-loaded aerogel formulations obtained by a technological combination of inkjet printing and superhydrophobic surfaces; Figure S3: Vancomycin release kinetics fitting to Ritger-Peppas equation from alginate aerogel matrices. Dotted lines representing theorical values estimated with the model. Standard deviations are not shown for the sake of clarity; Video S1: Manufacturing of the drop-by-demand aerogels implementing superhydrophobic surfaces.

Author Contributions: Conceptualisation, C.A.G.-G. and M.I.R.-H.; methodology, P.R.-P., C.L.-I., A.I.-M., J.F.M., C.A.G.-G. and M.I.R.-H.; validation, C.A.G.-G. and M.I.R.-H.; investigation, P.R.-P., C.L.-I., A.I.-M. and M.I.R.-H.; resources, C.A.G.-G.; data curation, P.R.-P., C.L.-I., A.I.-M., C.A.G.-G. and M.I.R.-H.; writing—original draft preparation, P.R.-P., C.A.G.-G. and M.I.R.-H.; writing—review and editing, C.A.G.-G. and M.I.R.-H.; supervision, C.A.G.-G. and M.I.R.-H.; project administration, C.A.G.-G.; funding acquisition, C.A.G.-G. All authors have read and agreed to the published version of the manuscript.

Funding: This research was funded by the Ministerio de Ciencia e Innovación, MICINN, Spain (PID2020-120010RB-I00), Xunta de Galicia, Spain (ED431C 2020/17) and Agencia Estatal de Investigación, Spain (AEI). Work carried out in the framework of the COST-Action "Advanced Engineering of aeroGels for Environment and Life Sciences" (AERoGELS, ref. CA18125) funded by the European Commission.

Data Availability Statement: Not applicable.

Acknowledgments: M.I.R.-H., C.L.-I. and A.I.-M. acknowledge Xunta de Galicia for their Postdoctoral contracts (ED481B 2018/009 and ED481B 2021/008) and Predoctoral grant [ED481A-2020/104], respectively. P.R.-P. acknowledges Banco Santander-USC for her Research grant.

Conflicts of Interest: The authors declare no conflict of interest. The funders had no role in the design of the study; in the collection, analyses or interpretation of data; in the writing of the manuscript, or in the decision to publish the results.

References

1. Wang, P.H.; Huang, B.S.; Horng, H.C.; Yeh, C.C.; Chen, Y.J. Wound healing. *J. Chin. Med. Assoc.* **2018**, *81*, 94–101. [CrossRef] [PubMed]
2. Han, G.; Ceilley, R. Chronic wound healing: A review of current management and treatments. *Adv. Ther.* **2017**, *34*, 599–610. [CrossRef] [PubMed]
3. Las Heras, K.; Igartua, M.; Santos-Vizcaino, E.; Hernandez, R.M. Chronic wounds: Current status, available strategies and emerging therapeutic solutions. *J. Control. Release* **2020**, *328*, 532–550. [CrossRef] [PubMed]
4. Sen, C.K. Human wound and its burden: Updated 2020 compendium of estimates. *Adv. Wound Care* **2021**, *10*, 281–292. [CrossRef]
5. Negut, I.; Dorcioman, G.; Grumezescu, V. Scaffolds for wound healing applications. *Polymers* **2020**, *12*, 2010. [CrossRef]
6. Esquivel-Castro, T.A.; Ibarra-Alonso, M.C.; Oliva, J.; Martínez-Luévanos, A. Porous aerogel and core/shell nanoparticles for controlled drug delivery: A review. *Mater. Sci. Eng. C* **2019**, *96*, 915–940. [CrossRef]
7. Bernardes, B.G.; Del Gaudio, P.; Alves, P.; Costa, R.; García-Gonzaléz, C.A.; Oliveira, A.L. Bioaerogels: Promising nanostructured materials in fluid management, healing and regeneration of wounds. *Molecules* **2021**, *26*, 3834. [CrossRef]
8. García-González, C.A.; Budtova, T.; Durães, L.; Erkey, C.; Del Gaudio, P.; Gurikov, P.; Koebel, M.; Liebner, F.; Neagu, M.; Smirnova, I. An opinion paper on aerogels for biomedical and environmental applications. *Molecules* **2019**, *24*, 1815. [CrossRef]
9. García-González, C.A.; Sosnik, A.; Kalmár, J.; De Marco, I.; Erkey, C.; Concheiro, A.; Alvarez-Lorenzo, C. Aerogels in drug delivery: From design to application. *J. Control. Release* **2021**, *332*, 40–63. [CrossRef]

10. López-Iglesias, C.; Barros, J.; Ardao, I.; Monteiro, F.J.; Alvarez-Lorenzo, C.; Gómez-Amoza, J.L.; García-González, C.A. Vancomycin-loaded chitosan aerogel particles for chronic wound applications. *Carbohydr. Polym.* **2019**, *204*, 223–231. [CrossRef]
11. López-Iglesias, C.; Barros, J.; Ardao, I.; Gurikov, P.; Monteiro, F.J.; Smirnova, I.; Alvarez-Lorenzo, C.; García-González, C.A. Jet cutting technique for the production of chitosan aerogel microparticles loaded with vancomycin. *Polymers* **2020**, *12*, 273. [CrossRef] [PubMed]
12. De Cicco, F.; Russo, P.; Reverchon, E.; García-González, C.A.; Aquino, R.P.; Del Gaudio, P. Prilling and supercritical drying: A successful duo to produce core-shell polysaccharide aerogel beads for wound healing. *Carbohydr. Polym.* **2016**, *147*, 482–489. [CrossRef] [PubMed]
13. Raman, S.P.; Keil, C.; Dieringer, P.; Hübner, C.; Bueno, A.; Gurikov, P.; Nissen, J.; Holtkamp, M.; Karst, U.; Haase, H.; et al. Alginate aerogels carrying calcium, zinc and silver cations for wound care: Fabrication and metal detection. *J. Supercrit. Fluids* **2019**, *153*, 104545. [CrossRef]
14. Lovskaya, D.; Menshutina, N.; Mochalova, M.; Nosov, A. Chitosan-based aerogel particles as highly effective local hemostatic agents production process and in vivo evaluations. *Polymers* **2020**, *12*, 2055. [CrossRef] [PubMed]
15. Delivery, A.; Yahya, E.B.; Jummaat, F.; Amirul, A.A.; Adnan, A.S.; Olaiya, N.G.; Abdullah, C.K.; Rizal, S.; Mohamad Haafiz, M.K.; Khalil, H.P.S. A review on revolutionary natural biopolymer-based aerogels for antibacterial delivery. *Antibiotics* **2020**, *9*, 648.
16. Rial-Hermida, M.I.; Rey-Rico, A.; Blanco-Fernandez, B.; Carballo-Pedrares, N.; Byrne, E.M.; Mano, J.F. Recent progress on polysaccharide-based hydrogels for controlled delivery of therapeutic biomolecules. *ACS Biomater. Sci. Eng.* **2021**, *7*, 4102–4127. [CrossRef] [PubMed]
17. Ferreira-Gonçalves, T.; Constantin, C.; Neagu, M.; Reis, C.P.; Sabri, F.; Simón-Vázquez, R. Safety and efficacy assessment of aerogels for biomedical applications. *Biomed. Pharmacother.* **2021**, *144*, 112356. [CrossRef] [PubMed]
18. Jose, J.; Pai, A.R.; Gopakumar, D.A.; Dalvi, Y.; Rubi, V.; Bhat, S.G.; Pasquini, D.; Kalarikkal, N.; Thomas, S. Novel 3D porous aerogels engineered at nano scale from cellulose nano fibers and curcumin: An effective treatment for chronic wounds. *Carbohydr. Polym.* **2022**, *287*, 119338. [CrossRef]
19. Zheng, L.; Zhang, S.; Ying, Z.; Liu, J.; Zhou, Y.; Chen, F. Engineering of aerogel-based biomaterials for biomedical applications. *Int. J. Nanomed.* **2020**, *15*, 2363–2378. [CrossRef]
20. Zhao, S.; Malfait, W.J.; Guerrero-Alburquerque, N.; Koebel, M.M.; Nyström, G. Biopolymer aerogels and foams: Chemistry, properties, and applications. *Angew. Chem. Int. Ed.* **2018**, *57*, 7580–7608. [CrossRef]
21. Puscaselu, R.G.; Lobiuc, A.; Dimian, M.; Covasa, M. Alginate: From food industry to biomedical applications and management of metabolic disorders. *Polymers* **2020**, *12*, 2417. [CrossRef]
22. Ahmed, A.; Getti, G.; Boateng, J. Ciprofloxacin-loaded calcium alginate wafers prepared by freeze-drying technique for potential healing of chronic diabetic foot ulcers. *Drug Deliv. Transl. Res.* **2018**, *8*, 1751–1768. [CrossRef] [PubMed]
23. Ganesan, K.; Budtova, T.; Ratke, L.; Gurikov, P.; Baudron, V.; Preibisch, I.; Niemeyer, P.; Smirnova, I.; Milow, B. Review on the production of polysaccharide aerogel particles. *Materials* **2018**, *11*, 2417. [CrossRef] [PubMed]
24. Yanniotis, S. Mass transfer by diffusion. In *Food Engineering Series*; Springer: New York, NY, USA, 2008; Volume I, pp. 141–153.
25. Mo, W.; Kong, F.; Chen, K.; Li, B. Relationship between freeze-drying and supercritical drying of cellulosic fibers with different moisture contents based on pore and crystallinity measurements. *Wood Sci. Technol.* **2022**, *56*, 867–882. [CrossRef]
26. Baudron, V.; Gurikov, P.; Smirnova, I.; Whitehouse, S. Porous starch materials via supercritical-and freeze-drying. *Gels* **2019**, *5*, 12. [CrossRef]
27. Li, X.; Liu, B.; Pei, B.; Chen, J.; Zhou, D.; Peng, J.; Zhang, X.; Jia, W.; Xu, T. Inkjet bioprinting of biomaterials. *Chem. Rev.* **2020**, *120*, 10793–10833. [CrossRef]
28. Feng, J.; Su, B.L.; Xia, H.; Zhao, S.; Gao, C.; Wang, L.; Ogbeide, O.; Feng, J.; Hasan, T. Printed aerogels: Chemistry, processing, and applications. *Chem. Soc. Rev.* **2021**, *50*, 3842–3888. [CrossRef]
29. López-Iglesias, C.; Casielles, A.M.; Altay, A.; Bettini, R.; Alvarez-Lorenzo, C.; García-González, C.A. From the printer to the lungs: Inkjet-printed aerogel particles for pulmonary delivery. *Chem. Eng. J.* **2019**, *357*, 559–566. [CrossRef]
30. Karakaya, E.; Bider, F.; Frank, A.; Teßmar, J.; Schöbel, L.; Forster, L.; Schrüfer, S.; Schmidt, H.-W.; Schubert, D.W.; Blaeser, A.; et al. Targeted printing of cells: Evaluation of ADA-PEG bioinks for drop on demand approaches. *Gels* **2022**, *8*, 206. [CrossRef]
31. Costa, A.M.S.; Alatorre-Meda, M.; Alvarez-Lorenzo, C.; Mano, J.F. Superhydrophobic surfaces as a tool for the fabrication of hierarchical spherical polymeric carriers. *Small* **2015**, *11*, 3648–3652. [CrossRef]
32. Rial-Hermida, M.I.; Oliveira, N.M.; Concheiro, A.; Alvarez-Lorenzo, C.; Mano, J.F. Bioinspired superamphiphobic surfaces as a tool for polymer- and solvent-independent preparation of drug-loaded spherical particles. *Acta Biomater.* **2014**, *10*, 4314–4322. [CrossRef] [PubMed]
33. Alcalá-Cerra, G.; Paternina-Caicedo, A.J.; Moscote-Salazar, L.R.; Gutiérrez-Paternina, J.J.; Niño-Hernández, L.M. Aplicación de vancomicina en polvo dentro de la herida quirúrgica durante cirugías de columna: Revisión sistemática y metaanálisis. *Rev. Esp. Cir. Ortop. Traumatol.* **2014**, *58*, 182–191. [CrossRef] [PubMed]
34. Lipsky, B.A.; Hoey, C. Topical antimicrobial therapy for treating chronic wounds. *Clin. Infect. Dis.* **2009**, *49*, 1541–1549. [CrossRef]
35. Martinez, L.R.; Han, G.; Chacko, M.; Mihu, M.R.; Jacobson, M.; Gialanella, P.; Friedman, A.J.; Nosanchuk, J.D.; Friedman, J.M. Antimicrobial and healing efficacy of sustained release nitric oxide nanoparticles against staphylococcus aureus skin infection. *J. Investig. Dermatol.* **2009**, *129*, 2463–2469. [CrossRef]
36. Gurikov, P.; Smirnova, I. Non-conventional methods for gelation of alginate. *Gels* **2018**, *4*, 14. [CrossRef] [PubMed]

37. Yarin, A.L. Drop impact dynamics: Splashing, spreading, receding, bouncing. *Annu. Rev. Fluid Mech.* **2006**, *38*, 159–192. [CrossRef]
38. Xu, Q.; Basaran, O.A. Computational analysis of drop-on-demand drop formation. *Phys. Fluids* **2007**, *19*, 102111. [CrossRef]
39. Rodríguez-Dorado, R.; López-Iglesias, C.; García-González, C.A.; Auriemma, G.; Aquino, R.P.; Del Gaudio, P. Design of aerogels, cryogels and xerogels of alginate: Effect of molecular weight, gelation conditions and drying method on particles' micromeritics. *Molecules* **2019**, *24*, 1049. [CrossRef]
40. Gonçalves, V.S.S.; Gurikov, P.; Poejo, J.; Matias, A.A.; Heinrich, S.; Duarte, C.M.M.; Smirnova, I. Alginate-based hybrid aerogel microparticles for mucosal drug delivery. *Eur. J. Pharm. Biopharm.* **2016**, *107*, 160–170. [CrossRef]
41. Fang, B.; Qiu, P.; Xia, C.; Cai, D.; Zhao, C.; Chen, Y.; Wang, H.; Liu, S.; Cheng, H.; Tang, Z.; et al. Extracellular matrix scaffold crosslinked with vancomycin for multifunctional antibacterial bone infection therapy. *Biomaterials* **2021**, *268*, 120603. [CrossRef]
42. Sing, K.S.; Everett, D.H.; Haul, R.A.W.; Moscou, L.; Pierotti, R.A.; Rouquerol, J. IUPAC recommendations reporting physisorption data for gas/solid systems. *Appl. Catal.* **1988**, *37*, 515–531.
43. Robitzer, M.; Tourrette, A.; Horga, R.; Valentin, R.; Boissire, M.; Devoisselle, J.M.; Di Renzo, F.; Quignard, F. Nitrogen sorption as a tool for the characterisation of polysaccharide aerogels. *Carbohydr. Polym.* **2011**, *85*, 44–53. [CrossRef]
44. Ritger, P.L.; Peppas, N.A. A simple equation for description of solute release II. Fickian and anomalous release from swellable devices. *J. Control. Release* **1987**, *5*, 37–42. [CrossRef]
45. Wang, C.; Okubayashi, S. 3D aerogel of cellulose triacetate with supercritical antisolvent process for drug delivery. *J. Supercrit. Fluids* **2019**, *148*, 33–41. [CrossRef]
46. Marin, M.A.; Mallepally, R.R.; McHugh, M.A. Silk fibroin aerogels for drug delivery applications. *J. Supercrit. Fluids* **2014**, *91*, 84–89. [CrossRef]
47. Antunes, J.; Gaspar, V.M.; Ferreira, L.; Monteiro, M.; Henrique, R.; Jerónimo, C.; Mano, J.F. In-air production of 3D co-culture tumor spheroid hydrogels for expedited drug screening. *Acta Biomater.* **2019**, *94*, 392–409. [CrossRef]
48. Lovskaya, D.D.; Lebedev, A.E.; Menshutina, N.V. Aerogels as drug delivery systems: In vitro and in vivo evaluations. *J. Supercrit. Fluids* **2015**, *106*, 115–121. [CrossRef]
49. Ghasemiyeh, P.; Vazin, A.; Zand, F.; Azadi, A.; Karimzadeh, I.; Mohammadi-Samani, S. A simple and validated HPLC method for vancomycin assay in plasma samples: The necessity of TDM center development in Southern Iran. *Res. Pharm. Sci.* **2020**, *15*, 529–540. [CrossRef]
50. De Jesús Valle, M.J.; López, F.G.; Navarro, A.S. Development and validation of an HPLC method for vancomycin and its application to a pharmacokinetic study. *J. Pharm. Biomed. Anal.* **2008**, *48*, 835–839. [CrossRef]
51. García-González, C.A.; Jin, M.; Gerth, J.; Alvarez-Lorenzo, C.; Smirnova, I. Polysaccharide-based aerogel microspheres for oral drug delivery. *Carbohydr. Polym.* **2015**, *117*, 797–806. [CrossRef]

gels MDPI

Article

Thermal Gelation for Synthesis of Surface-Modified Silica Aerogel Powders

Kyoung-Jin Lee, Jae Min Lee, Ki Sun Nam and Haejin Hwang *

Department of Materials Science and Engineering, Inha University, Incheon 22212, Korea; legna211@naver.com (K.-J.L.); 22201275@inha.edu (J.M.L.); skarltjs2@naver.com (K.S.N.)
* Correspondence: hjhwang@inha.ac.kr; Tel.: +82-32-860-7521

Abstract: A spherical silica aerogel powder with hydrophobic surfaces displaying a water contact angle of 147° was synthesized from a water glass-in-hexane emulsion through ambient pressure drying. Water glass droplets containing acetic acid and ethyl alcohol were stabilized in *n*-hexane with a surfactant. Gelation was performed by heating the droplets, followed by solvent exchange and surface modification using a hexamethyldisilazane (HMDS)/*n*-hexane solution. The pH of the silicic acid solution was crucial in obtaining a highly porous silica aerogel powder with a spherical morphology. The thermal conductivity, tapped density, pore volume, and BET surface area of the silica aerogel powder were 22.4 $mW \cdot m^{-1}K^{-1}$, 0.07 $g \cdot cm^{-3}$, 4.64 $cm^3 \cdot g^{-1}$, and 989 $m^2 \cdot g^{-1}$, respectively. Fourier transform infrared (FT–IR) spectroscopy analysis showed that the silica granule surface was modified by Si-CH_3 groups, producing a hydrophobic aerogel.

Keywords: silica aerogel; thermal gelation; porous; thermal conductivity; hydrophobicity

Citation: Lee, K.-J.; Lee, J.M.; Nam, K.S.; Hwang, H. Thermal Gelation for Synthesis of Surface-Modified Silica Aerogel Powders. *Gels* **2021**, *7*, 242. https://doi.org/10.3390/gels7040242

Academic Editors: István Lázár and Melita Menelaou

Received: 31 October 2021
Accepted: 25 November 2021
Published: 29 November 2021

1. Introduction

A silica aerogel is a mesoporous solid with outstanding properties, including low thermal conductivity, a low dielectric constant, a low refractive index, and high specific surface area. Silica aerogels are considered as promising materials for thermal insulation [1,2], anti-reflection coatings [3], low dielectrics [4], supports for cosmetics [5], adsorbents [6–8], and viscosity agents [9].

Silica aerogels can be fabricated in the form of a monolith or a powder. Silica aerogel monoliths are normally produced by supercritical or ambient pressure drying of a wet gel derived via hydrolysis, and polymerization of alkoxide- or water glass-based precursor solutions [10]. Wei et al. reported an ambient pressure-dried silica aerogel monolith with multiple surface modifications, low thermal conductivity (36 mW/mK), and high porosity (97%) [11]. However, the repeated modification process to transfer hydrophilicity to the hydrophobic surface of the silica aerogel monolith is tedious and requires an extremely long processing time. The reactions between the surface modification agent and silanol groups (Si OH) are diffusion-limited processes, suggesting that the processing time for surface modification increases in proportion to the size of the silica aerogel sample. In addition, silica aerogel monoliths are fragile [12].

Silica aerogel powders and granules are easily and inexpensively fabricated and have a short processing time compared with silica aerogel monoliths. Researchers have recently reported novel methods for rapid synthesis of hydrophobic silica aerogel powders and granules using ambient pressure drying. Bhagat et al. proposed a one-step process with simultaneous surface modification, solvent exchange, and sodium ion removal [13]. Huber et al. presented a one-pot synthesis method for silica aerogel granulates [14]. They argued that gelation after surface modification is crucial for reducing the amount of solvent and production time. In our previous study focusing on catalysts for hydrolysis and condensation of water glass, we proposed a novel fast synthesis technique for spherical silica aerogel powders with a narrow particle size distribution. This synthesis technique reduced the total processing time to less than 2 h [15]. However, silica aerogel powders using the aforementioned fast synthesis processes exhibit some drawbacks. The powder quality is somewhat poor. Tapped density and pore volume are lower than for supercritically dried aerogels [10,11,16]. In addition, the silica aerogels lack particle size homogeneity.

Silica aerogel powders can be prepared by crushing bulk dried silica aerogel or using emulsion polymerization techniques. Although it appears simple and straightforward, crushing silica aerogel bulk is cumbersome, and the resulting powder is bulky. In addition, size and shape control is challenging. Emulsion polymerization is a promising technique for controlling the size and shape of silica aerogel particles. Spherical silica aerogel powder is produced from a water glass-in-hexane emulsion. The particle size of the silica aerogel can be determined by the water glass droplet size, which depends on the force applied to the homogenizer and the emulsifier content.

The silica aerogel powder produced by emulsion polymerization in our previous study had a tapped density of 0.12 $g \cdot cm^{-3}$, a pore volume of 2.35 $cm^3 \cdot g^{-1}$, and a thermal conductivity of 26 $mW \cdot m^{-1} K^{-1}$, somewhat inferior to commercially available silica aerogel powders. We believe that the inferior properties are attributed to inhomogeneous hydrolysis and gelation in the water glass-in-hexane emulsion. In this study, a novel synthesis technique is proposed to produce high-quality spherical silica aerogel particles. We used a novel gelation process (thermal gelation), used for a sol-gel transition of natural polymers such as methylcellulose [17,18]. Water glass droplets containing acid and a gelation catalyst were stabilized in *n*-hexane with a surfactant, followed by thermal gelation, surface modification, and solvent exchange.

2. Materials and Methods

A water glass sodium silicate solution (silica content: 28–30 wt.%, SiO_2:Na_2O = 3.4:1, Young Il Chemical Co., Ltd., Incheon, Korea) was used as the starting material. Initially, the water glass solution was diluted to 5.3–8.7 wt.% with deionized water; 75 mL of water glass, 5 mL of acetic acid (99.5%, Samchun Pure Chemical, Pyeongtaek, Korea), and 5 mL of ethyl alcohol (95.0%, Samchun Pure Chemical) were mixed simultaneously. Ethyl alcohol was used as a condensation (gelation) catalyst, and as a con-solvent because a protic solvent such as ethyl alcohol can promote condensation. Next, 85 mL of *n*-hexane (95%, Samchun Pure Chemical) containing a surfactant, sorbitan monooleate (Span80, Junsei Chemical Co., Ltd., Tokyo, Japan), was added to the water glass/acetic acid/ethyl alcohol solution. The water glass solution to *n*-hexane ratio was fixed at 1. Water glass and *n*-hexane were emulsified using a homogenizer (UltraTurrax IKA T25:S25D-10G-KS, IKA Werke, Konigswinter, Germany) at 6000 rpm for 10 min. A stable water glass-in-hexane emulsion was obtained and heated at 60 °C for condensation (thermal gelation).

Most of the *n*-hexane was drained from the emulsion, and the silica wet gel spheres were immersed in 150 mL of ethyl alcohol. Silica wet gel spheres were solvent-exchanged with ethyl alcohol, which can induce hydrogel-to-alcogel transformation. The surfaces of the silica alcogel spheres were chemically modified in 150 mL of 20% hexamethyldisilazane (HMDS, 98%, Samchun Pure Chemical)/*n*-hexane solution at 60 °C for 3 h. The silylated silica wet gel spheres were washed using an ethyl alcohol/*n*-hexane solution to remove the remaining surface modification agents and reaction products. The surface modification process was repeated three times. The silica wet gel spheres were dried at 100 °C in ambient pressure for 1 h. A schematic of the spherical silica aerogel powder preparation procedure is shown in Figure 1.

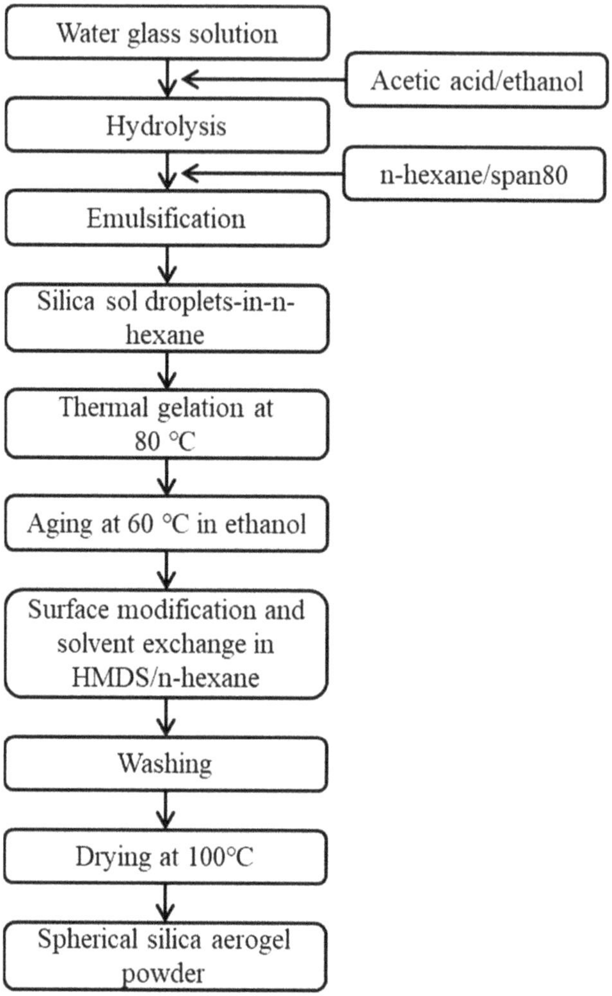

Figure 1. Experimental flow chart for synthesis of spherical silica aerogel powder.

The tapped density of the aerogel powders was determined using a tapping density tester (TAP-2S, Logan Instruments Co., Somerset, NJ, USA). The viscosity of the silicic acid solution was measured using a viscometer (LVT B, Brookfield, Chander, AZ) at 25 °C. The surface area was determined by BET analysis from the amount of N_2 gas adsorbed at different partial pressures ($0.01 < p/p0 < 1$, ASAP 2010; Micrometrics, Norcross, GA, USA). Fourier transform infrared (FT–IR) spectroscopy (FTS-165, Bio-Rad, Hercules, CA, USA) was used to confirm the surface chemical structure of the aerogels in the wavenumber range of 400–4000 cm^{-1}. The contact angle of the water droplet on the silica aerogel powder was calculated from the height and width of the water droplet [19].

The thermal conductivity of the silica aerogel powder was measured using the heat flow metering method with a heat flow meter (HFM 436 Lambda, NETZSCH, Selb, Germany). A silica aerogel powder was placed between two flat plates (25 cm × 25 cm), with the upper and lower plates set at 35 °C and 15 °C, respectively. When thermal equilibrium was reached, the thermal conductivity was estimated using Fourier's law. The thermal conductivity was calculated from the heat flux, the thickness of the silica aerogel powder, and the temperature gradient of the two plates. The microstructure of the silica aerogel powder was observed using field-emission scanning electron microscopy (FESEM, S-4200, Hitachi, Tokyo, Japan).

3. Results and Discussion

Both homogeneous gelation and surface modification are required to obtain high-quality hydrophobic silica aerogel powders (with low bulk density and high pore volume) [20]. Generally, hydrophobic silica aerogels are synthesized through gelation of silica sol or silicic acid solution and subsequent solvent exchange (SE) and surface modification (SM) using hydrophobizing agents. The SE and SM processes require a long processing time. To reduce the processing time, a one-pot process with simultaneous gelation, SE, and SM via the co-precursor method has been proposed [13, 14]. However, the one-pot process results in some reduced physical properties, including density and pore volume [21].

In this study, acetic acid and ethyl alcohol catalysts were used for the hydrolysis and condensation of the water glass solution. Generally, silica sol made with weak acids such as acetic acid is less stable than sol made with strong acids such as hydrochloric and nitric acids. In our previous study, it was deduced that acetic acid and isopropanol catalysts for hydrolysis and condensation of water glass were crucial for spherical silica aerogel powder synthesis with a narrow particle size distribution and a shorter production time. The pH of the silica sol made with acetic acid was 4.5–5.0; the silica sol exhibited adequate gelation time, neither long nor short, suggesting that silica sol catalyzed with acetic acid is stable.

When acetic acid and ethyl alcohol catalysts were added to the water glass solution, gelation did not occur in this step because the concentration of water glass was low (5–8%) and the silica sol made with acetic acid and ethyl alcohol was sufficiently stable at pH = 4.5–5.0. Gelation occurred when the silica sol in the *n*-hexane emulsion was heated to 60 °C. A silica wet gel with a spherical morphology was obtained.

Figure 2a,b show the tapped density of the obtained silica aerogel powders as a function of the water glass concentration and pH of the silicic acid solution. The lowest tapped density (0.067 $g \cdot cm^{-3}$) was obtained in a 6.6% water glass solution. Less than 6.6% water glass solution resulted in an increased tapped density. Above 6.6%, the increase in the tapped density was somewhat reasonable because the tapped density of the aerogel is proportional to the concentration of the starting water glass solution. In Figure 2a, below 6.6%, the tapped density increases again as a function of the water glass concentration. It can be inferred that this is associated with the pH of the silicic acid solution. The pH of the starting water glass solution was estimated to be 11.4 and does not depend on its concentration. In the silicic acid solution, a minimum tapped density was observed at pH = 4.5 (6.6% water glass solution).

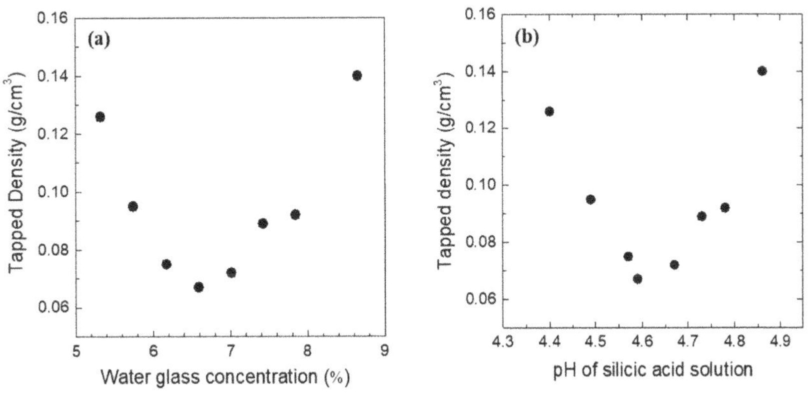

Figure 2. Tapped density of spherical silica aerogel powder as a function of water glass concentration (**a**) and pH of silicic acid solution (**b**).

Sarawade et al. addressed the mechanism of sol-gel polymerization from a water glass solution with an acetic acid catalyst. The gelation time decreased as the pH of the silicic acid solution increased; they showed that the minimum gelation time (5 min) was observed with a pH of 5–6 [22]. To obtain low-density silica aerogels, they used a silicic acid solution with a pH of 4. Above pH = 4, the gelation is too fast, leading to an extremely unstable silica sol. Below pH = 4, the hydrolysis and condensation reactions are not sufficient in the gelation process, which can affect the properties of the final silica aerogel products [23].

Figure 3 shows the change in viscosity as a function of aging time for silicic acid solutions at different pH values. The viscosity abruptly increased to 400 cP from 30–180 min at pH = 5.08 and 4.10, respectively. The rate of viscosity change increased as the pH increased (acetic acid content decreased). Table 1 summarizes the gelation times of the silicic acid solutions and the tapped densities of the synthesized silica aerogel powders as a function of acetic acid content and pH. The concentration of the water glass solution was 6.6%. Gelation time is defined as the time after which the viscosity deviates from linearity [24]. The gelation time strongly depends on the pH of the silicic acid solution. As the acetic acid content increases, the pH value of the silicic acid solution decreases, and the gelation time gradually increases. The condensation rate is generally faster than the hydrolysis rate and increases with increasing pH from 4–10 [25].

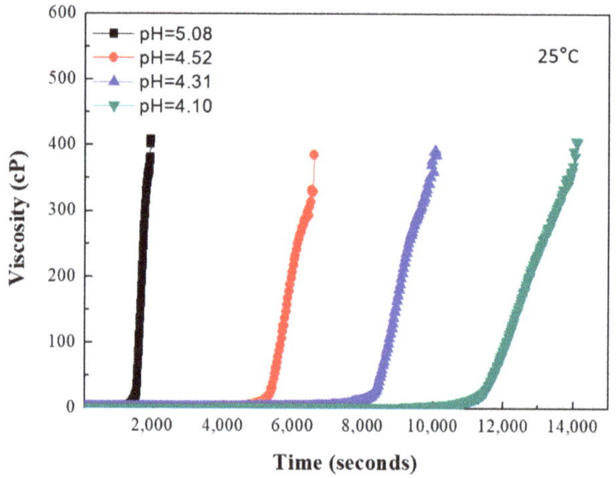

Figure 3. Viscosity of silicic acid solutions at different pH as a function of aging time.

Table 1. Gelation time of silicic acid solutions with different pH and tapped densities of silica aerogel powders (concentration of water glass solution: 6.6%).

Acetic Acid Content (mL)	pH of Silicic Acid Solution	Gelation Time (min)	Tapped Density ($g{\cdot}cm^{-3}$)
3	5.08	24	0.227
5	4.52	85	0.090
7	4.31	128	0.097
10	4.10	180	0.236

From Table 1, the silica aerogel powder sample prepared with a water glass concentration of 6.6% and a pH value of 4.52 exhibited the lowest tapped density (0.090 $g{\cdot}cm^{-3}$), which was also observed in the aerogel powder sample with a water glass concentration higher than 6.6%. This result suggests that an optimum pH exists at which the aerogel powder sample exhibits the lowest tapped density. When the pH of the silicic acid solution is higher than approximately 4.5, gelation is too fast, which can lead to an increased tapped density owing to insufficient time for solvent exchange and surface modification. In Table 1, the tapped density of the silica aerogel powder sample prepared using a silicic acid solution with a pH of 5.08 was significantly higher than those of silica aerogel powder samples with pH values of 4.31 and 4.51. The tapped density increased at pH = 4.1 (acetic acid content: 10 mL) owing to the slow hydrolysis and condensation rate at a low pH. If the pH of the silicic acid solution is too low, hydrolysis, condensation, and surface modification reactions occur simultaneously, which can have a detrimental effect on the mesoporous structure development of silica aerogel.

Figure 4a–f show SEM images of silica aerogel powders prepared with different water glass concentrations. Figure 4a–c show that the obtained silica aerogel powders have a spherical shape, with average estimated sizes of approximately 10–30 μm, regardless of the water glass concentration (Figure S1). This result suggests that the water glass concentration does not significantly affect the size and morphology (sphericity) of the silica aerogel powders. High-magnification SEM images (Figure 4d–f) show that the silica particles possess a typical highly porous three-dimensional network structure consisting of silica nanoparticles, although their tapped densities are slightly different.

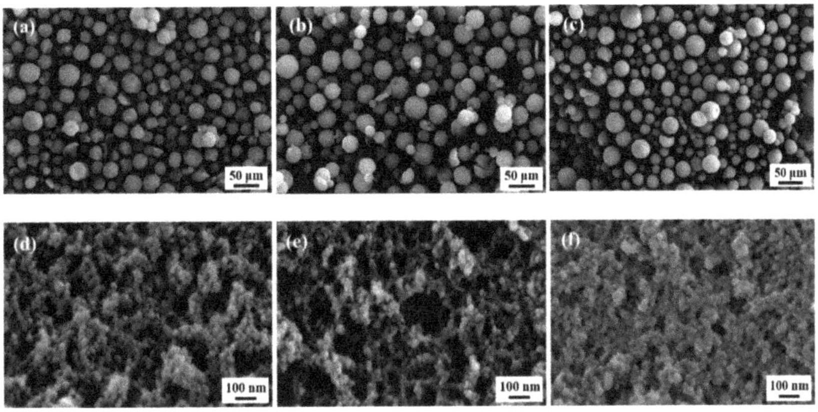

Figure 4. SEM images of silica aerogel powders prepared from (**a**,**d**) 5.7%; (**b**,**e**) 6.6%; (**c**,**f**) 8.7% water glass solutions.

N_2 adsorption–desorption isotherms of the silica aerogel powder are shown in Figure 5. N_2 absorption sharply increases near the high relative pressure area (Type IV adsorption–desorption isotherm with type H1 hysteresis loop), indicating that the silica aerogel is mesoporous [26,27]. It is known that hysteresis is attributed to capillary condensation and evaporation occurring in the mesopores [28]. The hysteresis loop in Figure 5b was found to be more significant, indicating many mesopores in the silica aerogel powder sample with a water glass concentration of 6.6%.

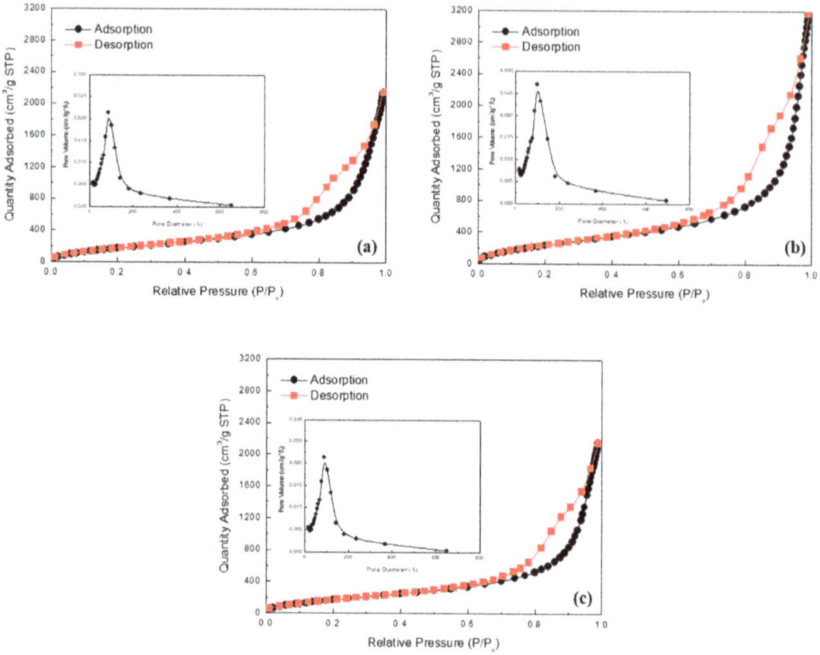

Figure 5. Adsorption–desorption isotherms of spherical silica aerogel powders synthesized from (**a**) 5.3%; (**b**) 6.6%; (**c**) 7.0% water glass solutions. Insets show pore size distribution based on BJH (adsorption).

Table 2 summarizes the physical properties of silica aerogel powders prepared with different water glass concentrations. Tapped density, BET specific surface area, and pore volume strongly depend on the water glass concentration. However, the water glass concentration does not affect the size of the mesopores. The mean pore diameter ranges from 17–19 nm. However, in Figure 4, it appears that the macropore size and distribution changes with the water glass concentration. This observed phenomenon is associated with the difference in agglomeration behavior between secondary silica particles. Depending on the pH of silicic acid solution with different water glass concentrations, silica aerogels had different macrostructures. The agglomeration between silica particles increases with an increase in the pH of the silicic acid solution [29], leading to the formation of macropores, as can be seen in Figure 4b. In the case of 8.7% of water glass concentration, the silica aerogel possessed denser macrostructure than 5.7% and 6.6% aerogel samples, which is due to the large shrinkage during ambient pressure drying.

Table 2. Physical properties of silica aerogel powders synthesized from different water glass concentrations.

Water Glass Concentration (%)	Specific Surface Area ($m^2 \cdot g^{-1}$)	Pore Volume ($cm^3 \cdot g^{-1}$)	Mean Pore Diameter (nm)
5.3	719	3.11	17.3
5.7	731	3.20	17.5
6.2	680	3.20	18.8
6.6	989	4.64	18.8
7.0	709	3.14	17.7

In Table 1, the silica aerogel powder prepared using a 6.6% water glass concentration exhibits the maximum BET specific surface area of 989 $m^2 \cdot g^{-1}$, and a pore volume of

4.64 $cm^3 \cdot g^{-1}$. The excellent properties of the 6.6% water glass solution sample are the result of a more porous microstructure and a slightly smaller particle size than the 5.7% and 8.7% samples. An adequate water glass concentration and pH are crucial for the silicic acid solution to obtain a silica aerogel powder with a low bulk density and a high pore volume.

Fourier transform infrared (FTIR) spectroscopy was used to confirm the hydrophobicity of the silica aerogel powder samples; the corresponding FTIR spectra are shown in Figure 6. According to previous studies [30,31], the absorption peaks near 1090 cm^{-1}, 760 cm^{-1}, and 460 cm^{-1} can be attributed to the asymmetric, symmetric and bending modes of Si-O-Si, respectively. These peaks are characteristic of typical silica aerogel network structures. The peaks at 1260 cm^{-1} and 850 cm^{-1} indicate the presence of a Si-C bond; the peaks at 2960 cm^{-1} and 1600 cm^{-1} correspond to C-H stretching. Peaks corresponding to the stretching vibration of O-H (3455 cm^{-1} and 1635 cm^{-1}) and Si-OH (~800 cm^{-1}) were not observed in any of the silica aerogel powder samples [32]. Thus, it can be inferred that the silica aerogel was modified into a hydrophobic form by the surface methyl groups ($-CH_3$). Photographs of water droplets on the silica aerogel powders prepared with water glass concentrations of 5.7%, 6.6%, and 8.7% are shown in Figure 7. The contact angles of the silica aerogel powder from the 6.6% water glass solution exhibited the highest contact angle of 147° (hydrophobicity); the contact angle decreased with increasing tapped density of the silica aerogel powder, suggesting that the contact angle (good hydrophobicity) is closely related to the tapped density of the silica aerogel powder. In general, the density of the silica aerogel depends on the degree of surface modification by functional groups. In other words, the high degree of modification results in low density because the pore water can be effectively exchanged by ethyl alcohol and *n*-hexane. Thus, it appears that the surface of the low-density silica aerogel is modified with large number of function groups, which can lead to good hydrophobicity.

The spherical silica aerogel powders prepared with water glass concentrations of 5.3% and 6.6% exhibited thermal conductivities of 24.2 $mWm^{-1} \cdot K^{-1}$ and 22.4 $mWm^{-1} \cdot K^{-1}$, respectively. Within our system, silica aerogel powder with a lower tapped density and higher pore volume exhibits a lower thermal conductivity. Although the thermal conductivity (22.4 $mW \cdot m^{-1} \cdot K^{-1}$) obtained in this study is somewhat lower than that of previously reported silica aerogel powder (~14 $mW \cdot m^{-1} \cdot K^{-1}$) [14], it is comparable to that of commercially available silica aerogel powder with a similar particle size [33]. Thus, the silica aerogel powder obtained in this study can be used commercially in the field of superinsulation.

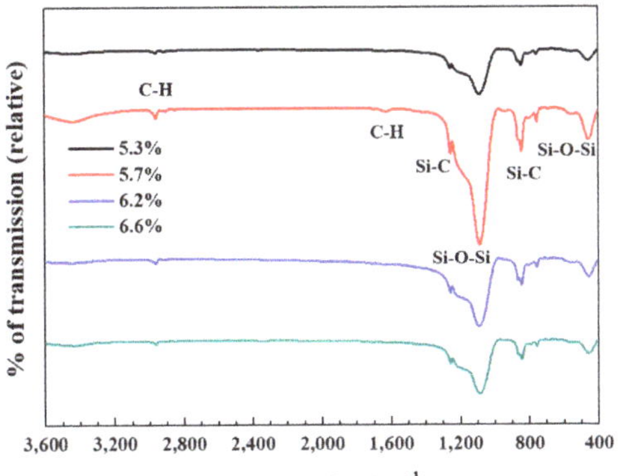

Figure 6. FT-IR spectra of silica aerogel powders prepared from water glass solutions with different concentrations.

(a) (b) (c)

Figure 7. Photographs of water droplets on silica aerogel powders prepared with water glass solution concentrations of (**a**) 5.7%; (**b**) 6.6%; (**c**) 8.7%.

4. Conclusions

Silica aerogel powders with spherical morphology and hydrophobic surfaces displaying a water contact angle of 147° can be synthesized from water glass solutions via emulsion polymerization, thermal gelation, and ambient pressure drying. In this study, acetic acid and ethyl alcohol were used as hydrolysis and condensation catalysts, respectively. The ratio of acetic acid content to water glass concentration, affecting the pH value of the silicic acid solution, was crucial in obtaining a highly porous spherical silica aerogel powder. The silica aerogel powder obtained from a silicic acid solution with a pH of approximately 4.5 exhibited the lowest thermal conductivity, and the greatest specific surface area and pore volume. As the acetic acid content was increased (as the pH of the silicic acid solution decreased), the gelation time was significantly increased, which led to an increased tapped density. If the pH of the silicic acid solution was above 4.5, the gelation rate was too fast and difficult to control, and physical properties of the silica aerogel powder such as pore volume and specific surface area deteriorated significantly.

Supplementary Materials: The following are available online at https://www.mdpi.com/article/10.3390/gels7040242/s1, Figure S1: Particle size distribu-tions of silica aerogel powder samples prepared with different water glass concentrations.

Author Contributions: Conceptualization: H.H. and K.-J.L.; methodology: J.M.L. and K.S.N.; formal analysis: K.-J.L.; investigation: K.-J.L.; writing—original draft preparation: K.-J.L.; writing—review and editing: H.H.; supervision: H.H.; project administration: H.H. All authors have read and agreed to the published version of the manuscript.

Funding: This work was supported by a National Research Foundation of Korea (NRF) grant funded by the Korean government (MSIT) (No. 2020R1A5A1019131), and a National Research Foundation of Korea (NRF) grant (NRF-2020M3D1A2102918).

Data Availability Statement: Not applicable.

Conflicts of Interest: The authors declare no conflict of interest.

References

1. Lei, Y.; Hu, Z.; Cao, B.; Chen, X.; Song, H. Enhancements of thermal insulation and mechanical property of silica aerogel monoliths by mixing graphene oxide. *Mater. Chem. Phys.* **2017**, *187*, 183–190. [CrossRef]
2. Lee, K.-Y.; Mahadik, D.B.; Parale, V.G.; Park, H.-H. Composites of silica aerogels with organics: A review of synthesis and mechanical properties. *J. Korean Ceram. Soc.* **2020**, *57*, 1–23. [CrossRef]
3. Zhang, M.; Liu, Q.; Tian, S.; Zhou, X.; Liu, B.; Zhao, X.; Tang, G.; Pang, A. Facile synthesis of silica composite films with good mechanical property for spectrally broad band antireflection coatings. *Colloids Surf. A* **2021**, *628*, 127255. [CrossRef]
4. Kim, Y.J.; Yoo, S.H.; Lee, H.G.; Won, Y.; Cho, J.; Kang, K. Structural analysis of silica aerogels for the interlayer dielectric in semiconductor devices. *Ceram. Int.* **2021**, *47*, 29722–29729. [CrossRef]
5. Yorov, K.E.; Kolesnik, I.V.; Romanova, I.P.; Mamaeva, Y.B.; Sipyagina, N.A.; Lermontov, S.A.; Kopitsa, G.P.; Baranchikov, A.E.; Ivanov, V.K. Engineering SiO_2–TiO_2 binary aerogels for sun protection and cosmetic applications. *J. Supercrit. Fluids* **2021**, *169*, 105099. [CrossRef]
6. Caputo, G.; Marco, I.D.; Reverchon, E. Silica aerogel–metal composites produced by supercritical adsorption. *J. Supercrit. Fluids* **2010**, *54*, 243–249. [CrossRef]

7. Matias, T.; Marques, J.; Quina, M.J.; Gando-Ferreira, L.; Valente, A.J.M.; Portugal, A.; Durães, L. Silica-based aerogels as adsorbents for phenol-derivative compounds. *Colloids Surf. A* **2015**, *480*, 260–269. [CrossRef]
8. Smirnova, I.; Suttiruengwong, S.; Arlt, W. Feasibility study of hydrophilic and hydrophobic silica aerogels as drug delivery systems. *J. Non-Cryst. Solids* **2004**, *350*, 54–60. [CrossRef]
9. Yamanaka, M.; Nakano, K.; Matsumura, Y.; Tanahashi, T. Perfluoropolyether Oil Diffusion Preventing Agent, and a Fluorine-Based Lubricant. JP6281084B2, 2 February 2018.
10. Mahadik, D.B.; Leea, Y.K.; Chavan, N.K.; Mahadik, S.A.; Park, H.-H. Monolithic and shrinkage-free hydrophobic silica aerogels via new rapid supercritical extraction process. *J. Supercrit. Fluids* **2016**, *107*, 84–91. [CrossRef]
11. Wei, T.-Y.; Chang, T.-F.; Lu, S.-Y.; Chang, Y.-C. Preparation of monolithic silica aerogel of low thermal conductivity by ambient pressure drying. *J. Am. Ceram. Soc.* **2007**, *90*, 2003–2007. [CrossRef]
12. Baetens, R.; Jelle, B.P.; Gustavsen, A. Aerogel insulation for building applications: A state-of-the-art review. *Energy Build.* **2011**, *43*, 761–769. [CrossRef]
13. Bhagat, S.D.; Kim, Y.-H.; Suh, K.-H.; Ahn, Y.-S.; Yeo, J.-G.; Han, J.-H. Superhydrophobic silica aerogel powders with simultaneous surface modification, solvent exchange and sodium ion removal from hydrogels. *Microporous Mesoporous Mater.* **2008**, *112*, 504–509. [CrossRef]
14. Huber, L.; Zhao, S.; Malfait, W.J.; Vares, S.; Koebel, M.M. Fast and minimal-solvent production of superinsulating silica aerogel granulate. *Angew. Chem. Int. Ed.* **2017**, *56*, 4753–4756. [CrossRef] [PubMed]
15. Lee, K.-J.; Kim, Y.H.; Lee, J.K.; Hwang, H. Fast synthesis of spherical silica aerogel powders by emulsion polymerization from water glass. *ChemistrySelect* **2018**, *3*, 1257–1261. [CrossRef]
16. Pan, Y.; He, S.; Cheng, X.; Li, Z.; Li, C.; Huang, Y.; Gong, L. A fast synthesis of silica aerogel powders-based on water glass via ambient drying. *J. Sol-Gel Sci. Technol.* **2017**, *82*, 594–601. [CrossRef]
17. Benslimane, A.; Bahlouli, I.M.; Bekkour, K.; Hammiche, D. Thermal gelation properties of carboxymethyl cellulose and bentonite-carboxymethyl cellulose dispersions: Rheological considerations. *Appl. Clay Sci.* **2016**, *132*, 702–710. [CrossRef]
18. Sarkar, N. Thermal gelation properties of methyl and hydroxypropyl methylcellulose. *J. Appl. Polym. Sci.* **1979**, *24*, 1073–1087. [CrossRef]
19. Ge, D.; Yang, L.; Li, Y.; Zhao, J. Hydrophobic and thermal insulation properties of silica aerogel/epoxy composite. *J. Non-Cryst. Solids* **2009**, *355*, 2610–2615. [CrossRef]
20. Piñero, M.; del Mar Mesa-Díaz, M.; de los Santos, D.; Reyes-Peces, M.V.; Díaz-Fraile, J.A.; de la Rosa-Fox, N.; Esquivias, L.; Morales-Florez, V. Reinforced silica-carbon nanotube monolithic aerogels synthesised by rapid controlled gelation. *J. Sol-Gel. Sci. Tech.* **2018**, *86*, 391–399. [CrossRef]
21. Wang, J.; Zhang, Y.; Wei, Y.; Zhang, X. Fast and one-pot synthesis of silica aerogels via a quasi-solvent-exchange-free ambient pressure drying process. *Microporous Mesoporous Mater.* **2015**, *218*, 192–198. [CrossRef]
22. Sarawade, P.B.; Kim, J.-K.; Hilonga, A.; Kim, H.T. Production of low-density sodium silicate-based hydrophobic silica aerogel beads by a novel fast gelation process and ambient pressure drying process. *Solid State Sci.* **2010**, *12*, 911–918. [CrossRef]
23. He, S.; Huang, D.; Bi, H.; Li, Z.; Yang, H.; Cheng, X. Synthesis and characterization of silica aerogels dried under ambient pressure bed on water glass. *J. Non-Cryst. Solids* **2015**, *410*, 58–64. [CrossRef]
24. Hamouda, A.A.; Amiri, H.A.A. Factors affecting alkaline sodium silicate gelation for in-depth reservoir profile modification. *Energies* **2014**, *7*, 568–590. [CrossRef]
25. Dorcheh, A.S.; Abbasi, M.H. Silica aerogel; synthesis, properties and characterization. *J. Mater. Process. Tech.* **2008**, *199*, 10–26. [CrossRef]
26. Iswar, S.; Malfait, W.J.; Balog, S.; Winnefeld, F.; Lattuada, M.; Koebel, M.M. Effect of aging on silica aerogel properties. *Microporous Mesoporous Mater.* **2017**, *241*, 293–302. [CrossRef]
27. Auniq, R.B.-Z.; Pakasri, N.; Boonyang, U. Synthesis and in vitro bioactivity of three-dimensionally ordered macroporous-mesoporous bioactive glasses; 45S5 and S53P4. *J. Korean Ceram. Soc.* **2020**, *57*, 305–313. [CrossRef]
28. Pan, Y.; He, S.; Gong, L.; Cheng, X.; Li, C.; Li, Z.; Liu, Z.; Zhang, H. Low thermal-conductivity and high thermal stable silica aerogel based on MTMS/Water-glass co-precursor prepared by freeze drying. *Mater. Des.* **2017**, *113*, 246–253. [CrossRef]
29. Alison, L.; Demirörs, A.F.; Tervoort, E.; Teleki, A.; Vermant, J.; Studart, A.R. Emulsions stabilized by chitosan-modified silica nanoparticles: PH control of structure–Property relations. *Langmuir* **2018**, *21*, 6147–6160. [CrossRef]
30. Shewale, P.M.; Rao, A.V.; Rao, A.P. Effect of different trimethyl silylating agents on the hydrophobic and physical properties of silica aerogels. *Appl. Surf. Sci.* **2008**, *254*, 6902–6907. [CrossRef]
31. He, S.; Huang, Y.; Chen, G.; Feng, M.; Dai, H.; Yuan, B.; Chen, X. Effect of heat treatment on hydrophobic silica aerogel. *J. Hazardou Mater.* **2019**, *362*, 294–302. [CrossRef]
32. Feng, Q.; Chen, K.; Ma, D.; Lin, H.; Liu, Z.; Qin, S.; Luo, Y. Synthesis of high specific surface area silica aerogel from rice husk ash via ambient pressure drying. *Colloids Surf. A* **2018**, *539*, 399–406. [CrossRef]
33. Hwang, H. The development of high performance silica aerogel spheres. Unpublished work, 2019.

gels

MDPI

Article

Chitosan-Silica Hybrid Biomaterials for Bone Tissue Engineering: A Comparative Study of Xerogels and Aerogels

Antonio Pérez-Moreno [1], Manuel Piñero [1,2,*], Rafael Fernández-Montesinos [3,4], Gonzalo Pinaglia-Tobaruela [3,4], María V. Reyes-Peces [1], María del Mar Mesa-Díaz [2,5], José Ignacio Vilches-Pérez [3,4], Luis Esquivias [6], Nicolás de la Rosa-Fox [1,2] and Mercedes Salido [3,4,*]

1 Departamento de Física de la Materia Condensada, Facultad de Ciencias, Universidad de Cádiz, 11510 Puerto Real, Spain; antoniopmoreno@gmail.com (A.P.-M.); maria.reyes@uca.es (M.V.R.-P.); nicolas.rosafox@uca.es (N.d.l.R.-F.)
2 Instituto de Microscopía Electrónica y Materiales (IMEYMAT), Universidad de Cadiz, 11510 Cádiz, Spain; mariadelmar.mesa@uca.es
3 Instituto de Biomedicina de Cádiz (INIBICA), Universidad de Cadiz, 11510 Cádiz, Spain; rafafdezmontesinos@gmail.com (R.F.-M.); gonzalo.pinaglia@uca.es (G.P.-T.); ignacio.vilches@uca.es (J.I.V.-P.)
4 Departamento de Histología, SCIBM, Facultad de Medicina, Universidad de Cádiz, 11004 Cádiz, Spain
5 Departamento de Ingeniería Química, Facultad de Ciencias, Universidad de Cádiz, 11510 Puerto Real, Spain
6 Departamento de Física de la Materia Condensada, Universidad de Sevilla, 41012 Sevilla, Spain; luisesquivias@us.es
* Correspondence: manolo.piniero@gm.uca.es (M.P.); mercedes.salido@uca.es (M.S.)

Abstract: Chitosan (CS) is a natural biopolymer that shows promise as a biomaterial for bone-tissue regeneration. However, because of their limited ability to induce cell differentiation and high degradation rate, among other drawbacks associated with its use, the creation of CS-based biomaterials remains a problem in bone tissue engineering research. Here we aimed to reduce these disadvantages while retaining the benefits of potential CS biomaterial by combining it with silica to provide sufficient additional structural support for bone regeneration. In this work, CS-silica xerogel and aerogel hybrids with 8 wt.% CS content, designated SCS8X and SCS8A, respectively, were prepared by sol-gel method, either by direct solvent evaporation at the atmospheric pressure or by supercritical drying in CO_2, respectively. As reported in previous studies, it was confirmed that both types of mesoporous materials exhibited large surface areas (821 m^2g^{-1}–858 m^2g^{-1}) and outstanding bioactivity, as well as osteoconductive properties. In addition to silica and chitosan, the inclusion of 10 wt.% of tricalcium phosphate (TCP), designated SCS8T10X, was also considered, which stimulates a fast bioactive response of the xerogel surface. The results here obtained also demonstrate that xerogels induced earlier cell differentiation than the aerogels with identical composition. In conclusion, our study shows that the sol-gel synthesis of CS-silica xerogels and aerogels enhances not only their bioactive response, but also osteoconduction and cell differentiation properties. Therefore, these new biomaterials should provide adequate secretion of the osteoid for a fast bone regeneration.

Keywords: xerogel; aerogel; chitosan; bioactivity; osteoconduction; focal adhesion; mechanotransduction; bone healing

Citation: Pérez-Moreno, A.; Piñero, M.; Fernández-Montesinos, R.; Pinaglia-Tobaruela, G.; Reyes-Peces, M.V.; Mesa-Díaz, M.d.M.; Vilches-Pérez, J.I.; Esquivias, L.; de la Rosa-Fox, N.; Salido, M. Chitosan-Silica Hybrid Biomaterials for Bone Tissue Engineering: A Comparative Study of Xerogels and Aerogels. *Gels* **2023**, *9*, 383. https://doi.org/10.3390/gels9050383

Academic Editors: István Lázár and Melita Menelaou

Received: 31 March 2023
Revised: 26 April 2023
Accepted: 29 April 2023
Published: 5 May 2023

1. Introduction

Natural polymers and composites for bone-tissue engineering (BTE) have gained considerable interest in recent years due to their good biocompatibility, biodegradability, and ability to mimic the bone extracellular matrix (ECM) [1–4]. Indeed, by combining biopolymers such as chitosan [5,6], collagen [7], alginate [8], silk fibroin [9], polycaprolactone [10], and gelatin [11,12] with inorganic materials, such as hydroxyapatite [13,14], other types of calcium phosphates [15], or silica hydrogels [16] and aerogels [17,18], hybrid composites with new material properties and applications in various areas of the biomedical field are made possible [4,19,20]. Several investigations on chitosan-based composites for BTE have recently been conducted in order to better understand the function of these novel biomaterials in overcoming the drawbacks of traditional bone-graft treatment. As a result, advancements in chitosan-based scaffolds have produced effective and efficient biological properties through material structural design, primarily through chitosan modification to address its drawbacks, such as poor mechanical properties, bad solubility in water, blood incompatibility and fast biodegradation in the body [21]. For example, CS-based hydrogels have been modified primarily through their main functional groups (OH and NH_2) with the addition of other compounds or biomaterials to achieve not only enhanced mechanical and physical properties through synergistic effects, but also to improve their biological behavior as bone-regeneration biomaterials. The result is that a lot of CS-based materials have been investigated and described in the literature. On this premise, hydrogel nanocomposites consisting of CS and polyhedral oligomeric silsesquioxanes (POSS) have demonstrated excellent in vitro biocompatibility and drug-release capabilities [22]. Additionally, novel hybrid organic–inorganic porous scaffolds with an interconnected pore network, excellent swelling properties, and adjustable degradation rate have been created from chitosan nanofibers modified with silane coupling agents and polyvinyl alcohol (PVA) [23]. In vitro cell culture investigations have also proven its cytocompatibility and prospective application in BTE.

Among these hybrid materials, the chitosan-silica system has received a lot of attention because of its outstanding bone regeneration characteristics, including osteoconductive and drug transport ability, which allows for the regulated release of pharmaceuticals such as growth factors or antibiotics [5,24–26]. As a result, several experiments using chitosan and inorganic silica gels—both with and without chemical alteration of its molecular structure—have been published. In both situations, CS has been demonstrated to be advantageous for BTE, regardless of the drying method (evaporative or supercritical drying) used to create porous CS-silica biomaterials for xerogel or aerogel [27–30]. The resulting specimens have demonstrated great mechanical strength, making them suitable for load-bearing applications. These hybrid biomaterials also have additional advantages for bone tissue repairing, such as increased porosity and specific surface area, which improve cell adhesion and proliferation [16,31].

To date, many methods such as the sol-gel process [32–34], electrospinning [35], spray drying [36], and freeze or supercritical drying [37,38], have been used to produce CS-silica biomaterials with varying microstructure and morphology. Among the numerous methods currently being utilized, the sol-gel method has been extensively used to produce a wide range of organic–inorganic hybrid materials due to its unique benefits, such as chemical homogeneity and purity, as well as the ability to regulate material structure at the nanoscale. So far, silane coupling agents such as aminopropyltriethoxysilane (APTES) or 3-glycidoxypropyltrimethoxysilane (GPTMS), have been used to crosslink the silica and the polymer organic network of chitosan to create mechanically strengthened hybrid composites via sol-gel [28,39]. However, studies have shown that these substances can be hydrolyzed in aqueous solutions, including bodily fluids. Depending on the concentration, this could result in the creation of cytotoxic degradation [40,41]. Incorporating calcium phosphates (CaP) into hybrid biomaterials, on the other hand, can enhance cell compatibility and encourage cell adhesion and proliferation, making these hybrids more appropriate for bone regeneration uses [1,42,43]. Various mesoporous CS–silica hybrids incorporating CaP have been prepared by sol-gel, including xerogels (dried by conventional evaporation of solvents) [27,33] and aerogels (dried in supercritical CO_2) [28] until now. Obviously, the textural properties of these hybrids are dependent on the drying processes performed. Indeed, they present large surface areas and also have a high potential as support matrices for replicating the structure of the native extracellular matrix (ECM) [27,29].

We are interested in these CS–SiO_2 hybrid materials as bone substitutes, because their large surface area provides support for other reactive species. However, we must develop a systematic understanding of how various material substrates may influence the bioactive and biological reactions. Thus, while some information exists about how the CS composition affects the textural properties and osteoconductive role of the final hybrid xerogels [27] and aerogels [29], there is little information about how initial cell differentiation, which is linked to mechanotransduction processes, is affected differently in the presence of these type of biomaterials.

Thus, we have created a CS–silica xerogel and aerogel monoliths using a previously described sol-gel approach [38,39] utilizing TEOS and CS powder as biomaterial precursors. For the study, we selected composition samples with well-known bioactive and osteoconductive characteristics, and we compared their effectiveness as bone-healing materials. In particular, we chose silica hybrid samples with an 8 wt.% CS content and no additional chemical additives or crosslinkers in order to avoid possible cytotoxic side effects. According to our previous results, (Perez-Moreno et al., 2020; Pérez-Moreno et al. 2021) this synthetic procedure guarantees that the resulting biomaterials will present high in vitro bioactivity response as well as osteoconduction behavior in cell culture media [27,29] Additionally, a 10 wt.% TCP was added prior to gelation to the original 8 wt.% CS–silica sol in order to evaluate the effects of CaP on cellular development. Because TCP leaching was observed during the CO_2 supercritical fluid extraction technique to produce the corresponding aerogels, the incorporation of TCP was only taken into account for samples in the xerogel stage. The resulting samples were characterized by thermogravimetric analysis (TGA) and Fourier transform infrared spectroscopy (FTIR). The microstructure and bioactivity were assessed by N_2-physisorption analysis and SEM, and correlated with actin cytoskeletal changes and focal adhesion maturation in order to identify the best platform for bone healing.

2. Results and Discussion

Our goal in this comparative research was to examine the similarities and differences between xerogels and aerogels based on the silica/chitosan hybrid system, with a focus on its use in bone regeneration. This was undertaken from the viewpoint of prior findings [27,29] taking into account the incorporation of TCP in the setting of xerogels. Prior to conventional evaporative drying, the wet gel samples were washed with ethanol solvent for a week to produce the corresponding xerogels. Alternatively, supercritical CO_2 was used in the drying process to produce the equivalent aerogels. The samples used in this study were chosen based on their in vitro biological performance, as well as well-known bioactive responses in simulated body fluid (SBF), and their synthesis process is discussed in more detail in Section 4.1. Aerogel samples were labeled SCS8A (aerogel), SCS8X (xerogel), and SCS8T10X (a xerogel containing 10 wt.% TCP). In addition, xerogel and aerogels of pure silica (SiO_2X and SiO_2A, respectively) were synthesized and included in this study as references. According to earlier research, this variety of compositions offers the materials with enhanced bioactivity as well as the necessary hybrid network stability to avoid undesirable fast biodegradation [27,29]. However, small differences between their biochemical and biophysical responses when exposed to simulated body fluid (SBF) and cell culture, can allow for the best performance of these types of biomaterials to be identified.

2.1. *Physical and Textural Properties of Xerogels and Aerogels*

Table 1 displays the physical properties of the samples under consideration. In all cases, homogeneous and elastic samples in the shape of monolithic cylinders were obtained, with bulk densities ranging from 0.18–0.19 gcm^{-3} in the case of aerogels and from 0.49 to 0.61 gcm^{-3} in xerogels. A significant volume shrinkage of up to 30–33% in aerogels and higher in xerogels (67–76%) was also observed (see Figure 1). Given the high hydrolysis ratio used in the synthesis (Rw = 30), these extremely large shrinkage values were attributed to both drying processes and conventional post-processing [44,45], including polycondensation continuity of the hybrid networks during washing and aging periods, an outcome which was particularly manifest in the samples undergoing solvent evaporation.

Table 1. Bulk density, volume shrinkage, and textural parameters calculated from N_2-physisorption experiments of the xerogel (X) and aerogel (A) samples involved in this study.

Sample	ρ * (gcm^{-3})	Volume Shrinkage * (%)	S_{BET} ** (m^2g^{-1})	V_p (cm^3g^{-1})	Pore Size (nm)
SiO_2A	0.19 ± 0.01	33.0 ± 2.3	978.2	4.1	16.9
SCS8A	0.18 ± 0.03	30.2 ± 3.1	857.7	3.9	17.3
SiO_2X	0.61 ± 0.05	75.7 ± 2.6	807.5	1.0	4.7
SCS8X	0.49 ± 0.10	72.3 ± 4.2	821.1	1.1	5.0
SCS8T10X	0.54 ± 0.03	67.4 ± 2.4	733.6	1.5	7.5

* Errors indicate the standard deviation computed from three replicate measurements. ** Correlation coefficient for BET surface area measurements was higher than 0.9996 in all cases.

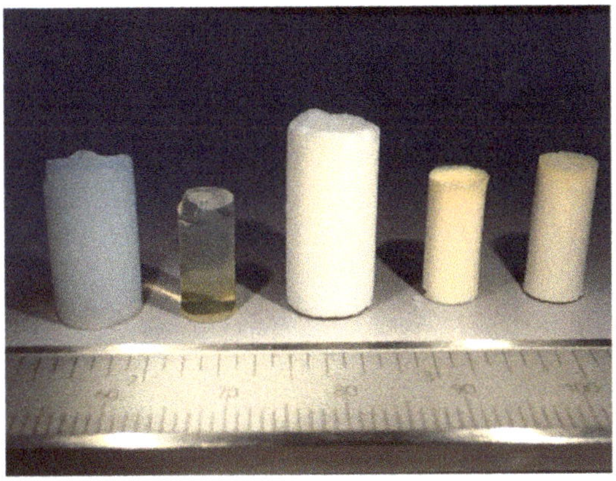

Figure 1. Representative sample set of the studied samples. From left to right: SiO_2A, SiO_2X, SCS8A, SCS8X and SCS8T10X.

Additionally, the textural parameters obtained from N_2-physisorption experiments are shown in Table 1 for the corresponding hybrid materials (SCS8A, SCS8X and SCS8T10X), and their respective pure silica matrices (SiO_2A and SiO_2X). In general, the xerogel samples showed rather high specific surface areas, varying from approximately 734 to 821 m^2g^{-1} for SCS8T10X and SCS8X samples, respectively. They also exhibited relatively low pore volumes of about 1.0 cm^3g^{-1} for both the pure silica xerogel and SCS8X sample. This volume increased by 50% when 10 wt.% TCP was incorporated with the SCS8 sample, allowing a xerogel sample with 1.5 gcm^{-3} of pore volume and also the highest pore size (7.5 nm) of all the three xerogel samples involved.

The before-mentioned results should be related to the presence of a certain amount of Ca^{2+} in dissolution after the TCP addition, providing the formation of calcium phosphate hydrates and resulting in a composite of interest in biomedical applications, particularly for bone-tissue engineering and repair [46,47]. As observed in Table 1, the xerogels displayed 70–75% smaller pore volumes and pore sizes than comparable aerogels, a result that can be attributed to the different drying process mechanisms involved. All these findings are consistent with aerogels of mesopore interconnected structures and high surface areas ranging from 857 m^2g^{-1} to approximately 978 m^2g^{-1} for the SCS8A and SiO_2A samples.

Figure 2a displays the N_2-physisorption isotherms for the pure silica aerogel and the xerogel samples, revealing that the aerogel SiO_2A adsorbs almost four times the volume of gas at high pressures as xerogel SiO_2X, which undergoes stabilization at high relative pressures (0.8–1.0). This is reflected by a reduction in surface area from 978.2 to 807.5 m^2g^{-1} and also pore size from 16.9 to 4.7 nm (Table 1). Additionally, type IV isotherms with H1 hysteresis loops were observed in both cases, indicating the existence of an interconnected mesopore network with widths between 2 and 50 nm [48], generated by the organic and inorganic precursors and the high hydrolysis ratio employed (30:1) [45]. The corresponding N_2-isotherms of the SCS8A, SCS8X, and SCS8T10X hybrid biomaterials are shown in Figure 2b, exhibiting characteristics similar to those of their respective inorganic matrices. Both SCS8X and SCS8T10X xerogel isotherms presented a well-defined horizontal plateau at relative pressures of 0.8–1.0 and 0.9–1.0, and a sharp step in the desorption branch at P/P0 values of 0.5–0.6 and 0.6–0.7, respectively. The shape of these isotherms and hysteresis loops is characteristic of cylindrical pores with a narrow pore size distribution and high pore size uniformity [49]. The pore size distribution (PSD) (see insets in Figure 2a,b) showed that the pore volume of the xerogels was mostly in the mesopore domain, with diameters in the 2–16 nm range, while its maximum was situated between 5.0–7.5 nm. Moreover, the t-plot analysis revealed the absence of micropores in all xerogel and aerogel samples [50]. Besides this, the isotherm curves of the aerogels have a steep adsorption branch, confirming their large surface area, followed by a large desorption hysteresis loop that closes at a lower relative pressure than the adsorption branch, which indicates a broad range of pore sizes from approximately 6 to 130 nm, noticeably invading the macropore region, but with the most abundant value of around 17 nm, in the mesopore domain (see Figure 2a,b inset). A cylindrical-like pore geometry with small openings was also suggested by the loop shape in this case, which could be caused by aggregates or bundles of cylindrical pores or interconnected channels [51].

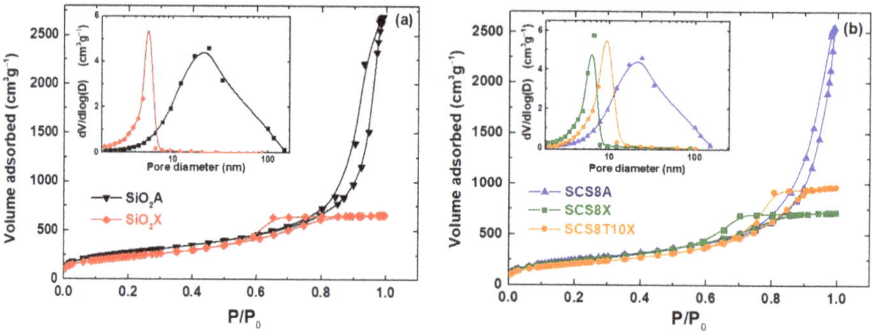

Figure 2. N_2-isotherms and pore size distribution (see insets) of (**a**) pure silica aerogel (SiO_2A) and xerogel (SiO_2X), and (**b**) of the three biomaterials under study (SCS8A, SCS8X and SCS8T10X).

2.2. Thermogravimetric Analysis

The thermal stability and decomposition behavior of the different materials was examined by thermogravimetric analysis. Figure 3a shows the thermogram curves for SiO_2A and SiO_2X silica matrices, while Figure 3b shows the thermogram curves for SCS8A, SCS8X, and SCS8T10X hybrid biomaterials. An initial weight loss in the range of 50–150 °C, accounting for the loss of adsorbed moisture, was observed in all cases (Figure 3a,b), suggesting the typical hydrophilic nature of all surfaces, most notably in the case of the SiO_2A aerogel specimen. The second weight loss occurred in the temperature range 150 °C–450 °C (Figure 3a). This is because of elimination of water molecules, which are produced by the dehydration of the silica network that develops during the hydrolysis and condensation of TEOS, with the silica aerogel losing additional weight. The degradation of any residual organic component contained in the silica aerogel caused a third weight loss that occurred between 450 and 600 °C (Figure 3a). This was mostly due to the decomposition of ethanol and unreacted precursors. Finally, the dehydroxylation of the silica network causes the fourth weight loss, which occurs between 600 and 900 °C, which is attributed to the elimination of hydroxyl groups from the silica network, which causes the creation of a more organized and denser silica structure [29]. Figure 3b shows comparable thermal processes for the SCS8A aerogel and SCS8X and SCS8T10 xerogel samples, including typical weight losses between 150 and

350 °C and 350–600 °C related to the degradation of chitosan [27,52]. In addition, a notable weight loss was observed in the thermogram of SCS8T10X between 350 and 600 °C, which was caused by the decomposition of TCP contained in the hybrid xerogel [27,53].

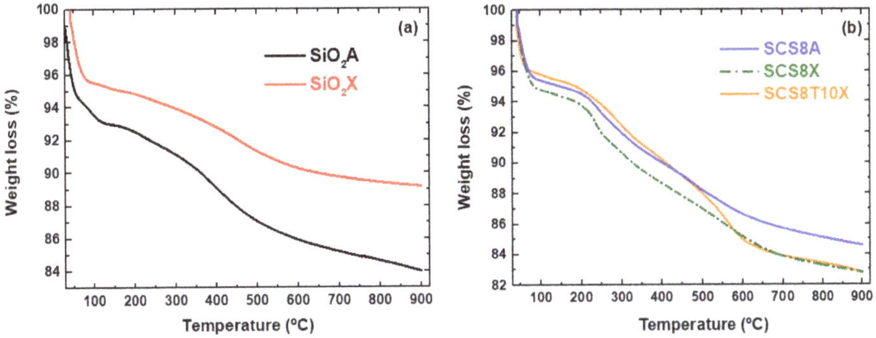

Figure 3. Thermograms of (**a**) (SiO_2A) and (SiO_2X) pure silica samples, and (**b**) of SCS8A, SCS8X, and SCS8T10X biomaterials.

2.3. FTIR Spectral Analysis

A few significant differences were observed in the FTIR spectra of the aerogels and the xerogels (Figure 4). First, aerogels (SiO_2S and SCS8A) often display larger and more intense bands at 450 cm^{-1} and 800 cm^{-1} (Si-O-Si bending vibration), and in the 1050–1200 cm^{-1} range (Si-O-Si stretching vibration), which are characteristics of the silica network [54]. This is because the supercritical drying technique used to create aerogels results in a more densely cross-linked network structure with a greater degree of polymerization, and therefore a more compact pore network structure. A shorter and narrower band, which denotes a lower level of polymerization and a more porous network structure, is generally observed in the same frequency region for the xerogels (SiO_2X, SCS8X, and SCS8T10X). The stretching vibrations of the OH groups, which correspond to the 3200–3700 cm^{-1} range, where aerogels also show more prominent peaks, suggest a larger number of residual silanol groups (-Si-OH) and adsorbed water owing to their typical hydrophilic character. In addition, the peaks at 950 cm^{-1} and at around 2350 cm^{-1} in the SiO_2A sample were attributed to the stretching and bending vibrations of the Si-OH bond, respectively, thus confirming the high -OH density at the aerogel surfaces [29]. Peaks at approximately 2900–3000 cm^{-1} were present only in the aerogels (SiO_2A and SCS8A) accounting for C-H stretching vibrations due to the use of ethanol as s M!"olvent to preserve the samples before supercritical drying [29].

However, the incorporation of chitosan into the hybrid gel structures of SCS8A, SCS8X, and SCS8T10X is difficult to discern, given that the majority of FTIR peaks overlap with those of silica [27]. The presence of chitosan in these hybrids provides the appearance of peaks at approximately 1150 cm^{-1} (stretching vibration of C-O-C), 1400 cm^{-1} (bending vibrations of C-H bond), 1560 cm^{-1}, and 1650 cm^{-1} (bending vibrations of the N-H and C=O groups, respectively). Both bands (amide I and amide II) may also be signals of the presence of chitosan. In addition, the broad peak in the range of 3200–3400 cm^{-1} can indicate stretching vibrations of -OH and N-H in chitosan [30].

Additionally, the FTIR spectrum of SCS8T10X includes a broad peak at approximately 1100–1200 cm^{-1}, which corresponds to the stretching vibration of the phosphate (PO_4) groups in TCP, while bending vibrations may appear between 600–900 cm^{-1}. Finally, a peak at approximately 400–600 cm^{-1} may be indicative of the stretching vibrations of Ca-O bonds in TCP [27].

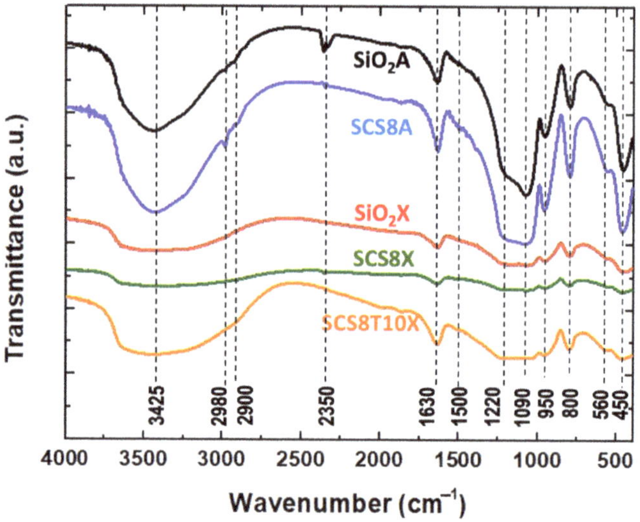

Figure 4. FTIR spectra of aerogel and xerogel pure and hybrid samples containing 8 wt.% chitosan and 10 wt.% TCP.

2.4. In Vitro Bioactivity in SBF

Figure 5 shows SEM micrographs of the hybrids after they were tested for bioactivity in SBF. The interaction of calcium and phosphate ions from the surrounding media formed an apatite-like structure on the surfaces of all samples after immersion for 28 days in SBF. As previously shown, the textural characteristics of the gels, along with the composition of the mineralized products, both affected the morphology and particle size of the HAp aggregates deposited on the three different surfaces [11,27,29]. As an illustration, Figure 3a,b show how the aerogel surface of SCS8A enables the formation of HAp spherulitic particles, which develop into an abundant accumulation with a nearly ideal spheric shape of uniform size (~10 μm). Instead, the corresponding xerogel (SCS8X) allows for the formation of HAp spherulites, which appear as agglomerates of varying sizes (~4–8 μm) and cover the surface to a lesser extent than the aerogel (see Figure 3c,d). Finally, as shown in Figure 3e,f, the TCP-containing xerogel sample (SCS8T10X) promoted the formation of an apatite layer that covered almost the entire surface of the substrate and had particle sizes of about 5 μm.

These results show the potential for the proposed biomimetic mineralization method to produce materials recovered by hydroxyapatite layers in all cases, with particle sizes ranging from 5 to 10 mm and a Ca/P ratio that is nearly stoichiometric (1.67) as described in previous work [27,29], with biocompatibility properties, which may be useful for a variety of biomedical applications as will be discussed below.

2.5. Osteoblast Behavior

The barrier that exists between a host and any implanted device that forms at the cell–biomaterial interface is considerably more than just a simple border; instead, it provides key cues for cell adhesion, subsequent induction, and tissue neogenesis as described by Biggs et al. [55]. A construct's function and cytocompatibility can be evaluated in vitro by examining cell viability and adherence at the substratum interface. The initial polarization of HOB cells was evident from 24 h onwards, and then followed by cell-adhesion protein expression and cell modifications identified as early osteoblast differentiation markers, which can be attributed to the biomaterial, as we and others have previously described [28,56–60].

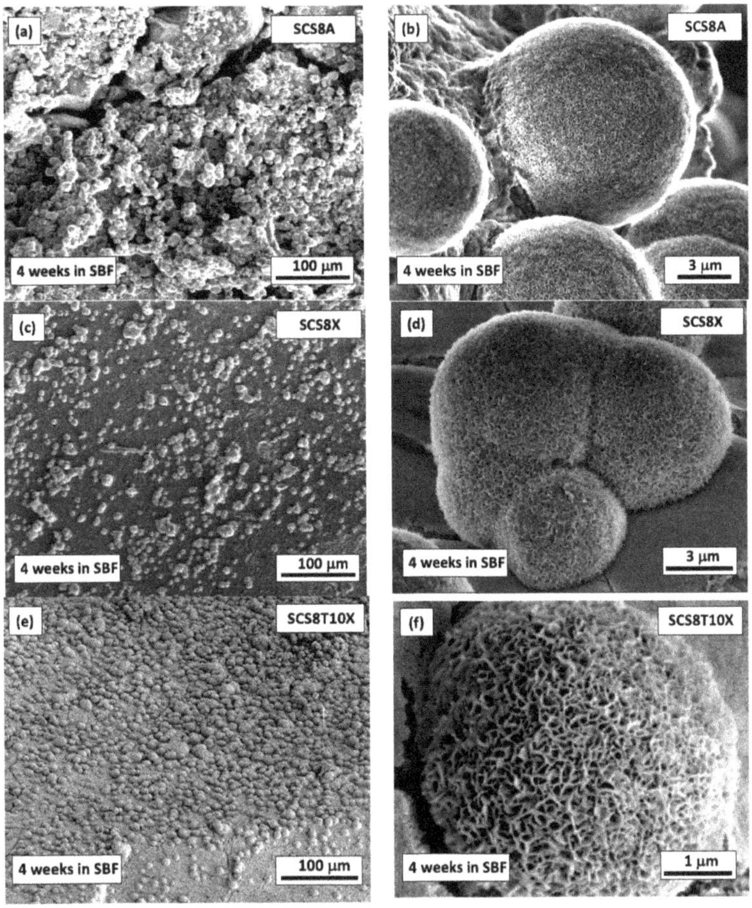

Figure 5. SEM micrographs of the surfaces of (**a**,**b**) SCS8A, (**c**,**d**) SCS8X, and (**c**,**f**) SCS8T10X hybrid gels after immersion in SBF for 28 days at 37 °C displaying strong bioactive response through the formation of a hydroxyapatite (HAp) layer.

At seeding, cell viability reached 98%. No appreciable apoptotic events were observed in either the experimental or control groups. The live dead assay was used in experimental groups to evaluate cell growth and viability, with a majority of osteoblasts, at any experimental time, were in a viable state (green), with only a small number of dead cells (red) (see Figure 6a,b).

In the first 48 and 72 hours of cell cultures, some of the materials cause variations in the cell proliferation of osteoblasts, although after one week of cell culture there is no difference between the materials, even when compared with the positive control. Therefore, the materials do not appear to affect the optimal growth of osteoblasts. (See Figure 7).

2.6. Cell Morphology, Cytoskeletal Organization, and Focal Adhesions

The importance and development of biochemical and biophysical features regulating cell–biomaterial interactions in the early stages are of capital importance in orchestrating the complex material–cytoskeleton crosstalk occurring at the interface. Although the precise underlying biomolecular mechanisms are not well defined, there is growing evidence that cytoskeleton-mediated signaling could prove to be sufficient to start and sustain differentiation programs [55,60]. In this milieu, focal adhesions (FAs) have a pivotal function in this process as a mechanical link between the cytoskeleton and the extracellular environment. They change in size, maturation stages, and in distribution according to the forces acting on them, as well as cytoskeleton changes according to FAs spatial distribution and size [61–64]. Two distinct forms of FAs have been identified that differ depending on cellular motility. In cells that retain their migratory capacity, small FAs are predominant, mostly composed of vinculin and talin, and appear on the leading edges of filopodia or lamellipodia.

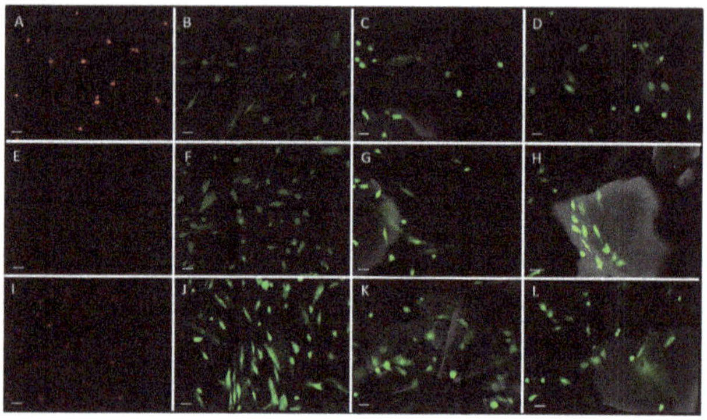

(**a**)

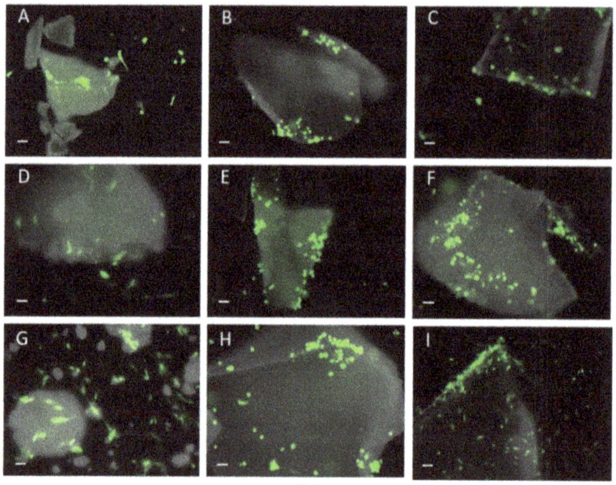

(**b**)

Figure 6. (**a**) Live/dead staining of HOB culture. Negative controls treated with 70% methanol for 30 min prior to staining: (**A**) 48 h HOB cultures, (**E**) 72 h HOB cultures and (**I**) one-week HOB cultures. The cultures of live control osteoblasts (positive control) were untreated cells grown underoptimal conditions at the described timepoints. Positive controls: (**B**) 48 h HOB cultures, (**F**) 72 h HOB cultures and (**J**) one-week HOB cultures. HOB cultures with SiO_2A: (**C**) 48 h HOB cultures, (**G**) 72 h HOB cultures and (**K**) one-week HOB cultures. HOB cultures with SiO_2X: (**D**) 48 h HOB cultures, (**H**) 72 h HOB cultures and (**L**) one-week HOB cultures. Green: living cells; red: dead cells and materials in gray. Scale: 20 µm. (**b**) Live/dead staining of HOB cells grown in the presence of SCS8A: (**A**) 48 h HOB cultures, (**D**) 72 h HOB cultures and (**G**) one-week HOB cultures. HOB cultures with SCS8X: (**B**) 48 h HOB cultures, (**E**) 72 h HOB cultures and (**H**) one-week HOB cultures. HOB cultures with SCS8T10X: (**C**) 48 h HOB cultures, (**F**) 72 h HOB cultures and (**I**) one-week HOB cultures. Live cells flouresce green; dead cells are imaged in red and materials in gray. Scale bar: 20 µm.

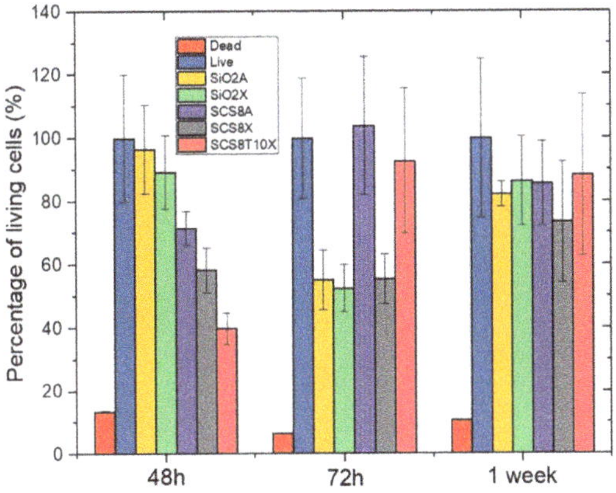

Figure 7. Histogram representations as mean ± SEM.of the percentage of living HOB cells. Negative controls (dead l) were treated for 30 min before labelling with 70% methanol. Live represent the reference value i.e., 100% of living cells obtained from untreated cultures grown under optimal conditions at the experimental times. Quantification analyzed in 10 images per experiment, in triplicate.

These are also called immature punctate nanoscale (0.2 μm^2) FAs and can undergo rapid turnover within the lamellipodium or mature into larger FAs influenced by cytoskeletal tension and motion state of the cell. Mature FAs are large (1.0–10 mm^2) and predominantly composed of paxillin, vinculin, focal adhesion kinases and α actinin. During the migration phase they anchor cells, to maintain cellular morphology and tensional homeostasis. In resting and non-migrating osteoblasts, they can be localized or even formed throughout the whole cytoplasm [11,65–67].

After osteoblast immunolabelling with rhodamine phalloidin and antivinculin antibodies, we observed focal adhesion development and cytoskeletal changes in both the aerogel and xerogel groups. Once the distribution of vinculin-positive events was assessed, we used image analysis to measure the size of FAs in the proposed silica–chitosan biomaterials (Figure 8).

Our findings showed that predominantly small- and medium-sized FAs, as a sign of cell migration, were identified after 48 h in culture in SCS8X and SCS8T10X with accompanying changes in the cytoskeleton. In the silica–chitosan xerogel groups (Figure 8), the development of stress fibers was observed from 48 h, increased with time, and were arranged in the cell periphery and tipped with FAs. After 1 week (Figure 8 I–L), actin stress fibers appear clearly organized on the entire mobile surface of the cell and reinforced by a large number of FAs on the cell's leading edges

Comparative images of osteoblasts cultured in the presence of SiO_2X or SiO_2A samples are shown in Figure 9 and revealed quite a different pattern both for focal adhesion distribution and also for cytoskeletal changes. Scarce and poorly developed FAs appeared in the first 48 and 72 h, and even after one week, especially in the SiO_2A groups, which can be associated with cells that are still in the migratory phase. In the SiO_2X groups, a greater differentiation of actin filaments towards stress fibers was observed after 1 week in culture, although at all experimental times, the cellular response was lower than that observed in the hybrid groups.

According to our data and the literature, osteoblasts grown in the presence of SCS8A seem to maintain their migration capability, according to focal adhesion patterns with an elongated morphology over time. In the aerogel groups (Figure 10A–C), the appearance of stress fibers occurred somewhat later, being already evident after 72 h, in which they presented a peripheral distribution associated with lamellipodia and filopodia and were very evident after one week of culture. Finally, in the control groups on glass (Figure 10D–F), elongated and narrow cells with few filopodia, some lamellipodia, few non-polarized FAs, and a cytoskeleton pattern with no obvious stress fibers were observed.

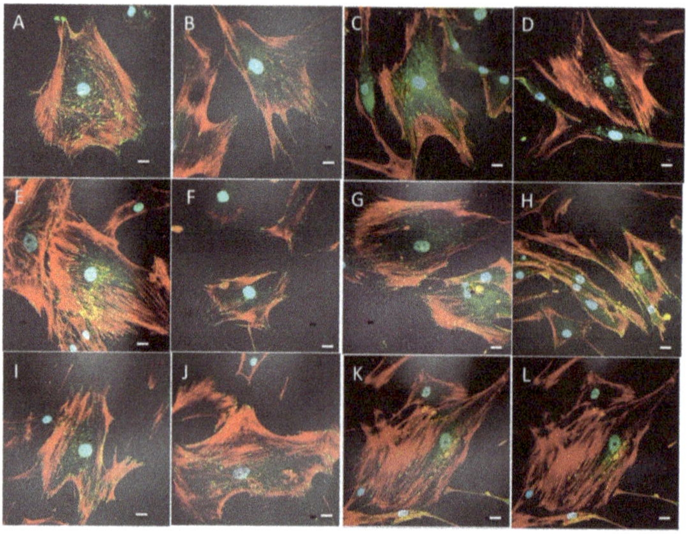

Figure 8. Immunolabelling and confocal examination of actin cytoskeleton with rhodamine phalloidin (red) and vinculin for focal adhesions (green) of HOB® osteoblasts cultured in the presence of: (**A**,**B**) xerogel SCS8X; (**C**,**D**) SCS8T10X, 48 hours; (**E**,**F**) xerogel SCS8X after 72 h,(**G**,**H**) SCS8T10X after 72 h in culture; (**I**,**J**) xerogel SCS8X, 1 week, and (**K**,**L**) SCS8T10X after 1 week in culture. Blue, DAPI-labelled nuclei. Scale bar: 20 µm.

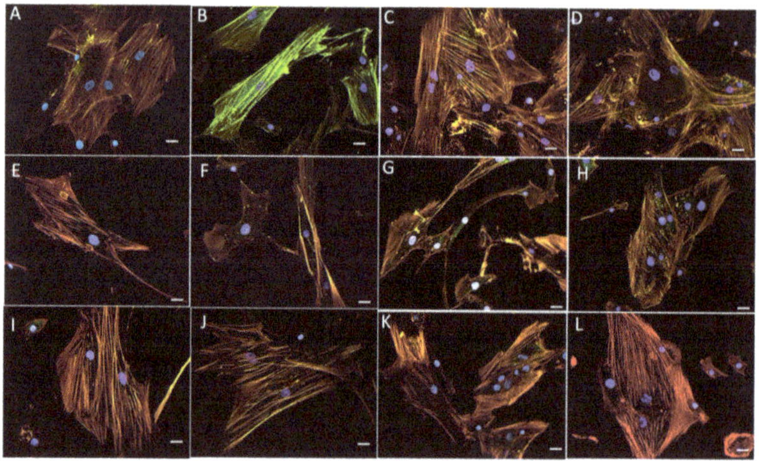

Figure 9. Immunolabelling with rhodamine phalloidin and confocal examination of actin cytoskeleton (red) and vinculin (green) for FAs of HOB® osteoblasts grown in the presence of: (**A**,**B**) SiO_2A, 48 h culture; (**C**,**D**) SiO_2X, 48 h culture; (**E**,**F**) SiO_2A 72 h culture; (**G**,**H**) SiO_2X 72 h culture; (**I**,**J**) SiO_2A 1 week culture; (**K**,**L**) SiO_2X after 1 week in culture. Blue, DAPI-labelled nuclei. Scale bar: 20 µm.

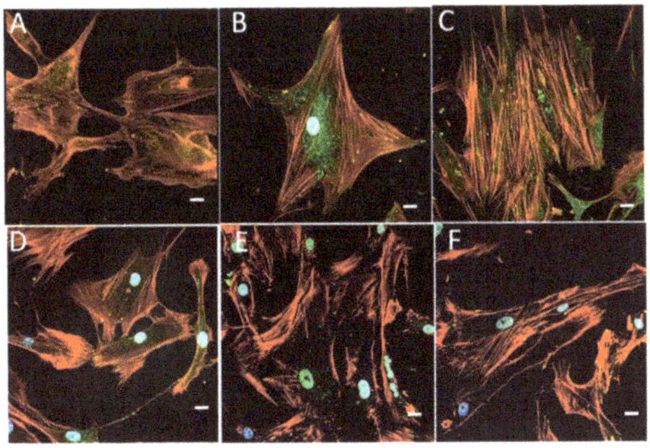

Figure 10. Immunolabelling and confocal examination of actin cytoskeleton with rhodamine phalloidin (red) and vinculin (green) for FAs of HOB® osteoblasts growing in culture in the presence of: SCS8A (**A**), 48 h culture; (**B**) 72 h culture; (**C**) 1 week culture and osteoblasts growing in glass (controls) (**D**), 48 h culture; (**E**) 72 h culture; (**F**) 1 week culture. Blue, DAPI-labelled nuclei. Scale bar: 20 μm.

Image analysis confirmed the data described above, with predominantly small- and medium-sized FAs (Figure 11), as expression of cell migration, after 48 h in culture in SCS8X and SCS8T10X. Concordantly, minor morphological changes were identified in these groups (Figure 12).

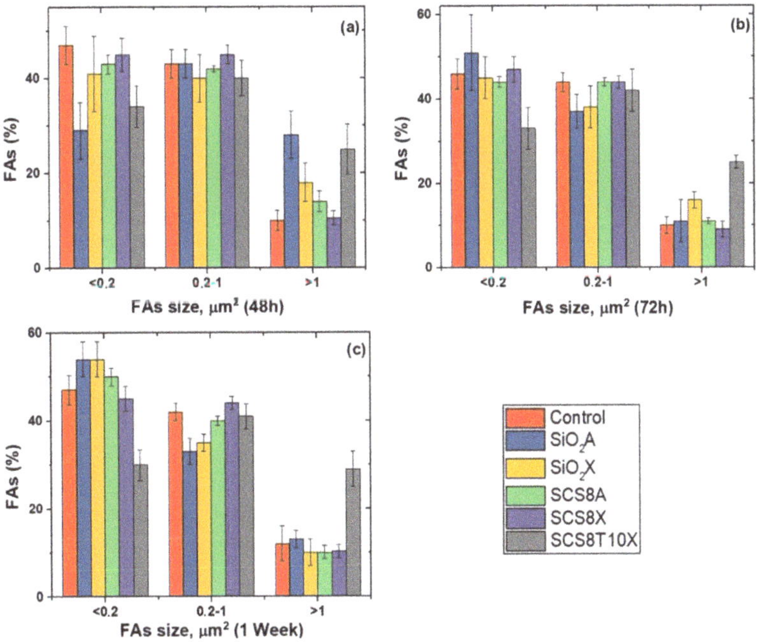

Figure 11. FA size in HOB® cultured on the xerogels. Time-dependent percentage after (**a**) 48 h, (**b**) 72 h, (**c**) 1 week. One way analysis of variance. Statistical significance was defined as $p < 0.05$.

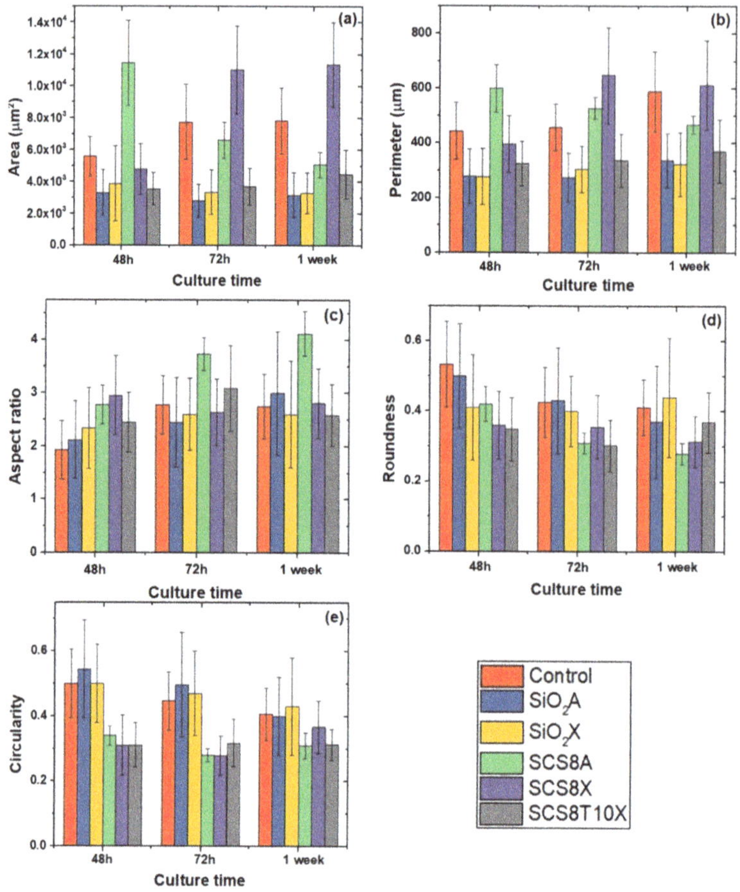

Figure 12. Quantification of shape variables. (**a**) Cell spread area, (**b**) cell aspect ratio, (**c**) circularity, (**d**) cell perimeter, and (**e**) cell roundness. One way ANOVA. Significance: $p < 0.05$.

In SiO_2A groups, as well as in SiO_2X groups, small FAs predominated and the percentage increased after 72 h and 1 week, while medium- and large-sized FAs percentages decreased with time (Figure 11). This tendency persisted for 72 h and 1 week in culture. Nevertheless, in the SCS8T10 X groups, the percentage of mature FAs, sized > 1 square micron, remain stable for 48 h onwards and significantly increased with respect to the rest of the experimental groups after one week of culture. In control cells, small- and medium-sized FAs predominate at any experimental times.

Morphological analysis of shape parameters revealed changes in cell area, mainly in SCS8A groups, which were the largest at initial times and elongated with time, as confirmed by the perimeter and aspect ratio data shown in Figure 12a–c. In contrast, cells grown on SCS8X increased both in area and perimeter, and consequently in aspect ratio, from 48 h to 72 h and finally 1 week. These data were consistent with FA expression, with small- and medium-sized FAs during all experimental times. Osteoblasts grown in the control groups and in biomaterials composed only of SiO_2 presented a less complex cell morphology with the highest circularity values.

Bone tissue capability of self-regeneration in response to a variety of diseases and injuries is mainly mediated by osteoblasts, cells from the mesenchymal lineage that can divide and generate both collagen fibers and extracellular bone matrix (EBM) components. During both osteogenesis and regeneration processes, osteoblasts differentiate into a second type of bony cell called osteocytes, which are in the final differentiation stage of the osteoblastic lineage. During the process, osteoblast morphology changes, and long branched osteocytic processes emerge from a smaller cell body, while the extracellular matrix osteoid is formed. Osteocytic channels within progressively calcified EBM are then created, while osteocytic processes grow to maintain cells in contact with nearby capillaries, osteocytes, and border cells. Osteocytic processes also play an important role in mechanotransduction events by keeping the cytoskeleton in deep contact with focal adhesion contacts, thus allowing mechanical forces from the surrounding environment to modulate cell differentiation [66,68].

Proteins, primarily serum vitronectin and fibronectin, begin to cling to the biomaterial surface when exposed to a supplemented cell culture medium in vitro. Cell membrane integrins can recognize specific binding sites on these proteins. Following this, via linker proteins such as vinculin, the integrins assemble into what are known as FA complexes and bind to F-actin. Osteogenic cells have a highly developed actin cytoskeleton, and it is well known that mechanical stress in their microenvironment plays a role in determining their development [56–58,69].

The role of FAs during initial cell adhesion couples their role as mechanosensors for biophysical and biochemical properties of the material with continuous assembly and disassembly as the cell moves, leading to focal adhesion changes in number and size, as described in the results section. Concordant cytoskeletal changes in actin networks leading to stress fiber development and changes in cell morphology are observed when osteoblasts are in the presence of an optimal scaffold [70].

We previously demonstrated that FAs are involved in the surface recognition process, which leads to cell adhesion and mechanotransduction-induced alterations in osteoblasts. Additional cues for cell placement, spreading, and differentiation can be achieved when the nanoscale biomaterial surface is chemically optimal. As we and others have previously described, chitosan is a natural polymer that offers adequate biocompatibility, biodegradability, hydrophilicity, and nontoxicity.

3. Conclusions

Nanoscale chitosan-based biomaterial surfaces described in the paper offer an adequate set of characteristics for BTE. In addition, they are chemically attractive and provide additional cues for cell positioning, spreading, and differentiation. The best osteoblastic response was observed in the presence of SCS8T10X with indicators that would hopefully favor adequate secretion of osteoid. SCS8A and SCS8X samples required more time to induce an osteoblastic differentiation response comparable to that of TCP-enriched samples, and osteoblasts grown on these samples retained a good migration capacity for quite some time. All xerogel samples appeared to induce cell differentiation earlier than the aerogels with identical compositions.

Finally, the data obtained for the samples composed only of SiO_2 showed the least influence on the differentiation indicators used in the time periods studied, and the cellular response was much slower. We can conclude that nanoscale chitosan-based biomaterial surfaces offer an adequate set of characteristics, such as biocompatibility, biodegradability, hydrophilicity, and non-toxicity. In addition, they are chemically appealing, providing additional cues for cell placement, osteoblast spreading, and cell differentiation, which can be improved by the addition of chitosan, TCP, or both. Therefore, the adequacy of the different parameters influencing synthesis (including drying strategies) must be considered to optimize the osteoblastic response.

4. Materials and Methods

4.1. Materials

Chitosan (CS; 50,000–190,000 Da; 75–85% deacetylation degree) was provided by Sigma Aldrich (St. Louis, MI, USA). Tetraethyl-orthosilicate (TEOS, 99%) and hydrochloric acid (37%, Pharma grade) were supplied for Alfa Aesar (Haverhill, MA, USA). Tri-calcium phosphate (TCP, pure, pharma grade) and absolute ethanol (99.5%) were obtained from Panreac (Barcelona, Spain), HOB® human osteoblasts, fetal calf serum, and osteoblast growing medium (Promocell, Heidelberg, Germany) paraformaldehyde, PBS, Triton x-100, bovine serum albumin, methanol, rhodamine phalloidin, and monoclonal anti-vinculin FITC conjugate were all purchased from Sigma Aldrich, (St. Louis, MI, USA) along with Hard Set Vectashield with DAPI ® (Vector, Burlingame, CA, USA).

4.2. Gel Synthesis

Chitosan–silica hybrid gels were synthesized using the sol-gel technique based on previously reported procedures [27,29] and schematized in Figure 13. First, silica sol was made by combining TEOS, water, and HCl in a molar ratio of 1:4:0.05, while exposed to 4.5 $kJcm^{-3}$ of ultrasound energy in a glass reactor. A CS solution was prepared separately by adding 2.5 g of low molecular weight CS (50,000–190,000 Da) and 0.6M HCl to a 250 mL solution. A homogeneous mixture was obtained under the ultrasound exposure with an energy of 10 $kJcm^{-3}$. The final solution was then obtained by combining both sols and adding an extra dose of ultrasound (0.5 $kJcm^{-3}$), resulting in totally transparent solutions. All samples included 8% chitosan and the water/TEOS molar ratio was maintained at 30:1. TCP was then added to the TEOS/CS sol and completely dissolved to produce silica-8 wt.% CS, 10 wt.% TCP sols that were kept at 50 °C on a stove to gel for 24–48 h. The hybrid samples were then aged in an ethanol bath for 28 days to strengthen them and achieve the greatest shrinkage before drying. Finally, residual water in the hybrid hydrogel was removed after another week of soaking in ethanol with daily exchange. SCS8X and SCS8T10X xerogels were thus obtained by direct liquid evaporation at 50 °C, while supercritical drying in CO_2 was employed to produce SCS8A aerogels. Samples were found in both cases as fracture-free monoliths with cylindrical geometry and well-differentiated volume contraction (see Figure 11).

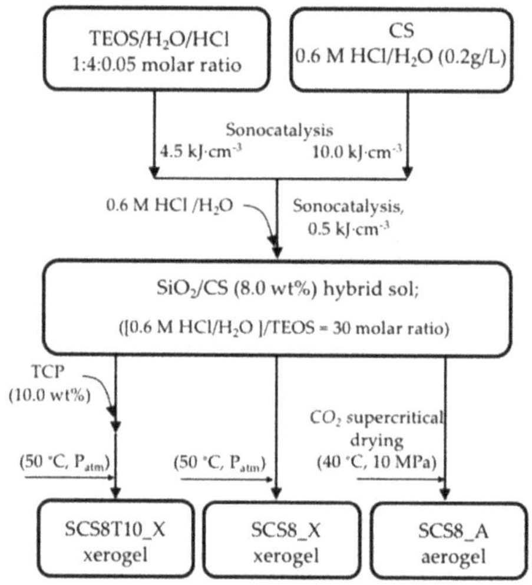

Figure 13. Process flow schematic for the production of SiO_2/CS aerogels and SiO_2/CS/TCP xerogels.

According to our previous results (Perez-Moreno et al., 2020; Pérez-Moreno et al., 2021), the above synthetic procedure ensures that the resulting biomaterials exhibit high in vitro bioactivity responses as well as osteoconduction behavior in cell culture media [27,29]. In addition, the inclusion of 10 wt.% TCP has impact on the production of osteoinductive surfaces [27,71]. However, this TCP incorporation was unsuccessful for aerogels, because total TCP leaching was observed during the CO_2 supercritical procedure.

4.3. Physical Characterization and Structural Properties

The mass of the samples was determined using a microbalance with an accuracy of 0.1 mg, and the sizes of monolithic specimens were measured by a Vernier caliper. Nitrogen physisorption tests (Micromeritics ASAP2010, Norcross, GA, USA) at 77 K with a pressure transducer resolution of 10^{-4} mm Hg were used to examine the textural characteristics of the hybrid aerogels. In order to conduct the study, specific surface area, pore volume, and pore size distribution were calculated using the BET and BJH standard models. The samples were degassed at 120 °C for 4 h before the tests.

4.4. Thermal Characterization

The thermal analysis of the materials was performed by thermogravimetric analysis (TGA) by measuring weight changes in the temperature range 50–900 °C in an air environment at a constant heating rate of 10 °C/min with a TGA Discovery instrument (TA Instruments, Ndw Castle, DE, USA).

4.5. Fourier Transform Infrared Spectroscopy

Fourier-transform infrared spectroscopy (FTIR) was employed to obtain information about the surface chemical structure of the samples. Experiments were recorded with a Bruker Alpha System Spectrophotometer (KBr wafer technique), using the same quantity of sample in all measurements. Signals were obtained in transmittance mode and the spectral range was set at 4000 to 500 cm^{-1} with a resolution of 4 cm^{-1}.

4.6. Evaluation of the Bioactivity in SBF

Aerogel pellets measuring 5 mm in length and 8 mm in diameter were submerged in 20 mL of simulated body fluid (SBF) [61] in polyethylene flasks to study bioactivity. Hydroxyapatite (HAp) was then measured at the top after 28 days of soaking at 37 °C. The samples were meticulously removed from the buffer solution on a weekly basis, washed with Milli-Q water (Millipore Sigma, Burlington, MA, USA), and then dried once more at 50 °C and ambient pressure. The test was conducted with weekly fluid exchange. The surface morphologies of the SBF-treated samples were examined using an FEI Nova NanoSEM 450 (FEI, Morristown, NJ, USA) with a precision of 1.4 nm after various soaking times. Additionally, using a Bruker SDD-EDS analyzer, the Ca/P composition of the specimen surface was determined.

4.7. Cell Culture

Under sterile conditions, HOB cells were seeded onto the scaffolds. Once at optimal confluence, osteoblasts were extracted and counted to determine the optimal seeding density, and osteoblasts viability was determined using an automatic Luna® cell counter (Invitrogen, San Diego, CA 92121 USA). The cell population doubling did not surpass ten. Both aerogels and xerogels were sterilized in a clinically standardized autoclave in accordance with the European standard DIN EN ISO 13,060 recommendations for class B autoclaves. After sterilization, the samples were placed in a laminar flow chamber in Mattek® glass bottom wells under sterile conditions. Samples with a surface area of 1 square cm were used. A drop of 50 µL of cell suspension at a density of 15,000 HOB® cells /cm^2 was added to each sample and kept for 30 min, to avoid dispersion and ensure optimal cell attachment, in humid conditions and under incubation at 37 °C and 5% CO_2. Following this, wells were refilled with OGM® supplemented to a final concentration of 0.1 mL/mL of fetal calf serum and incubated during experimental times. Every two days, the media were replaced, and the degradation products were determined. The test groups were: SCS8A, SCS8X, SCS8T10X. Control groups consisted of HOB® cells grown on glass.

4.8. Live/Dead Cell Assay

The cell viability and cytotoxicity of were assessed using a live/dead cell assay at experimental times of 24 h, 48 h, and 7 days. The samples were rinsed twice with PBS and exposed to calcein-AM (0.5 μL/mL) in PBS and ethidium homodimer-1 (EthD-1) (2 μL/mL) diluted in PBS. A confocal laser scanning microscope (CSLM) was employed to identify live and dead cells.

4.9. Cell Morphology and Spreading

Cell changes in morphology, alignment and spreading were assessed using a phase-contrast microscope prior to immunolabelling for fluorescence and confocal laser scanning microscopy. Both the fluorescence and confocal modes were combined, with the Nomarski mode for both material and cell imaging.

4.10. Actin Cytoskeletal Organization

Osteoblasts were immunolabeled with rhodamine-phalloidin and anti-vinculin antibodies after 48 h, 72 h, and 7 days of incubation. After washing with pH 7.4 prewarmed phosphate-buffered saline (PBS), samples were fixed with 3.7% paraformaldehyde at RT and permeabilized with 0.1% Triton x-100 after washing. Finally, preincubation with 1% bovine serum albumin in PBS for 20 min, after PBS rinsing, was performed in order to remove background prior to cell immunolabelling with rhodamine phalloidin and then rinsed with prewarmed PBS prior to mounting with Vectashield ®. At least five samples of each type were seeded and analyzed for each experiment. All experiments were repeated in triplicates unless otherwise stated. The test groups were: SCS8A, SCS8X, SCS8T10X. HOB cells grown on glass under the conditions described above were used as controls.

4.11. Confocal Examination

At least five samples were analyzed for each group using an Olympus confocal microscope to assess the influence of the surface on focal adhesion number and development, cytoskeletal organization, and cell morphology. At least 50 cells per sample were analyzed. Samples were exposed to the lowest laser power necessary to generate a fluorescent signal for a time interval not higher than 5 min to avoid photobleaching. Images were acquired at a resolution of 1024 × 1024, and processed.

4.12. Image Analysis

Sample images were collected as frames obtained at 40x magnification and processed using Image J software (NIH, http://rsb.info.nih.gov/ij (accessed on 26 June 2021). The perimeter, area, circularity, roundness, and aspect ratio were analyzed as shape variables. At least 40 regions of interest (ROIs) were measured for quantitative analysis. ROIs were selected under the following criteria: well-defined limits, clear identification of the nucleus, and absence of intersection with neighboring cells. Data were analyzed using SPSS and expressed as the mean ± standard deviation. Once normality and homoscedasticity were confirmed, the difference between the mean values was analyzed using one-way analysis of variance and the Brown–Forsythe and Games–Howell tests. Statistical significance was defined as $p < 0.05$.

Author Contributions: Conceptualization, A.P.-M., M.P. and M.S.; methodology, A.P.-M., M.P., R.F.-M., G.P.-T., M.V.R.-P., M.d.M.M.-D. and M.S.; software, R.F.-M., G.P.-T. and A.P.-M.; validation, R.F.-M., M.V.R.-P., L.E. and M.S.; formal analysis, A.P.-M., R.F.-M., M.d.M.M.-D. and M.S.; investigation, A.P.-M., M.P., R.F.-M., M.V.R.-P., J.I.V.-P., M.d.M.M.-D., L.E. and M.S.; resources, J.I.V.-P., M.d.M.M.-D., N.d.l.R.-F. and M.S.; data curation, A.P.-M., R.F.-M., G.P.-T., M.P. and M.S.; writing—original draft preparation, A.P.-M., M.P. and M.S.; writing—review and editing, A.P.-M., M.S., N.d.l.R.-F. and M.P.; visualization, A.P.-M., M.P., N.d.l.R.-F. and M.S.; supervision, M.S., M.P., L.E. and N.d.l.R.-F.; project administration, M.P. and M.S.; funding acquisition, M.S. All authors have read and agreed to the published version of the manuscript.

Funding: This research was funded by Andalucía FEDER /ITI 2014-2020 Grant for PI 013/017, Junta de Andalucía TEP115, and CTS 253 PAIDI research groups (Spain). The work has also been co-financed by the Junta de Andalucía, FEDER-UCA18_106598.

Institutional Review Board Statement: Not applicable.

Informed Consent Statement: Not applicable.

Data Availability Statement: Not applicable.

Acknowledgments: Authors acknowledge the use of instrumentation as well as the technical advice provided by SCCYT (UCA) for SEM divisions at University of Cadiz. The authors would also like to thank J. Vilches-Troya, retired Professor of Histology and Pathology of the University of Cadiz, for his expert advice and supervision.

Conflicts of Interest: The authors declare no conflict of interest.

References

1. Jeong, J.E.; Park, S.Y.; Shin, J.Y.; Seok, J.M.; Byun, J.H.; Oh, S.H.; Kim, W.D.; Lee, J.H.; Park, W.H.; Park, S.A. 3D Printing of Bone-Mimetic Scaffold Composed of Gelatin/β-Tri-Calcium Phosphate for Bone Tissue Engineering. *Macromol. Biosci.* **2020**, *20*, 2000256. [CrossRef] [PubMed]
2. Fernandez-Yague, M.A.; Abbah, S.A.; McNamara, L.; Zeugolis, D.I.; Pandit, A.; Biggs, M.J. Biomimetic approaches in bone tissue engineering: Integrating biological and physicomechanical strategies. *Adv. Drug Deliv. Rev.* **2015**, *84*, 1–29. [CrossRef] [PubMed]
3. Veres, P.; Király, G.; Nagy, G.; Lázár, I.; Fábián, I.; Kalmár, J. Biocompatible silica-gelatin hybrid aerogels covalently labeled with fluorescein. *J. Non-Cryst. Solids* **2017**, *473*, 17–25. [CrossRef]
4. Rezwan, K.; Chen, Q.Z.; Blaker, J.J.; Bocaccini, A.R. Biodegradable and bioactive porous polymer/inorganic composite scaffolds for bone tissue engineering. *Biomaterials* **2006**, *27*, 3413–3431. [CrossRef] [PubMed]
5. Wang, D.; Romer, F.; Connell, L.; Walter, C.; Saiz, E.; Yue, S.; Lee, P.D.; McPhail, D.S.; Hanna, J.V.; Jones, J.R. Highly flexible silica/chitosan hybrid scaffolds with oriented pores for tissue regeneration. *J. Mater. Chem. B* **2015**, *3*, 7560–7576. [CrossRef] [PubMed]
6. Jerome, C.; Croisier, F. Chitosan-based biomaterials for tissue engineering. *Eur. Polym. J.* **2013**, *49*, 780–792. [CrossRef]
7. Kechagias, S.; Moschogiannaki, F.; Stratakis, E.; Tzeranis, D.S.; Vosniakos, G.C. Porous collagen scaffold micro-fabrication: Feature-based process planning for computer numerically controlled laser systems. *Int. J. Adv. Manuf. Technol.* **2020**, *111*, 749–763. [CrossRef]
8. Lucía, T.; Hern, A.C.; Rodríguez-lorenzo, L.M. Preparation of covalently bonded silica-alginate hybrid hydrogels by SCHIFF base and sol-gel reactions. *Carbohydr. Polym.* **2021**, *267*, 118186. [CrossRef]
9. Wu, J.; Zheng, K.; Huang, X.; Liu, J.; Liu, H.; Boccaccini, A.R.; Wan, Y.; Guo, X.; Shao, Z. Thermally triggered injectable chitosan/silk fibroin/bioactive glass nanoparticle hydrogels for in-situ bone formation in rat calvarial bone defects. *Acta Biomater.* **2019**, *91*, 60–71. [CrossRef]
10. Hollister, S.J. Porous scaffold design for tissue engineering. *Nat. Mater.* **2005**, *4*, 518–524. [CrossRef]
11. Reyes-Peces, M.V.; Fernández-Montesinos, R.; del Mar Mesa-Díaz, M.; Vilches-Pérez, J.I.; Cárdenas-Leal, J.L.; de la Rosa-Fox, N.; Salido, M.; Piñero, M. Structure-Related Mechanical Properties and Bioactivity of Silica–Gelatin Hybrid Aerogels for Bone Regeneration. *Gels* **2023**, *9*, 67. [CrossRef] [PubMed]
12. Mahony, O.; Tsigkou, O., Ionescu, C.; Minelli, C.; Ling, L.; Hanly, R.; Smith, M.E.; Stevens, M.M.; Jones, J.R. Silica-gelatin hybrids with tailorable degradation and mechanical properties for tissue regeneration. *Adv. Funct. Mater.* **2010**, *20*, 3835–3845. [CrossRef]
13. Dong, Y.; Liang, J.; Cui, Y.; Xu, S.; Zhao, N. Fabrication of novel bioactive hydroxyapatite-chitosan-silica hybrid scaffolds: Combined the sol-gel method with 3D plotting technique. *Carbohydr. Polym.* **2018**, *197*, 183–193. [CrossRef]
14. Martínez-vázquez, F.J.; Cabañas, M.V.; Paris, J.L.; Lozano, D.; Vallet-regí, M. Fabrication of novel Si-doped hydroxyapatite/gelatine scaffolds by rapid prototyping for drug delivery and bone regeneration. *Acta Biomater.* **2015**, *15*, 200–209. [CrossRef] [PubMed]
15. Heinemann, S.; Heinemann, C.; Wenisch, S.; Alt, V.; Worch, H.; Hanke, T. Calcium phosphate phases integrated in silica/collagen nanocomposite xerogels enhance the bioactivity and ultimately manipulate the osteoblast/osteoclast ratio in a human co-culture model. *Acta Biomater.* **2013**, *9*, 4878–4888. [CrossRef] [PubMed]
16. Jayash, S.N.; Cooper, P.R.; Shelton, R.M.; Kuehne, S.A.; Poologasundarampillai, G. Novel chitosan-silica hybrid hydrogels for cell encapsulation and drug delivery. *Int. J. Mol. Sci.* **2021**, *22*, 12267. [CrossRef] [PubMed]
17. Demilecamps, A.; Beauger, C.; Hildenbrand, C.; Rigacci, A.; Budtova, T. Cellulose-silica aerogels. *Carbohydr. Polym.* **2015**, *122*, 293–300. [CrossRef]
18. García-González, C.A.; Budtova, T.; Dur, L.; Erkey, C.; Del Gaudio, P.; Gurikov, P.; Koebel, M.; Liebner, F.; Neagu, M.; Smirnova, I. An Opinion Paper on Aerogels for Biomedical and Environmental Applications. *Molecules* **2019**, *24*, 1815. [CrossRef]
19. Bharadwaz, A.; Jayasuriya, A.C. Recent Trends in the Application of Widely Used Natural and Synthetic Polymer Nanocomposites in Bone Tissue Regeneration. *Mater. Sci. Eng. C* **2021**, *110*, 110698. [CrossRef]
20. Roseti, L.; Parisi, V.; Petretta, M.; Cavallo, C.; Desando, G.; Bartolotti, I.; Grigolo, B. Scaffolds for Bone Tissue Engineering: State of the art and new perspectives. *Mater. Sci. Eng. C* **2017**, *78*, 1246–1262. [CrossRef]
21. Zhu, Y.; Zhang, Y.; Zhou, Y. Application Progress of Modified Chitosan and Its Composite Biomaterials for Bone Tissue Engineering. *Int. J. Mol. Sci.* **2022**, *23*, 6574. [CrossRef] [PubMed]

22. Celesti, C.; Iannazzo, D.; Espro, C.; Visco, A.; Legnani, L.; Veltri, L.; Visalli, G.; Di Pietro, A.; Bottino, P.; Chiacchio, M.A. Chitosan/POSS Hybrid Hydrogels for Bone Tissue Engineering. *Materials* **2022**, *15*, 8208. [CrossRef] [PubMed]
23. Ma, W.; Zhang, S.; Xie, C.; Wan, X.; Li, X.; Chen, K.; Zhao, G. Preparation of High Mechanical Strength Chitosan Nanofiber/NanoSiO$_2$/PVA Composite Scaffolds for Bone Tissue Engineering Using Sol–Gel Method. *Polymers* **2022**, *14*, 2083. [CrossRef] [PubMed]
24. Seo, S.; Kim, J.; Kim, J.; Lee, J.; Sang, U.; Lee, E.; Kim, H. Enhanced mechanical properties and bone bioactivity of chitosan/silica membrane by functionalized-carbon nanotube incorporation. *Compos. Sci. Technol.* **2014**, *96*, 31–37. [CrossRef]
25. Logithkumar, R.; Keshavnarayan, A.; Dhivya, S.; Chawla, A.; Saravanan, S.; Selvamurugan, N. A review of chitosan and its derivatives in bone tissue engineering. *Carbohydr. Polym.* **2016**, *151*, 172–188. [CrossRef]
26. Rinki, K.; Dutta, P.K.; Hunt, A.J.; MacQuarrie, D.J.; Clark, J.H. Chitosan aerogels exhibiting high surface area for biomedical application: Preparation, characterization, and antibacterial study. *Int. J. Polym. Mater. Polym. Biomater.* **2011**, *60*, 988–999. [CrossRef]
27. Perez_Moreno, A.; Reyes-Peces, M.V.; Vilches-Pérez, J.I.; Fernández-Montesinos, R.; Pinaglia-Tobaruela, G.; Salido, M.; de la Rosa-fox, N.; Piñero, M. Effect of Washing Treatment on the Textural Properties and Bioactivity of Silica/Chitosan/TCP Xerogels for Bone Regeneration. *Int. J. Mol. Sci.* **2021**, *22*, 8321. [CrossRef]
28. Reyes-Peces, M.V.; Pérez-Moreno, A.; De-los-Santos, D.M.; del Mar Mesa-Díaz, M.; Pinaglia-Tobaruela, G.; Vilches-Pérez, J.I.; Fernández-Montesinos, R.; Salido, M.; de la Rosa-Fox, N.; Piñero, M. Chitosan-GPTMS-silica hybrid mesoporous aerogels for bone tissue engineering. *Polymer* **2020**, *12*, 2723. [CrossRef]
29. Perez-Moreno, A.; Reyes-Peces, M.V.; de los Santos, D.M.; Pinaglia-Tobaruela, G.; de la Orden, E.; Vilches-Pérez, J.I.; Salido, M.; Piñero, M.; de la Rosa-Fox, N. Hydroxyl groups induce bioactivity in silica/chitosan aerogels designed for bone tissue engineering. In vitro model for the assessment of osteoblasts behavior. *Polymers* **2020**, *12*, 2802. [CrossRef]
30. Trujillo, S.; Pérez-Román, E.; Kyritsis, A.; Gõmez Ribelles, J.L.; Pandis, C. Organic-inorganic bonding in chitosan-silica hybrid networks: Physical properties. *J. Polym. Sci. Part B Polym. Phys.* **2015**, *53*, 1391–1400. [CrossRef]
31. Palla-rubio, B.; Araújo-gomes, N.; Fernández-gutiérrez, M.; Rojo, L.; Suay, J.; Gurruchaga, M. Synthesis and characterization of silica-chitosan hybrid materials as antibacterial coatings for titanium implants. *Carbohydr. Polym.* **2019**, *203*, 331–341. [CrossRef] [PubMed]
32. Pipattanawarothai, A.; Suksai, C.; Srisook, K.; Trakulsujaritchok, T. Non-cytotoxic hybrid bioscaffolds of chitosan-silica: Sol-gel synthesis, characterization and proposed application. *Carbohydr. Polym.* **2017**, *178*, 190–199. [CrossRef] [PubMed]
33. Budnyak, T.M.; Pylypchuk, I.V.; Tertykh, V.A.; Yanovska, E.S.; Kolodynska, D. Synthesis and adsorption properties of chitosan-silica nanocomposite prepared by sol-gel method. *Nanoscale Res.Lett.* **2015**, *10*, 87–96. [CrossRef] [PubMed]
34. Alvarez Echazú, M.I.; Renou, S.J.; Alvarez, G.S.; Desimone, M.F.; Olmedo, D.G. Synthesis and Evaluation of a Chitosan–Silica-Based Bone Substitute for Tissue Engineering. *Int. J. Mol. Sci.* **2022**, *23*, 13379. [CrossRef] [PubMed]
35. Toskas, G.; Cherif, C.; Hund, R.D.; Laourine, E.; Mahltig, B.; Fahmi, A.; Heinemann, C.; Hanke, T. Chitosan(PEO)/silica hybrid nanofibers as a potential biomaterial for bone regeneration. *Carbohydr. Polym.* **2013**, *94*, 713–722. [CrossRef]
36. da Costa Neto, B.P.; da Mata, A.L.M.L.; Lopes, M.V.; Rossi-Bergmann, B.; Ré, M.I. Preparation and evaluation of chitosan-hydrophobic silica composite microspheres: Role of hydrophobic silica in modifying their properties. *Powder Technol.* **2014**, *255*, 109–119. [CrossRef]
37. Connell, L.S.; Romer, F.; Suárez, M.; Valliant, E.M.; Zhang, Z.; Lee, P.D.; Smith, M.E.; Hanna, J.V.; Jones, J.R. Chemical characterisation and fabrication of chitosan-silica hybrid scaffolds with 3-glycidoxypropyl trimethoxysilane. *J. Mater. Chem. B* **2014**, *2*, 668–680. [CrossRef]
38. Ayers, M.R.; Hunt, A.J. Synthesis and properties of chitosan-silica hybrid aerogels. *J. Non-Cryst. Solids* **2001**, *285*, 123–127. [CrossRef]
39. Connell, L.S.; Gabrielli, L.; Mahony, O.; Russo, L.; Cipolla, L.; Jones, J.R. Functionalizing natural polymers with alkoxysilane coupling agents: Reacting 3-glycidoxypropyl trimethoxysilane with poly(γ-glutamic acid) and gelatin. *Polym. Chem.* **2017**, *8*, 1095–1103. [CrossRef]
40. Balavigneswaran, C.K.; Venkatesan, R.; Karuppiah, P.S.; Kumar, G.; Paliwal, P.; Krishnamurthy, S.; Kadalmani, B.; Mahto, S.K.; Misra, N. Silica release from silane cross-linked gelatin based hybrid scaffold affects cell proliferation. *ACS Appl. Bio Mater.* **2020**, *3*, 197–207. [CrossRef]
41. Fuentes, C.; Ruiz-Rico, M.; Fuentes, A.; Barat, J.M.; Ruiz, M.J. Comparative cytotoxic study of silica materials functionalised with essential oil components in HepG2 cells. *Food Chem. Toxicol.* **2021**, *147*, 111858. [CrossRef] [PubMed]
42. Houmard, M.; Fu, Q.; Genet, M.; Saiz, E.; Tomsia, A.P. On the structural, mechanical, and biodegradation properties of HA/b -TCP robocast scaffolds. *J. Biomed. Mater. Res. Part B* **2013**, *101*, 1233–1242. [CrossRef] [PubMed]
43. Cao, H.; Kuboyama, N. A biodegradable porous composite scaffold of PGA/β-TCP for bone tissue engineering. *Bone* **2010**, *46*, 386–395. [CrossRef]
44. Zhao, X.; Wang, Y.; Luo, J.; Wang, P.; Xiao, P.; Jiang, B. The Influence of Water Content on the Growth of the Hybrid-Silica Particles by Sol-Gel Method. *Silicon* **2021**, *13*, 3413–3421. [CrossRef]
45. Buckley, A.M.; Greenblatt, M. A comparison of the microstructural properties of silica aerogels and xerogels. *J. Non-Cryst. Solids* **1992**, *143*, 1–13. [CrossRef]

46. Canillas, M.; Pena, P.; De Aza, A.H.; Rodríguez, M.A. Calcium phosphates for biomedical applications. *Bol. Soc. Esp. Ceram. Vidr.* **2017**, *56*, 91–112. [CrossRef]
47. Safronova, T.V.; Selezneva, I.I.; Tikhonova, S.A.; Kiselev, A.S.; Davydova, G.A.; Shatalova, T.B.; Larionov, D.S.; Rau, J.V. Biocompatibility of biphasic α,β-tricalcium phosphate ceramics in vitro. *Bioact. Mater.* **2020**, *5*, 423–427. [CrossRef]
48. Gregg, S.J.; Sing, K.S.W. *Adsorption, Surface Area and Porosity*, 2nd ed.; Gregg, S.J., Ed.; Academic Press: London, UK, 1982.
49. Sing, K. Reporting physisorption data for gas/solid systems with Special Reference to the Determination of Surface Area and Porosity. *Pure Appl. Chem.* **1982**, *54*, 2201–2218. [CrossRef]
50. Galarneau, A.; Mehlhorn, D.; Guenneau, F.; Coasne, B.; Villemot, F.; Minoux, D.; Aquino, C.; Dath, J. Specific Surface Area Determination for Microporous/Mesoporous Materials: The Case of Mesoporous FAU-Y Zeolites. *Langmuir* **2018**, *34*, 14134–14142. [CrossRef]
51. Sing, K.S.W.; Williams, R.T. Physisorption hysteresis loops and the characterization of nanoporous materials. *Adsorpt. Sci. Technol.* **2004**, *22*, 773–782. [CrossRef]
52. Smitha, S.; Shajesh, P.; Mukundan, P.; Warrier, K.G.K. Sol-gel synthesis of biocompatible silica-chitosan hybrids and hydrophobic coatings. *J. Mater. Res.* **2008**, *23*, 2053–2060. [CrossRef]
53. Jinlong, N.; Zhenxi, Z.; Dazong, J. Investigation of Phase Evolution During the Thermochemical Synthesis of Tricalcium Phosphate. *J. Mater. Synth. Process.* **2001**, *9*, 235–240. [CrossRef]
54. Venkateswara Rao, A.; Kalesh, R.R. Comparative studies of the physical and hydrophobic properties of TEOS based silica aerogels using different co-precursors. *Sci. Technol. Adv. Mater.* **2003**, *4*, 509–515. [CrossRef]
55. Biggs, M.J.P.; Dalby, M.J. Focal adhesions in osteoneogenesis. *Proc. Inst. Mech. Eng. H* **2010**, *224*, 1441–1453. [CrossRef] [PubMed]
56. Bačáková, L.; Filová, E.; Rypáček, F.; Švorčík, V.; Starý, V. Cell Adhesion on Artificial Materials for Tissue Engineering. *Physiol. Res.* **2004**, *53*, S35–S45. [CrossRef] [PubMed]
57. Terriza, A.; Vilches-Pérez, J.I.; González-Caballero, J.L.; de la Orden, E.; Yubero, F.; Barranco, A.; Gonzalez-Elipe, A.R.; Vilches, J.; Salido, M. Osteoblasts Interaction with PLGA Membranes Functionalized with Titanium Film Nanolayer by PECVD. *Materials* **2014**, *7*, 1687–1708. [CrossRef] [PubMed]
58. Coyer, S.R.; Singh, A.; Dumbauld, D.W.; Calderwood, D.A.; Craig, S.W.; Delamarche, E.; García, A.J. Nanopatterning reveals an ECM area threshold for focal adhesion assembly and force transmission that is regulated by integrin activation and cytoskeleton tension. *J. Cell Sci.* **2012**, *125*, 5110–5123. [CrossRef]
59. Bays, J.L.; DeMali, K.A. Vinculin in cell–cell and cell–matrix adhesions. *Cell. Mol. Life Sci.* **2017**, *74*, 2999–3009. [CrossRef]
60. Natale, C.F.; Ventre, M.; Netti, P.A. Tuning the material-cytoskeleton crosstalk via nanocon fi nement of focal adhesions. *Biomaterials* **2014**, *35*, 2743–2751. [CrossRef]
61. Salido, M.; Vilches, J.I.; Gutiérrez, J.L.; Vilches, J. Actin cytoskeletal organization in human osteoblasts grown on different dental titanium implant surfaces. *Histol. Histopathol.* **2007**, *22*, 1355–1364. [CrossRef]
62. Salido, M.; Vilches-perez, J.I.; Gonzalez, J.L.; Vilches, J. Mitochondrial bioenergetics and distribution in living human osteoblasts grown on implant surfaces. *Histol. Histopathol.* **2009**, *24*, 1275–1286. [PubMed]
63. Lamers, E.; van Horssen, R.; te Riet, J.; van Delft, F.C.M.J.M.; Luttge, R.; Walboomers, X.F.; Jansen, J.A. The influence of nanoscale topographical cues on initial osteoblast morphology and migration. *Eur. Cells Mater.* **2010**, *20*, 329–343. [CrossRef] [PubMed]
64. Jonathan, M.; Biggs, P.; Richards, R.G.; Dalby, M.J. Nanotopographical modification: A regulator of cellular function through focal adhesions. *Nanomedicine* **2010**, *6*, 619–633. [CrossRef]
65. Vallet-Regí, M.; Feito, M.J.; Arcos, D.; Oñaderra, M., Matesanz, M.C., Martínez-Vázquez, F.J., Sánchez-Salcedo, S.; Portolés, M.T.; Linares, J. Response of osteoblasts and preosteoblasts to calcium deficient and Si substituted hydroxyapatites treated at different temperatures. *Colloids Surf. B Biointerfaces* **2015**, *133*, 304–313. [CrossRef]
66. Owens, G.J.; Singh, R.K.; Foroutan, F.; Alqaysi, M.; Han, C.M.; Mahapatra, C.; Kim, H.W.; Knowles, J.C. Sol-gel based materials for biomedical applications. *Prog. Mater. Sci.* **2016**, *77*, 1–79. [CrossRef]
67. Serra, I.R.; Fradique, R.; Vallejo, M.C.S.; Correia, T.R.; Miguel, S.P.; Correia, I.J. Production and characterization of chitosan/gelatin/β-TCP scaffolds for improved bone tissue regeneration. *Mater. Sci. Eng. C* **2015**, *55*, 592–604. [CrossRef]
68. Arcos, D.; Vallet-regí, M. Acta Biomaterialia Sol-gel silica-based biomaterials and bone tissue regeneration. *Acta Biomater.* **2010**, *6*, 2874–2888. [CrossRef]
69. Bittig, A.T.; Matschegewski, C.; Nebe, J.B.; Stählke, S.; Uhrmacher, A.M. Membrane related dynamics and the formation of actin in cells growing on micro-topographies: A spatial computational model. *BNC Syst. Biol.* **2014**, *8*, 106–125. [CrossRef]
70. Mullen, C.A.; Vaughan, T.J.; Voisin, M.C.; Brennan, M.A.; Layrolle, P.; McNamara, L.M. Cell morphology and focal adhesion location alters internal cell stress. *J. R. Soc. Interface* **2014**, *11*, 1–12. [CrossRef]
71. Tang, Z.; Li, X.; Tan, Y.; Fan, H.; Zhang, X. The material and biological characteristics of osteoinductive calcium phosphate ceramics. *Regen. Biomater.* **2018**, *5*, 43–59. [CrossRef]

gels

Article

Robust Silica-Bacterial Cellulose Composite Aerogel Fibers for Thermal Insulation Textile

Huazheng Sai [1,2,3,*], Meijuan Wang [1,2,3], Changqing Miao [1,2,3], Qiqi Song [1,2,3], Yutong Wang [1,2,3], Rui Fu [1,2,3,*], Yaxiong Wang [1,2,3], Litong Ma [1,2,3] and Yan Hao [1,2,3]

1 School of Chemistry and Chemical Engineering, Inner Mongolia University of Science & Technology, Baotou 014010, China; wmjbest1014@163.com (M.W.); qingmc@163.com (C.M.); songqiqiaa@163.com (Q.S.); wangyut@163.com (Y.W.); wangyaxiong2021@126.com (Y.W.); mlt0916@126.com (L.M.); haoyannk@163.com (Y.H.)
2 Inner Mongolia Engineering Research Center of Comprehensive Utilization of Bio-Coal Chemical Industry, Inner Mongolia University of Science & Technology, Baotou 014010, China
3 Inner Mongolia Key Laboratory of Coal Chemical Engineering & Comprehensive Utilization, Inner Mongolia University of Science & Technology, Baotou 014010, China
* Correspondence: shz15@tsinghua.org.cn (H.S.); furui14@mails.ucas.edu.cn (R.F.)

Abstract: Aerogels are nanoporous materials with excellent properties, especially super thermal insulation. However, owing to their serious high brittleness, the macroscopic forms of aerogels are not sufficiently rich for the application in some fields, such as thermal insulation clothing fabric. Recently, freeze spinning and wet spinning have been attempted for the synthesis of aerogel fibers. In this study, robust fibrous silica-bacterial cellulose (BC) composite aerogels with high performance were synthesized in a novel way. Silica sol was diffused into a fiber-like matrix, which was obtained by cutting the BC hydrogel and followed by secondary shaping to form a composite wet gel fiber with a nanoscale interpenetrating network structure. The tensile strength of the resulting aerogel fibers reached up to 5.4 MPa because the quantity of BC nanofibers in the unit volume of the matrix was improved significantly by the secondary shaping process. In addition, the composite aerogel fibers had a high specific area (up to 606.9 m^2/g), low density (less than 0.164 g/cm^3), and outstanding hydrophobicity. Most notably, they exhibited excellent thermal insulation performance in high-temperature (210 °C) or low-temperature (−72 °C) environments. Moreover, the thermal stability of CAFs (decomposition temperature was about 330 °C) was higher than that of natural polymer fiber. A novel method was proposed herein to prepare aerogel fibers with excellent performance to meet the requirements of wearable applications.

Keywords: fibrous aerogel; nanoscale interpenetrating network; secondary shaping; strength; thermal properties

Citation: Sai, H.; Wang, M.; Miao, C.; Song, Q.; Wang, Y.; Fu, R.; Wang, Y.; Ma, L.; Hao, Y. Robust Silica-Bacterial Cellulose Composite Aerogel Fibers for Thermal Insulation Textile. *Gels* **2021**, *7*, 145. https://doi.org/10.3390/gels7030145

Academic Editors: István Lázár and Melita Menelaou

Received: 13 August 2021
Accepted: 14 September 2021
Published: 17 September 2021

1. Introduction

Aerogels are materials with excellent features, such as large specific surface area (500–1200 m^2/g), high porosity (80–99.8%), and low density (0.003–0.5 g/cm^3) [1], which make them readily applicable in adsorption [2,3], heat preservation [4,5], and catalysis [6,7]. However, as ultra-porous materials, aerogels are often highly brittle, especially in the case of a three-dimensional gel skeleton composed of nanoparticles, e.g., silica and other inorganic oxide aerogels [8]. This poor mechanical property is mainly due to the very small connection area between the nanoparticles that make up the gel skeleton [9]. Hence, the utilization and promotion of aerogels are severely restricted by their low mechanical strength.

Researchers worldwide have mainly focused on improving the mechanical properties of aerogel materials through precursor regulation and external doping. Methods of regulating gel precursors include increasing the quantity of precursors [10], using precursors containing inert groups [11–13], using precursors with flexible chains [14–16], and using precursors containing polymer monomers to construct bimolecular chain cross-linking network structures [17,18] to reduce the content of rigid –Si–O–Si–, giving the gel skeleton excellent flexibility to resist the impact of external forces. The external doping strategy includes a mixed extrusion of aerogel powder and long fiber [19], dispersion fiber doping [20], reinforcement by fiber felt [21,22] and isometric growth of polymers on the aerogel skeleton [23,24] to connect different areas of the skeleton or expend the connection area of the nanoparticles. These two strategies improve the mechanical properties to a great extent, but they are mainly applied to aerogel blocks, sheets, and blankets.

At present, the excellent thermal insulation performance of aerogel materials makes them suitable for application in thermal insulation clothing and other fabrics, which would require the processing and manufacture of aerogels as high-strength fibers. Aerogel fibers not only enrich the morphology of aerogels but also expand the applicability of aerogels in different fields and contexts. In addition to thermal insulation applications [25,26], aerogel fibers could also be used in other fields, such as adsorption [27–29], biological sensing [30], and supercapacitors [31]. As a result of these potential advantages, it is essential to develop different strategies to obtain high-strength aerogel fibers.

Freeze spinning and wet spinning are the most commonly reported methods used to prepare aerogel fibers [32]. The raw materials of the aerogel fibers are typically synthetic polymers [33], natural polymers [34], graphene oxide [35], or their composites [36]. Although some studies have reported on inorganic oxide (e.g., SiO_2) aerogels, their tensile strength generally does not exceed 0.5 MPa [27,37], which is much lower than that of the previously mentioned fibrous aerogels. This is because there are significant differences between their microstructures. As a result of the extremely limited particle connection area, the "pearl-necklace" network formed by inorganic oxide nanoparticles is weaker than the gel skeleton composite of nanofiber-like or nanosheet-like nanostructure units connected to each other [38]. Unlike direct freeze spinning or wet spinning, a hollow polymer fiber is prepared by coaxial wet spinning, and then silk fibroin, graphene oxide, etc. are poured into it. Subsequently, the fiber being freeze dried to construct a gel skeleton was achieved recently [26,36]. The tensile strength of the obtained sheath–core coaxial aerogel fibers is significantly improved compared with that of fibrous aerogels without an outer polymer layer. This preparation process, which uses coaxial wet spinning to construct the hollow polymer fiber as a protective layer, is complicated and time consuming. Therefore, it is critical to continuously explore new strategies to prepare high-strength aerogel fibers, especially inorganic oxide-based aerogel fibers with excellent thermal insulation properties.

In this paper, a novel method that does not use spinning is proposed for fabricating aerogel fibers with a nanoscale interpenetrating network (as shown in Figure 1), which is a structure that has been proven to significantly improve the mechanical properties of aerogel blocks or films. Bacterial cellulose (BC) hydrogel was processed into a long fiber as the matrix, and then, the silica sol was diffused into the matrix. After secondary shaping to regulate the matrix morphology at both macroscopic and microscopic levels, the nanoscale interpenetrating network structure was obtained by an in situ sol-gel reaction. Finally, silica-BC composite aerogel fibers (CAFs) were obtained by hydrophobic modification and atmospheric pressure drying. The secondary shaping process, which increases the content of BC nanofibers per unit volume of the matrix, combined with the excellent mechanical properties of BC [39], significantly improves the tensile strength of the silica aerogel fibers.

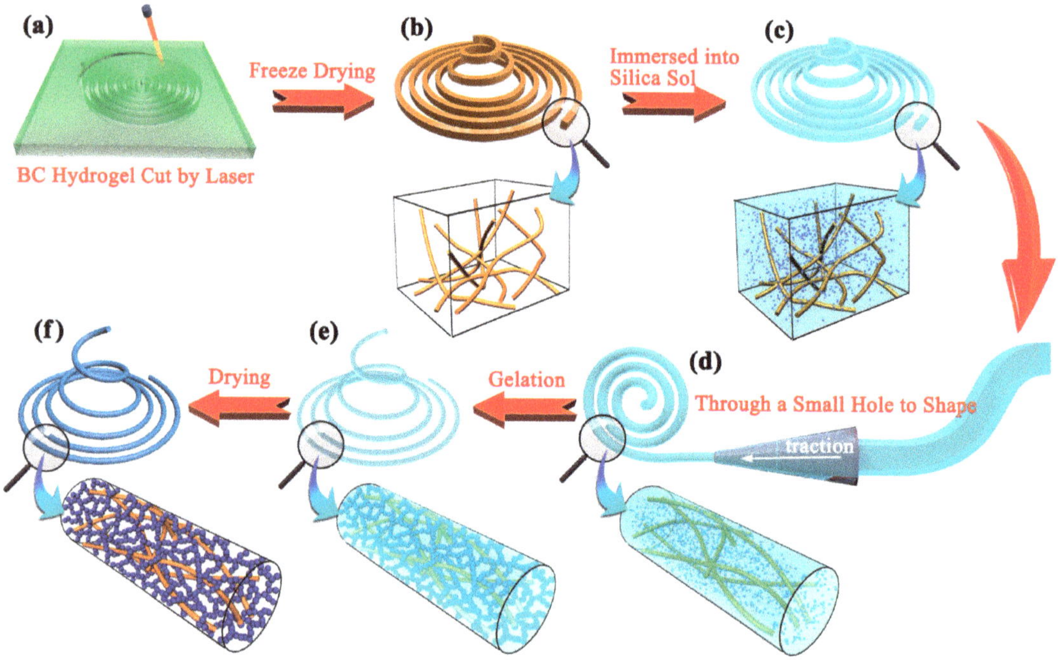

Figure 1. Schematic of the preparation process of CAFs. A *nata-de-coco* slice was cut by a laser (**a**), followed by freeze drying to obtain a fiber-like matrix consisting of BC nanofibers (**b**). The matrix containing silica sols (**c**) was reshaped by a small hole mold (**d**). After the silica gel skeleton was formed in the matrix (**e**), the fiber-like composite wet gel was dried at ambient pressure after hydrophobization to obtain CAFs (**f**).

2. Results and Discussion

2.1. Synthesis of CAFs and Their Macroscopic Characteristics

The detailed preparation method was described in the section "Materials and Methods part (Section 4). The diameter of the obtained CAFs was approximately 0.7 mm, as obtained from scanning electron microscopy (SEM) images; this diameter was close to the inner diameter of the mold. This means that the sample was able to almost completely spring back in the process of atmospheric drying owing to the effective hydrophobic modification. The hydroxyl groups (including –C–OH and –Si–OH) on the surface of BC nanofibers and silica gel skeleton were replaced with the inert and hydrophobic methyl groups through the hydrophobic modification process, which inhibited the formation of new –Si–O–Si– on the surface of the gel skeleton, allowing it to spring back during the drying process. Moreover, the hydrophobic modification also endowed the CAFs with excellent hydrophobicity. As shown in Figure 2 and Video S1, the water droplet maintained its spherical shape even when the fiber was pushed against the droplet. The excellent hydrophobicity could prevent the nanoporous structure of the CAFs from being destroyed by water vapor or water droplets.

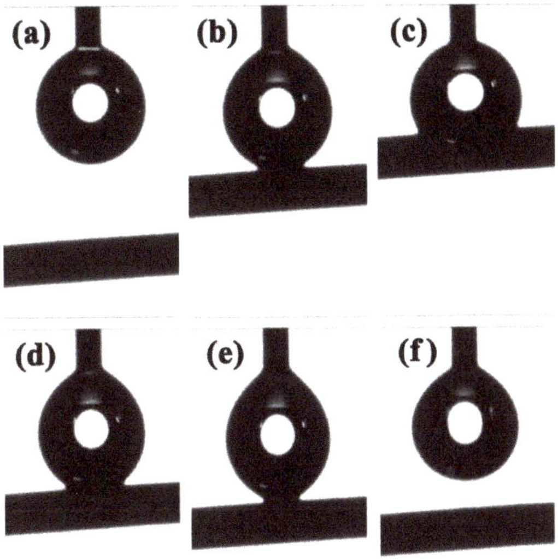

Figure 2. Wettability of CAF-3. (**a–c**) The gradual contact process between water droplets and the CAF. (**d–f**) The process of water droplets leaving the CAF.

2.2. Microstructures

As shown in the SEM images (Figure 3), when the precursor concentration was low, the CAF-1 was mainly composed of BC nanofiber aggregates, with only a small amount of silica nanoparticles attached to the BC by hydrogen bonding. At larger precursor concentrations, as in CAF-2 and CAF-3, the silica gel skeleton was gradually formed in the BC nanofiber network. Thus, the nanoscale interpenetrating network structure was constructed. When the precursor concentration was improved further, the silica gel skeleton became more compact, as shown in CAF-4. It can be seen from the cross-section that the diameter of the sample is approximately 0.7 mm, which is close to the inner diameter of the secondary shaping mold, indicating that the sample did not significant shrink during the atmospheric pressure drying process. The diameter of CAF-1 was slightly smaller than those of the other samples because of the lack of a rigid silica gel skeleton, which caused the sample to shrink by a certain extent during the drying process. As the gel skeleton first shrinks and then springs back during the drying process, the gel skeleton must reach a certain strength to ensure that it can effectively spring back. As a result, a certain concentration of precursors (TEOS) to construct a gel skeleton with sufficient strength is very important for the preparation of composite aerogel fibers.

In addition, the spatial distribution of BC nanofibers in CAFs was obviously denser than that of BC nanofibers without secondary shaping (Figure S2), which is beneficial for further improving the strength of the composite aerogels. Compared with our previous study, the higher density of the obtained BC nanofibers enables the samples to better retain their morphology without shrinkage, even at low concentrations of silica precursors [40]. This could be because denser BC nanofibers could more effectively resist the shrinkage caused by capillary forces during the ambient pressure drying process.

The nitrogen adsorption–desorption isotherms of the prepared CAFs (Figure 4a) were type IV isotherms with hysteresis loops, confirming the formation of mesoporous structures. The hysteresis loops of CAF-3 and CAF-4 were more obvious than those of CAF-1 and CAF-2. Moreover, the pore size distribution (Figure 4b) showed that CAF-3 and CAF-4 had more significant mesoporous structure characteristics than CAF-1 and CAF-2. This suggests that sufficient silica precursors are required to form a complete gel skeleton. When the concentration of silica precursor was relatively low, there was a lack of sufficient silica nanoparticles among the nanofiber network to form a gel skeleton due to the attachment of silica nanoparticles to BC nanofibers. In particular, the number of BC nanofibers per unit volume was enhanced by secondary shaping, which further enhanced the adhesion of silica particles to the BC nanofibers. Based on the results of a BET test, the specific surface areas of the samples grew from 367.9 to 606.9 m^2/g (Table 1) with an increase in TEOS concentration, which is higher than that of PAO@ANF symbiotic aerogel fiber [28] and Kevlar aerogel fibers [25]. This is because a gel skeleton composed of inorganic oxide nanoparticles has a rougher surface compared to a gel skeleton composed of nanofibers or nanosheets.

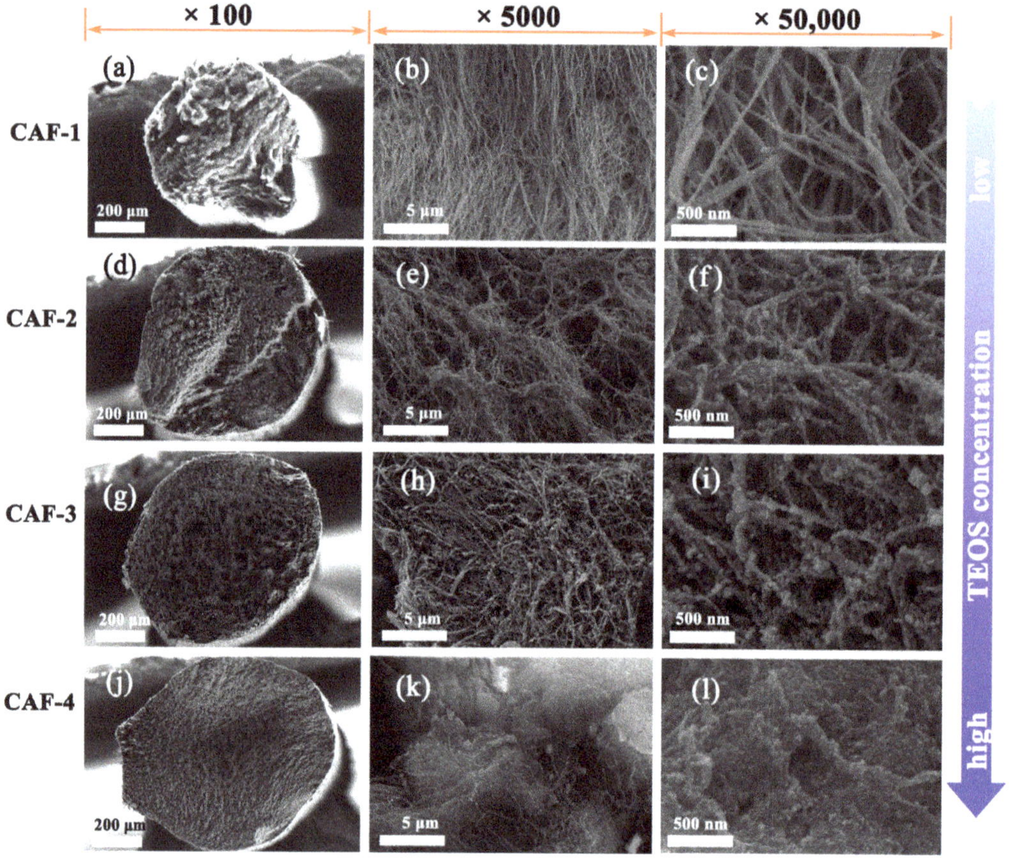

Figure 3. The SEM images of CAF-1 (**a**–**c**), CAF-2 (**d**–**f**), CAF-3 (**g**–**i**), and CAF-4 (**j**–**l**) with different magnifications.

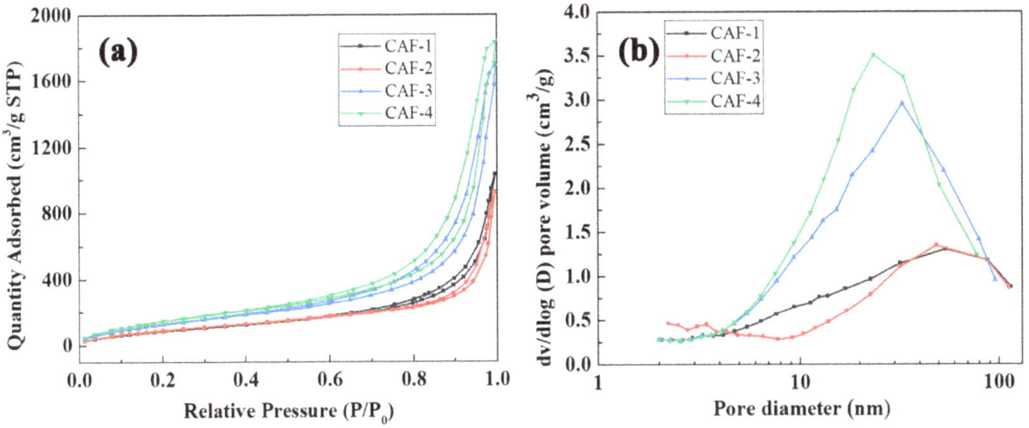

Figure 4. Nitrogen adsorption–desorption isotherms (**a**) and pore size distribution (**b**) of CAFs.

Table 1. Physical properties of CAFs.

Samples	SiO_2 in Aerogels [% *w*/*w*]	Bulk Density [g cm^{-3}]	S_{BET} [m^2 g^{-1}]	Pore Size [nm]	Porosity [a] [%]
CAF-1	27	0.110	367.9	14.2	93.6
CAF-2	40	0.121	387.6	13.7	93.3
CAF-3	49	0.143	541.1	15.5	92.2
CAF-4	55	0.164	606.9	15.1	91.2

[a] The porosity includes the voids caused by crystal growth among gel skeletons during gel freezing.

2.3. Mechanical Properties

High mechanical performance is crucial for a wearable thermal insulation material. The mechanical performances of CAFs prepared with different concentrations of TEOS precursor are presented in Figure 5. The stress–strain curves of CAFs obtained from tensile tests showed that the breaking stress was in the range of 4.5–5.4 MPa. All the samples showed good tensile strength, which was much higher than that of native silica aerogel fibers, and it was comparable or even superior to those of CA/PAA-SF aerogel fibers [41] (3 MPa), SF/GO aerogel fibers [36] (3.2 MPa), and PAO@ANF aerogel fibers [28] (4.56 MPa). Moreover, the curves also showed that the elongation at break of the sample decreased from 6.8% to 1.1% with increasing precursor concentration. This could be because when more and denser gel skeletons formed between the BC nanofibers, the free movement of the nanofibers was restricted, resulting in the free deformation space of the nanofiber network being compressed. At the same time, the sample density increased slightly with increasing precursor concentration because of the enhancement of silica that forms the gel skeleton in the obtained CAFs, but it was generally in a low range (Table 1).

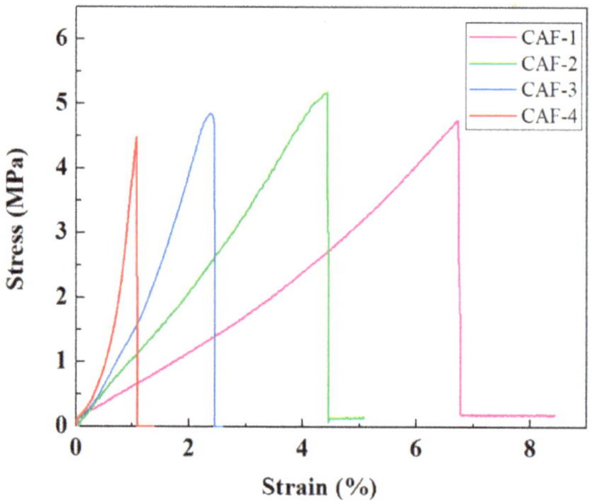

Figure 5. Stress–strain curves of CAFs.

2.4. Thermal Insulation

The insulation performance of CAFs obtained from different concentrations of TEOS was tested under hot and cold conditions. Cotton threads and silk fabric, with similar diameters or thicknesses as those of the CAFs, were also tested under the same conditions. First, several CAFs were packed tightly and aligned in one direction to form a single-layer mat with a thickness of approximately 0.7 mm and placed on a hot plate. The thermocouple was connected to the fiber surface and the hot plate, and the temperature of the surface of CAFs (T_f) was recorded during the heating of the hot plate from 30 to 200 °C. The absolute value of the temperature difference ($|\Delta T|$) of the CAF surface (T_f) and the hot plate were plotted against the hot plate temperature (T_h). As shown in Figure 6a, while heating the hot plate from 30 to 200 °C, the temperature difference of the one-layer CAF mat was always higher than those of the one-layer silk fabric mat and cotton fabric mat. When T_h was 120 °C, the CAF mat temperature reached 79 °C, and the temperatures of the silk and cotton fabric mats reached 101.3 °C and 107 °C, respectively. The higher the temperature difference ($|\Delta T|$), the better the thermal insulation of the studied materials. Hence, the thermal insulation properties of CAFs are superior to those of silk fabrics and cotton threads. Meanwhile, for visually comparing the thermal insulation difference between the CAF fabric and cotton threads at the same heat source temperature (80 °C), a layer of CAF-3 formed by approximately 10 samples and a layer of 10 cotton threads of the same length were heated at the same time. After the temperature was stable, an infrared camera was used to capture photographs (Figure 6b), which showed that the temperature of the cotton fabric was higher than that of the CAF fabric, intuitively proving that the CAFs have excellent thermal insulation properties.

As seen in Figure 6a, with an increase in the amounts of TEOS in CAF-1, CAF-2, and CAF-3, the thermal insulation performances of the corresponding CAFs were gradually enhanced, owing to the complete gel skeleton gradually forming with increasing solid concentration in the aerogel. However, when the concentration of the silica precursor increased further (CAF-4), the thermal insulation performance of CAF-4 decreased because of the high content of the solid phase and because heat is more easily transmitted within the solid phase. At the same time, when the temperature was high, the slope of $|\Delta T|$ to T_h decreased slightly with an increase in temperature, implying that the proportion of thermal radiation on heat transfer rose gradually at high temperatures. In fact, the thermal insulation mechanism of aerogels mainly depended on the blocking of thermal convection and heat conduction, not thermal radiation. Figure 6c shows the precise values of temperature change on the surface of the hot plate (T_h) and aerogel fibers (CAF-3) during the heating process of the hot plate. The surface temperature of CAF-3 varied from 28 to 131 °C as the temperature of the hot plate rose from 30 to 210 °C. When T_h was stable at 210 °C, the $|\Delta T|$ of the hot plate and CAF-3 was approximately 80 °C. When CAF-3 was heated again after a heating-cooling process, the $|\Delta T|$ showed no obvious change, indicating that the thermal insulation performance of CAF-3 was stable. The temperature–time curves for the other samples are shown in Figure S3 (in the Supporting Information) and indicate the stability of the thermal insulation.

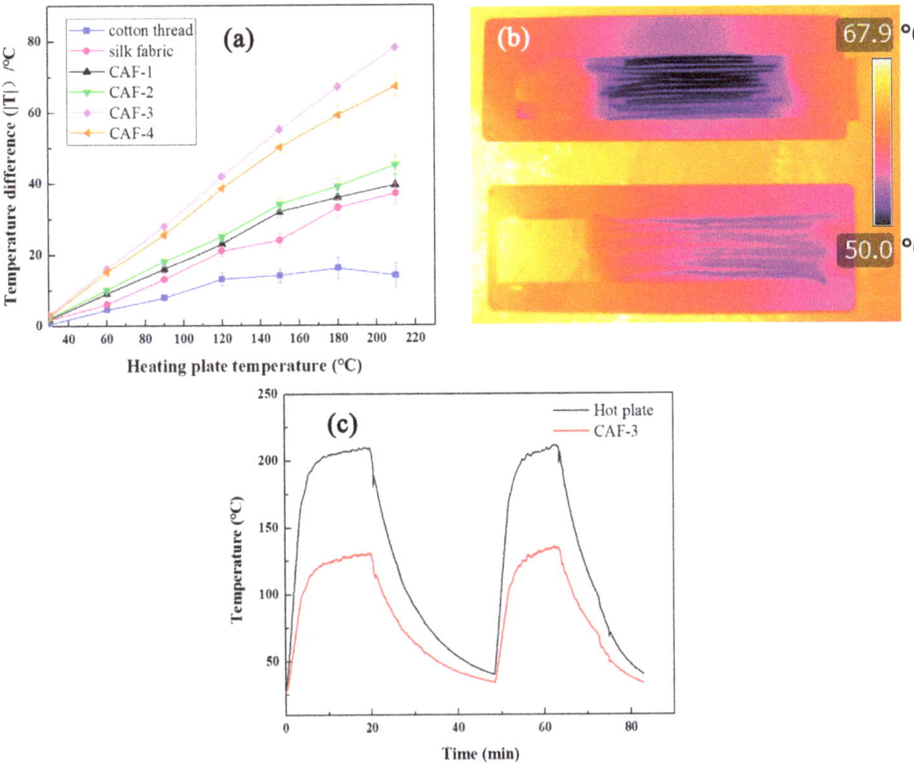

Figure 6. Thermal insulation properties of CAFs, silk fabric, and cotton threads. Temperature difference between the fiber surface and hot plate versus the hot plate for the single-layer mat of CAFs, silk fabric, and cotton threads (**a**). Infrared photo of one-layer mats of CAF-3 and cotton threads at high temperatures (**b**). Temperature–time curves of CAF-3 and hot plate (**c**).

To test the thermal insulation performance of the sample at lower temperatures, the three single-layer fiber mats of CAF-3, silk fabric, and cotton threads were placed on a sheet of iron with 2.5 cm of dry ice underneath. The temperatures of the fiber surface and iron sheet were monitored simultaneously. When the sheet temperature was −72 °C, the absolute temperature difference $|\Delta T|$ values of the one-layer aerogel fiber mat, silk fabric mat, and cotton thread mat were 59 °C, 22 °C, and 46 °C, respectively. This demonstrates that CAF-3 has excellent insulation performance in cold environments.

2.5. Thermal Stability

As shown in Figure 7a, the BC matrix demonstrates thermal stability similar to that of cotton thread and silk fabric, which are also composites of natural polymers. The thermogravimetric analysis curves showed that the temperature of decomposition of the cellulose (BC matrix) in composite aerogel fibers gradually shifted from about 280 °C to about 330 °C (Figure 7b) as the silica content increased. Moreover, when the mass gradually decreased to 88% of the initial mass, as shown by the blue arrow in Figure 7b, the temperature corresponding to the sample gradually rose from 300 °C to about 370 °C as the silica content increased. Hence, silica improved the thermal stability of the polymer matrix. This allows the CAFs to be used at higher temperatures than the traditional polymer fiber, owing to the stabilizing effect of silica on BC matrix.

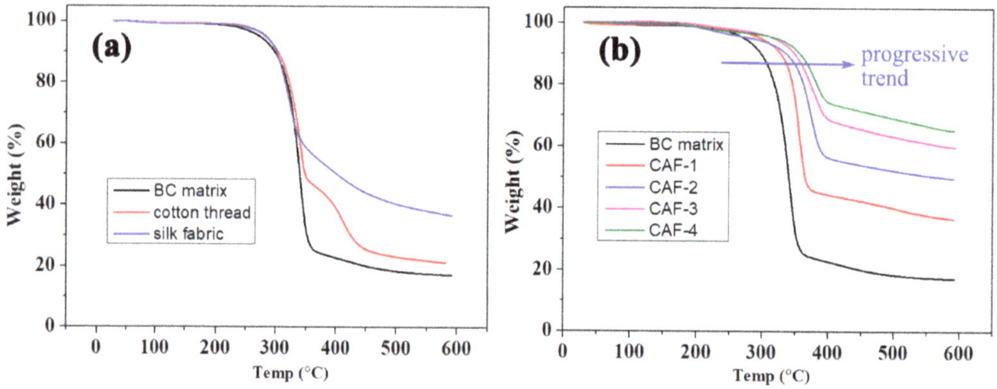

Figure 7. (**a**) Thermogravimetry analysis (TGA, 10 °C min^{-1} heating) curves of BC matrix, cotton thread, and silk fabric. (**b**) Thermogravimetry analysis (TGA, 10 °C min^{-1} heating) curves of BC matrix and CAFs. The blue arrow shows the temperature changes of each material at the same change in weight proportion (88%).

3. Conclusions

In summary, a novel and simple method for preparing aerogel fibers with excellent mechanical properties and thermal insulation has been demonstrated here. The silica precursor TEOS and BC matrix were used to obtain the CAFs through in situ sol–gel reaction. The mechanical properties of CAFs were significantly improved by increasing the content of BC nanofibers per unit volume via a secondary shaping process. Efficient silica gel skeleton formed in the BC matrix endowed the CAFs with excellent thermal insulation performance and large specific surface area (606.9 m^2/g). Moreover, the stabilizing effect of silica on the BC matrix makes the CAFs usable at relatively high temperatures (up to about 330 °C). In addition, the outstanding hydrophobicity of CAFs enables them to resist erosion by water vapor and have good weather resistance. Consequently, these aerogel fibers are promising materials for wearable thermal insulation. This work introduced the nanoscale interpenetrating network structure into aerogel fibers and provided a new way for the preparation and toughening of aerogel fibers, especially the inorganic oxide-based aerogel fibers.

4. Materials and Methods

4.1. Materials

Nata-de-coco slices were purchased from Wenchang Baocheng Industry and Trade Co., Ltd. (Hainan, China). Tetraethoxysilane (TEOS), *n*-hexane, triethylamine (TEA), trimethylchlorosilane (TMCS), and tert-butanol were obtained from Aladdin Reagent Co., Ltd. (Shanghai, China). Ethanol, hydrochloric acid (HCl), sodium hydroxide (NaOH, 4 wt %), and ammonium hydroxide ($NH_3 \cdot H_2O$) were purchased from Beijing Chemical Reagent Co. (Beijing, China), LTD. All the chemicals were of analytical grade and were used as received without any further purification.

4.2. Preparation of BC Fibers

Nata-de-coco slices (i.e., bacterial cellulose hydrogel, BC hydrogel) of thicknesses of 3 mm were first soaked in deionized water for 4 h. The water was replaced several times to remove the added sucrose; after this process, the slices were boiled in NaOH for 6 h at 90 °C. Subsequently, clean BC hydrogel was obtained by rinsing it with deionized water until it became neutral. The *nata-de-coco* slice (Figure S1a) was placed on a glass plate, squeezed to remove about 80% of the water, and then cut using a laser (15 W power, 80W6040, Liaocheng Julong Laser Equipment Co., Ltd., Liaocheng, China) to obtain fibers of uniform width (2 mm) and length (about 500 mm) (Figure 1a and Figure S1b). Finally, the dried fiber-like BC matrix (Figure 1b and Figure S1c) was obtained by freeze drying for 24 h after the solvent replacement of the mixed liquid of water and tert-butanol (V_{water}:$V_{tert\text{-}butanol}$ = 3:2). This step was shown on the left side of Figure 8.

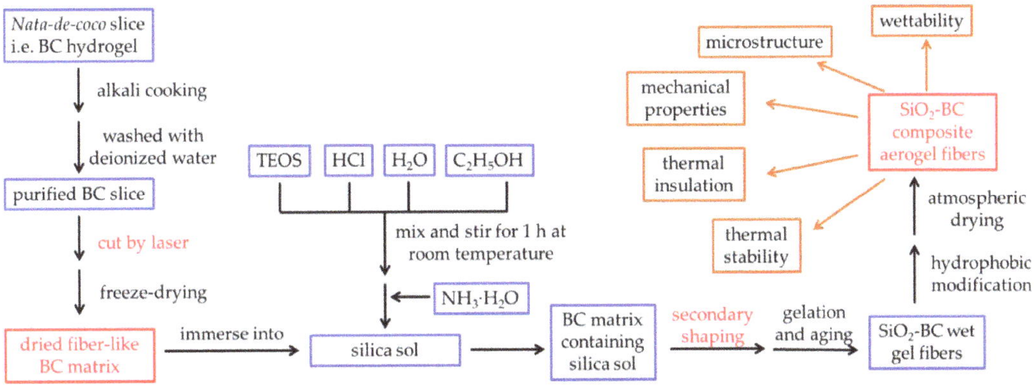

Figure 8. Flowchart illustrating the overall processes used in this work.

4.3. Preparation of Silica Sols

The preparation of silica sols with different concentrations of precursor TEOS was carried out by following the previously reported method [40]. The dosages of deionized water, HCl (*w*/*w* 1%), and ethanol in the process of sol preparation were 2, 0.2, and 9.2 mL, respectively. Different amounts (Table 2) of TEOS, deionized water, ethanol, and HCl were mixed in a beaker and stirred for 1 h at room temperature. Dilute ammonia (0.1 mol L^{-1}, 1 mL) was added to this system to form silica alcosols as shown in the middle of Figure 8.

Table 2. CAFs formed by varying concentration of precursor TEOS.

Sample	CAF-1	CAF-2	CAF-3	CAF-4
TEOS (mL)	0.85	1.7	2.55	3.4

4.4. Preparation of Silica–BC Composite Wet Gel Fibers

The dried fiber-like BC matrix was immersed in silica sol and then stirred (Figure 1c and Figure S1d). The silica sol with the matrix was placed in an ice-water bath to inhibit the gelation process, ensuring that the silica sol could fully diffuse into the BC matrix. After sufficient diffusion for 2 h, the BC matrix containing silica sol was removed and passed through a tapered mold (the front end of 1000 µL pipette gun head with an inner diameter of 0.8 mm was cut as mold, produced by Nantong Hairui Experimental Equipment Co., Ltd., Nantong, China) immediately to make the fiber-like BC matrix finer and more uniform (Figure 1d and Figure S1e); this was the secondary shaping process. Then, the fiber-like BC matrix was placed in a sealed container filled with ethanol vapor for approximately 25 min to obtain silica–BC composite wet gel fibers (Figure 1e). The ethanol vapor was used to ensure that the composite wet gel did not shrink.

4.5. Hydrophobic Modification and Atmospheric Drying of CAFs

The wet gel fibers were soaked in ethanol and heated at 70 °C for 1 h to age the silica gel skeleton. Then, ethanol was replaced with *n*-hexane for 3 h for solvent replacement. *N*-hexane (50 mL), TEA (4 mL), and TMCS (3 mL) were added to the glass reactor, and the wet gel fibers were immersed in the liquid mixture. The glass reactor was heated in an oil bath and refluxed for 2 h. Then, the wet gel fibers were dipped into a 100 mL beaker containing 50 mL ethanol, which was replaced every 30 min; this process was repeated twice. Subsequently, ethanol was replaced with *n*-hexane, and the same operation as above was repeated. Subsequently, the wet gel fibers after hydrophobic modification were removed and heated in an oven at 80 °C for 20 min. Finally, the hydrophobic CAFs were obtained (Figure 1f and Figure S1f).

4.6. Characterization

As shown in the upper right corner of Figure 8, the morphology of the CAFs was determined, and the specific surface area, porosity, pore-size distribution, mechanical properties, wettability, density, and content of silica in the CAFs were examined. Thermal insulation performance and thermal stability were also evaluated. Detailed characterization methods are provided in the Supporting Information.

Supplementary Materials: The following are available online at https://www.mdpi.com/article/10.3390/gels7030145/s1, Video S1: video of the wettability of the silica–bacterial cellulose composite aerogel fibers (CAFs), Figure S1: Photos of the preparation process of CAFs; Figure S2: SEM images of BC without secondary shaping; Figure S3: Temperature–time curves of CAFs at high temperature; Characterization and Reference.

Author Contributions: Conceptualization, project administration, validation, funding acquisition, writing—review and editing, H.S.; investigation, data curation, writing—original draft, M.W.; methodology, C.M.; validation, Q.S.; data curation, Y.W. (Yutong Wang); supervision, conceptualization, R.F.; project administration, Y.W. (Yaxiong Wang); supervision, L.M.; funding acquisition, Y.H. All authors have read and agreed to the published version of the manuscript.

Funding: This research was funded by the Innovation Fund of Inner Mongolia University of Science & Technology (2018QDL-B01), the Natural Science Foundation of Inner Mongolia (2019BS05022, 2019LH02004), the Scientific Research Projects of Colleges and Universities of Inner Mongolia (NJZZ20087), Science and Technology Plan Project of Inner Mongolia Autonomous Region (2020GG0152), National Natural Science Foundation of China (NNSFC, No. 52164013 and No. 11965015) and Outstanding Youth Fund Project of Innovation Fund of Inner Mongolia University of Science and Technology (No. 2019YQL05). We also appreciate the financial support from Human Resources and Social Security Department of Inner Mongolia.

Institutional Review Board Statement: Not applicable.

Informed Consent Statement: Not applicable.

Conflicts of Interest: The authors declare no conflict of interest.

References

1. Ziegler, C.; Wolf, A.; Liu, W.; Herrmann, A.-K.; Gaponik, N.; Eychmüller, A. Modern Inorganic Aerogels. *Angew. Chem. Int. Ed.* **2017**, *56*, 13200–13221. [CrossRef]
2. Sai, H.; Fu, R.; Xing, L.; Xiang, J.; Li, Z.; Li, F.; Zhang, T. Surface Modification of Bacterial Cellulose Aerogels' Web-like Skeleton for Oil/Water Separation. *ACS Appl. Mater. Interfaces* **2015**, *7*, 7373–7381. [CrossRef] [PubMed]
3. Peydayesh, M.; Suter, M.K.; Bolisetty, S.; Boulos, S.; Handschin, S.; Nyström, L.; Mezzenga, R. Amyloid Fibrils Aerogel for Sustainable Removal of Organic Contaminants from Water. *Adv. Mater.* **2020**, *32*, 1907932. [CrossRef]
4. Adhikary, S.K.; Ashish, D.K.; Rudžionis, Ž. Aerogel based thermal insulating cementitious composites: A review. *Energy Build.* **2021**, *245*, 111058. [CrossRef]
5. Wu, K.; Zhang, L.; Yuan, Y.; Zhong, L.; Chen, Z.; Chi, X.; Lu, H.; Chen, Z.; Zou, R.; Li, T.; et al. An Iron-Decorated Carbon Aerogel for Rechargeable Flow and Flexible Zn-Air Batteries. *Adv. Mater.* **2020**, *32*, 2002292. [CrossRef] [PubMed]
6. Cai, B.; Eychmüller, A. Promoting Electrocatalysis upon Aerogels. *Adv. Mater.* **2019**, *31*, 1804881. [CrossRef]

7. Fu, G.; Yan, X.; Chen, Y.; Xu, L.; Sun, D.; Lee, J.-M.; Tang, Y. Boosting Bifunctional Oxygen Electrocatalysis with 3D Graphene Aerogel-Supported Ni/MnO Particles. *Adv. Mater.* **2018**, *30*, 1704609. [CrossRef] [PubMed]
8. Adhikary, S.K.; Rudžionis, Ž.; Tučkutė, S.; Ashish, D.K. Effects of carbon nanotubes on expanded glass and silica aerogel based lightweight concrete. *Sci. Rep.* **2021**, *11*, 2104. [CrossRef]
9. Mohite, D.P.; Larimore, Z.J.; Lu, H.; Mang, J.T.; Sotiriou-Leventis, C.; Leventis, N. Monolithic Hierarchical Fractal Assemblies of Silica Nanoparticles Cross-Linked with Polynorbornene via ROMP: A Structure–Property Correlation from Molecular to Bulk through Nano. *Chem. Mater.* **2012**, *24*, 3434–3448. [CrossRef]
10. Fricke, J. Aerogels—Highly tenuous solids with fascinating properties. *J. Non-Cryst. Solids* **1988**, *100*, 169–173. [CrossRef]
11. Cai, L.; Shan, G. Elastic silica aerogel using methyltrimethoxysilane precusor via ambient pressure drying. *J. Porous Mat.* **2015**, *22*, 1455–1463. [CrossRef]
12. Zhong, L.; Chen, X.; Song, H.; Guo, K.; Hu, Z. Highly flexible silica aerogels derived from methyltriethoxysilane and polydimethylsiloxane. *New J. Chem.* **2015**, *39*, 7832–7838. [CrossRef]
13. Rao, A.V.; Bhagat, S.D.; Hirashima, H.; Pajonk, G.M. Synthesis of flexible silica aerogels using methyltrimethoxysilane (MTMS) precursor. *J. Colloid Interface Sci.* **2006**, *300*, 279–285. [PubMed]
14. Wang, Z.; Dai, Z.; Wu, J.; Zhao, N.; Xu, J. Vacuum-Dried Robust Bridged Silsesquioxane Aerogels. *Adv. Mater.* **2013**, *25*, 4494–4497. [CrossRef] [PubMed]
15. Wang, Z.; Wang, D.; Qian, Z.; Guo, J.; Dong, H.; Zhao, N.; Xu, J. Robust Superhydrophobic Bridged Silsesquioxane Aerogels with Tunable Performances and Their Applications. *ACS Appl. Mater. Interfaces* **2015**, *7*, 2016–2024. [CrossRef] [PubMed]
16. Yun, S.; Luo, H.; Gao, Y. Low-density, hydrophobic, highly flexible ambient-pressure-dried monolithic bridged silsesquioxane aerogels. *J. Mater. Chem. A* **2015**, *3*, 3390–3398. [CrossRef]
17. Zu, G.; Kanamori, K.; Maeno, A.; Kaji, H.; Nakanishi, K. Superflexible Multifunctional Polyvinylpolydimethylsiloxane-Based Aerogels as Efficient Absorbents, Thermal Superinsulators, and Strain Sensors. *Angew. Chem. Int. Ed.* **2018**, *57*, 9722–9727. [CrossRef]
18. Zu, G.; Shimizu, T.; Kanamori, K.; Zhu, Y.; Maeno, A.; Kaji, H.; Shen, J.; Nakanishi, K. Transparent, Superflexible Doubly Cross-Linked Polyvinylpolymethylsiloxane Aerogel Superinsulators via Ambient Pressure Drying. *ACS Nano* **2018**, *12*, 521–532. [CrossRef]
19. Yuan, B.; Ding, S.; Wang, D.; Wang, G.; Li, H. Heat insulation properties of silica aerogel/glass fiber composites fabricated by press forming. *Mater. Lett.* **2012**, *75*, 204–206. [CrossRef]
20. Xu, L.; Jiang, Y.; Feng, J.; Feng, J.; Yue, C. Infrared-opacified Al_2O_3–SiO_2 aerogel composites reinforced by SiC-coated mullite fibers for thermal insulations. *Ceram. Int.* **2015**, *41*, 437–442. [CrossRef]
21. Chandradass, J.; Kang, S.; Bae, D.-S. Synthesis of silica aerogel blanket by ambient drying method using water glass based precursor and glass wool modified by alumina sol. *J. Non-Cryst. Solids* **2008**, *354*, 4115–4119. [CrossRef]
22. Oh, K.; Kim, D.; Kim, S. Ultra-porous flexible PET/Aerogel blanket for sound absorption and thermal insulation. *Fiber. Polym.* **2009**, *10*, 731–737. [CrossRef]
23. Nguyen, B.N.; Meador, M.A.B.; Tousley, M.E.; Shonkwiler, B.; McCorkle, L.; Scheiman, D.A.; Palczer, A. Tailoring Elastic Properties of Silica Aerogels Cross-Linked with Polystyrene. *ACS Appl. Mater. Interfaces* **2009**, *1*, 621–630. [CrossRef] [PubMed]
24. Randall, J.P.; Meador, M.A.B.; Jana, S.C. Tailoring Mechanical Properties of Aerogels for Aerospace Applications. *ACS Appl. Mater. Interfaces* **2011**, *3*, 613–626. [CrossRef]
25. Liu, Z.; Lyu, J.; Fang, D.; Zhang, X. Nanofibrous Kevlar Aerogel Threads for Thermal Insulation in Harsh Environments. *ACS Nano* **2019**, *13*, 5703–5711. [CrossRef] [PubMed]
26. Zhou, J.; Hsieh, Y.-L. Nanocellulose aerogel-based porous coaxial fibers for thermal insulation. *Nano Energy* **2020**, *68*, 104305. [CrossRef]
27. Meng, S.; Zhang, J.; Chen, W.; Wang, X.; Zhu, M. Construction of continuous hollow silica aerogel fibers with hierarchical pores and excellent adsorption performance. *Microporous Mesoporous Mat.* **2019**, *273*, 294–296. [CrossRef]
28. Li, J.; Wang, J.; Wang, W.; Zhang, X. Symbiotic Aerogel Fibers Made via In-Situ Gelation of Aramid Nanofibers with Polyamidoxime for Uranium Extraction. *Molecules* **2019**, *24*, 1821. [CrossRef]
29. Meng, S.; Zhang, J.; Xu, W.; Chen, W.; Zhu, L.; Zhou, Z.; Zhu, M. Structural control of silica aerogel fibers for methylene blue removal. *Sci. China Technol. Sci.* **2019**, *62*, 958–964. [CrossRef]
30. Mitropoulos, A.N.; Burpo, F.J.; Nguyen, C.K.; Nagelli, E.A.; Ryu, M.Y.; Wang, J.; Sims, R.K.; Woronowicz, K.; Wickiser, J.K. Noble Metal Composite Porous Silk Fibroin Aerogel Fibers. *Materials* **2019**, *12*, 894. [CrossRef]
31. Lai, H.; Wang, Y.; Wang, Y.; Liu, W.; Bao, X.; Liu, F.; Li, X.; Lei, Z.; Jiao, H.; Fan, Z. Macroscale amphiphilic aerogel fibers made from nonwoven nanofibers for large active mass loading. *J. Power Sources* **2020**, *474*, 228612. [CrossRef]
32. Liu, Y.; Zhang, Y.; Xiong, X.; Ge, P.; Wu, J.; Sun, J.; Wang, J.; Zhuo, Q.; Qin, C.; Dai, L. Strategies for Preparing Continuous Ultraflexible and Ultrastrong Poly(Vinyl Alcohol) Aerogel Fibers with Excellent Thermal Insulation. *Macromol. Mater. Eng.* **2021**, *306*, 2100399. [CrossRef]
33. Li, M.; Gan, F.; Dong, J.; Fang, Y.; Zhao, X.; Zhang, Q. Facile Preparation of Continuous and Porous Polyimide Aerogel Fibers for Multifunctional Applications. *ACS Appl. Mater. Interfaces* **2021**, *13*, 10416–10427. [CrossRef]
34. Ying, C.; Huaxin, G.; Yujie, W.; Dewen, L.; Hao, B. A Thermally Insulating Textile Inspired by Polar Bear Hair. *Adv. Mater.* **2018**, *30*, 1706807.

35. Xu, Z.; Zhang, Y.; Li, P.; Gao, C. Strong, Conductive, Lightweight, Neat Graphene Aerogel Fibers with Aligned Pores. *ACS Nano* **2012**, *6*, 7103–7113. [CrossRef]
36. Wang, Z.; Yang, H.; Li, Y.; Zheng, X. Robust Silk Fibroin/Graphene Oxide Aerogel Fiber for Radiative Heating Textiles. *ACS Appl. Mater. Interfaces* **2020**, *12*, 15726–15736. [CrossRef]
37. Du, Y.; Zhang, X.; Wang, J.; Liu, Z.; Zhang, K.; Ji, X.; You, Y.; Zhang, X. Reaction-Spun Transparent Silica Aerogel Fibers. *ACS Nano* **2020**, *14*, 11919–11928. [CrossRef] [PubMed]
38. Zhang, G.H.; Dass, A.; Rawashdeh, A.M.M.; Thomas, J.; Counsil, J.A.; Sotiriou-Leventis, C.; Fabrizio, E.F.; Ilhan, F.; Vassilaras, P.; Scheiman, D.A.; et al. Isocyanate-crosslinked silica aerogel monoliths: Preparation and characterization. *J. Non-Cryst. Solids* **2004**, *350*, 152–164. [CrossRef]
39. Wang, S.; Jiang, F.; Xu, X.; Kuang, Y.; Fu, K.; Hitz, E.; Hu, L. Super-Strong, Super-Stiff Macrofibers with Aligned, Long Bacterial Cellulose Nanofibers. *Adv. Mater.* **2017**, *29*, 1702498. [CrossRef] [PubMed]
40. Sai, H.; Fu, R.; Xiang, J.; Guan, Y.; Zhang, F. Fabrication of elastic silica-bacterial cellulose composite aerogels with nanoscale interpenetrating network by ultrafast evaporative drying. *Compos. Sci. Technol.* **2018**, *155*, 72–80. [CrossRef]
41. Yang, H.; Wang, Z.; Liu, Z.; Cheng, H.; Li, C. Continuous, Strong, Porous Silk Firoin-Based Aerogel Fibers toward Textile Thermal Insulation. *Polymers* **2019**, *11*, 1899. [CrossRef] [PubMed]

gels

MDPI

Article

Environment-Friendly Catalytic Mineralization of Phenol and Chlorophenols with Cu- and Fe- Tetrakis(4-aminophenyl)-porphyrin—Silica Hybrid Aerogels

Enikő Győri [1], Ádám Kecskeméti [1], István Fábián [1,2], Máté Szarka [3,4] and István Lázár [1,*]

[1] Department of Inorganic and Analytical Chemistry, University of Debrecen, Egyetem tér 1, H-4032 Debrecen, Hungary; gyori.eniko@science.unideb.hu (E.G.); adam.kecskemeti@teva.hu (Á.K.); ifabian@science.unideb.hu (I.F.)
[2] MTA-DE Redox and Homogeneous Catalytic Reaction Mechanisms Research Group, University of Debrecen, Debrecen Egyetem tér 1, H-4032 Debrecen, Hungary
[3] Institute for Nuclear Research, Debrecen Bem tér 18/c, H-4026 Debrecen, Hungary; szarka.mate@atomki.hu
[4] Vitrolink Kft., Debrecen, Sárosi utca 9, H-4033 Debrecen, Hungary
* Correspondence: lazar@science.unideb.hu; Tel.: +36-52-512-900-22376

Abstract: Fenton reactions with metal complexes of substituted porphyrins and hydrogen peroxide are useful tools for the mineralization of environmentally dangerous substances. In the homogeneous phase, autooxidation of the prophyrin ring may also occur. Covalent binding of porphyrins to a solid support may increase the lifetime of the catalysts and might change its activity. In this study, highly water-insoluble copper and iron complexes of 5,10,15,20-tetrakis(4-aminophenyl)porphyrin were synthesized and bonded covalently to a very hydrophilic silica aerogel matrix prepared by co-gelation of the propyl triethoxysilyl-functionalized porphyrin complex precursors with tetramethoxysilane, followed by a supercritical carbon dioxide drying. In contrast to the insoluble nature of the porphyrin complexes, the as-prepared aerogel catalysts were highly compatible with the aqueous phase. Their catalytic activities were tested in the mineralization reaction of phenol, 3-chlorophenol, and 2,4-dichlorophenol with hydrogen peroxide. The results show that both aerogels catalyzed the oxidation of phenol and chlorophenols to harmless short-chained carboxylic acids under neutral conditions. In batch experiments, and also in a miniature continuous-flow tubular reactor, the aerogel catalysts gradually reduced their activity, due to the slow oxidation of the porphyrin ring. However, the rate and extent of the degradation was moderate and did not exclude the possibility that the as-prepared catalysts, as well as their more stable derivatives, might find practical applications in environment protection.

Keywords: silica aerogel; aerogel hybrid; covalent immobilization; porphyrin complexes; heterogeneous catalyst; phenol mineralization; chlorophenol mineralization

Citation: Győri, E.; Kecskeméti, Á.; Fábián, I.; Szarka, M.; Lázár, I. Environment-Friendly Catalytic Mineralization of Phenol and Chlorophenols with Cu- and Fe-Tetrakis(4-aminophenyl)-porphyrin—Silica Hybrid Aerogels. *Gels* **2022**, *8*, 202. https://doi.org/10.3390/gels8040202

Academic Editor: Miguel Sanchez-Soto

Received: 17 February 2022
Accepted: 21 March 2022
Published: 23 March 2022

1. Introduction

Phenol is an extensively used reagent in the production of phenolic resins, bisphenol A, caprolactam, and other chemicals. As an unfortunate consequence, similarly to many compounds used in the industry, phenol can be found in wastewater or in the soil. Due to its high toxicity, complete elimination of phenol is an important environmental goal.

Chlorophenols are produced by the chlorination of phenol, resulting in 19 different compounds including isomers. All of them show significant antiseptic activity and higher toxicity than that of phenol. That property is utilized when they are applied as disinfectants, preservative agents, herbicides, or insecticides [1]. However, these compounds are not biodegradable at all. By bioaccumulation in plant and animal species, they can affect the food chain. Due to their persistence in the environment, they are listed as first-priority pollutants, so the necessity of their effective mineralization is indisputable [2]. Since chlorinated organic compounds are resistant to biodegradation, processes other than microbial processes are needed to eliminate them [3]. Thus far, many techniques have been developed for the removal of phenols from the environment [4,5]. The most efficient methods are the advanced oxidation processes (AOPs), in which hydroxyl or sulfate radicals act as active agents. In practice, several reactions and materials are used to generate them, such as UV light combined with either hydrogen peroxide or ozone, the Fenton reaction, or salts of the peroxy monosulfate ion, for example [6–9].

In the Fenton and Fenton-like reactions, hydroxyl radical (OH^-) is generated from hydrogen peroxide, although the exact mechanism is still not clarified. H_2O_2 can oxidize Fe^{2+} ion to produce Fe^{3+} ion, hydroxide ion and highly reactive radicals (see Equations (1) and (2)) and these free radicals can easily oxidize the organic pollutants [10–12].

$$Fe^{2+} + H_2O_2 \rightarrow Fe^{3+} + OH^- + OH^\bullet \quad (1)$$

$$OH^\bullet + H_2O_2 \rightarrow HO_2^\bullet + H_2O \quad (2)$$

The Fenton-like metal ion catalysts are used mostly in the homogeneous phase. Their main drawbacks are the difficulty of separation and regeneration of the catalyst, as well as the increasing concentration of metal ions in water and the soil.

The most common catalytically active materials are metals, oxides and sulfides. The efficiency of the heterogeneous catalysts may be increased by applying them on a solid support. Although the turnover rate is in general lower compared to the homogeneous phase reactions, heterogeneous catalysis has several advantages. Easy separation from the reaction mixture, better tolerance towards extreme reaction parameters, larger surface area and a higher number of active sites make them valuable materials [13,14].

Numerous heterogeneous catalysts have been studied, such as, the Cu- or Fe-containing zeolites due to their high catalytic activity and selectivity [15–17]. Additional examples are the porphyrins, which are tetradentate macrocyclic ligands and form complexes of extremely high stability. Various porphyrin derivatives containing zinc, copper, iron, manganese, palladium, vanadium or other transition metal ions are extensively used as catalysts [18–20]. In the homogeneous phase catalytic processes, porphyrins are susceptible to self-degradation and loss of activity [21,22]. Therefore, it is an important goal to develop heterogeneous phase porphyrin catalysts possessing high activity and increased stability.

Aerogels are extremely light solid materials exhibiting unique physical and surface properties. Some of the most versatile ones are silica aerogels, which are prepared by the sol-gel technique from an organic silane precursor, and dried to a solid under supercritical conditions. The process results in a substance with specific properties, such as high and open porosity, large specific surface area, extremely low bulk density, high insulating capacity, to name a few. Thanks to these features, aerogels can be applied in many different industrial fields, for example as insulating materials, Cherenkov radiators, biomaterials, or catalysts [23–27]. The siloxane network can be functionalized by covalently binding organic moieties to the skeleton, embedding guest particles in the structure, or adsorbing metal ions on the surface. Such aerogel-based materials are widely used as catalysts in hydrogen production [28,29], dye degradation in wastewaters [30], or methanol electrooxidation [31,32].

There are several methods for the immobilization of porphyrins on a carrier. The simplest technique is adsorption of the molecules on the surface [33]. Although the process is straightforward, the chance of leaching from the matrix is rather high. Another method is the "ship in a bottle" process [34], which embeds the molecules in narrow necked cavities. A major disadvantage of such a catalyst is the limited access of substrates to the catalytically active centers. The most sophisticated way is the covalent immobilization, which prevents leaching from the solid phase, and provides a good contact with the substrates. However, the technique may require special knowledge of synthetic chemistry [35–37].

In an earlier study it was demonstrated that highly water-soluble porphyrin complexes may undergo decomposition in Fenton reactions due to their autocatalytic oxidation [38].

It was supposed that immobilization of catalytically active porphyrin complexes may decrease the rate of autooxidation. Recently, silica aerogels covalently functionalized with tetraaza macrocyclic copper complexes were prepared and tested in our laboratory for catalytic oxidation of phenols with hydrogen peroxide [26]. In this paper we report the synthesis and characterization of silica aerogels which are covalently functionalized with water-insoluble 5,10,15,20-tetrakis(4-aminophenyl)porphyrin complexes. Their catalytic activities were tested in mineralization of environment-polluting materials, phenol and chlorophenols, in batch mode and in continuous-flow microreactor and the life expectances of the catalysts were determined at different temperatures and molar ratios.

2. Results and Discussion

2.1. Preparation of the Heterogeneous Catalysts

Scheme 1 shows the synthetic steps of the preparation. First, the porphyrin ring was functionalized with 3-isocyanatopropyl triethoxysilane (3-IPTES), which acted as a bifunctional spacer and coupling agent. A 40.4 mg (5.99×10^{-5} mmol) portion of TAPP was dissolved in 8.00 mL anhydrous DMF under an argon atmosphere. To that 0.67 cm^3 (656 mg, 2.39×10^{-3} mmol) of 3-IPTES was added, and the solution was heated at 70 °C for 96 h. Metal complexes were prepared then by carefully reacting the functionalized porphyrin rings with substoichiometric portions of either copper(2+) acetate or crystalline iron(2+) sulfate heptahydrate under anhydrous conditions in DMF or DMSO until the UV fluorescence of the free porphyrine ring disappeared. The aerogel catalysts were obtained by the ammonia-catalyzed co-hydrolysis and co-condensation of the triethoxypropylsilyl-functionalized porphyrin complexes with tetramethoxysilane (TMOS) in a methanol-water mixture using the sol-gel technique, as described earlier. After the necessary ageing and solvent exchange process, the supercritical carbon dioxide drying process resulted in aerogel monoliths [39]. The photographs of the as-prepared catalysts are shown in Scheme 1. Although it might be difficult to see their true colour in the cylindrical form, microscopic images of thin fragments revealed the red colour of the copper-containing (denoted as CuPA), and greenish brown colour of the iron-containing (denoted as FePA) aerogels.

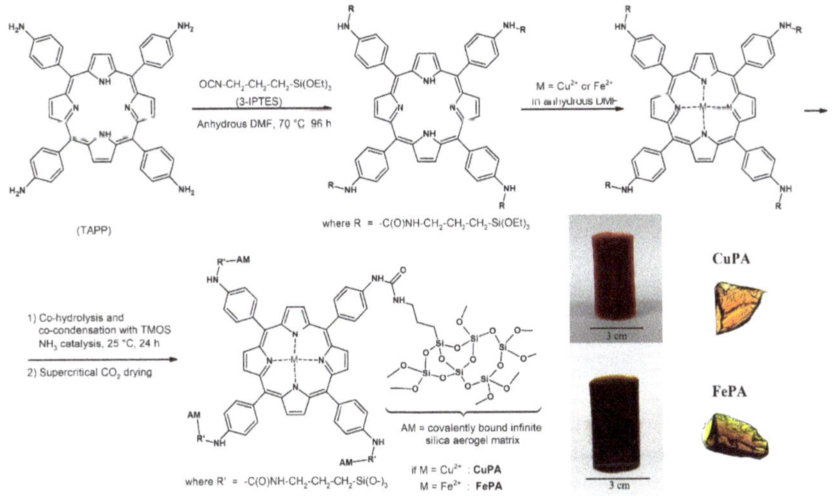

Scheme 1. Synthetic pathway and reaction conditions of the complex formation, covalent binding and co-gelation of the porphyrin complexes with tetramethoxysilane (TMOS) leading to hybrid metal complex–aerogel catalysts FePA and CuPA. On the right side: Photographs (left) and microscopic images (40×) (right) of monolithic and fragmented pieces of CuPA and FePA catalysts.

2.2. Characterization of the Catalysts

The functionalization of the porphyrin ring with 3-isocyanatopropyltriethoxysilane (3-IPTES) was monitored by NMR spectroscopy (See in Figures S1 and S2). The differences between the two spectra can be clearly seen. The significant change in the aromatic region of the spectra (Figure S2) indicates that the coupling was successful. The complexation of the functionalized porphyrin ring with Cu(2+) and Fe(2+) ion was monitored with a 366 nm UV lamp. The non-complexed porphyrin ring had a strong red fluorescence, which disappears when the complex is formed. We could observe the vanishing of the fluorescence in the case of both metal ions.

The FT-IR spectra of the complex solutions were recorded as well, they can be found in the Supplementary Materials (Figure S3). Figure 1 shows the fingerprint regions of the spectra compared with that of the non-complexed porphyrin ring. The differences prove the change in the structure of the porphyrin ring and thus the formation of the complexes. Figure 2 shows the FT-IR spectra of an aerogel functionalized with porphyrin complex. The following peaks can be assigned to silica aerogels: The O–H stretching vibration at ~3400 cm^{-1}, the Si–O–Si asymmetric stretching vibration at around 1050 cm^{-1} and the asymmetric stretching vibration at approximately 950 cm^{-1}. The vibrations of C–H bonds appearing between 2800 and 3000 cm^{-1} indicate the successful functionalization.

The Raman spectra of the porphyrin ring, the complexes and the functionalized aerogels were also recorded. Figure 3 shows the difference between the spectrum of the empty porphyrin ring and the complexes. In the high-frequency region the Raman bands are sensitive to the electron density, the axial ligation and to the core size of the central metal ion. The band at 1543 cm^{-1} of the porphyrin ligand can be assigned to the $C_\beta C_\beta$ stretch, which was upshifted to 1546 cm^{-1} in the Fe and 1576 cm^{-1} in the Cu complexes, respectively. This band appears as one of the most intense bands in the spectra for the Fe and Cu complexes. The band at 1487 cm^{-1} of the porphyrin ligand could be assigned to the phenyl ring vibration, which was practically the same in the Fe complex but was shifted to 1497 cm^{-1} in the Cu complex, indicating a higher effect of the Cu ion on the phenyl at the meso-positions. The bands between 1300 cm^{-1} and 1450 cm^{-1} are most likely the out-of-phase coupled $C_\alpha C_\beta/C_\alpha N$ stretching modes. The 1323 and 1360 cm^{-1} bands of the porphyrin ligand were most probably the pyrrole quarter ring stretching and the $C_\alpha C_\beta/C_\alpha N$ stretching modes, respectively. The $C_\alpha C_\beta/C_\alpha N$ stretching appeared at 1333 cm^{-1} for both the Cu and the Fe complexes too. The pyrrole stretching was found at 1360 cm^{-1} for the Cu and Fe complexes as well. The 1236 cm^{-1} band of the 5,10,15,20-tetrakis(4-aminophenyl)-porphyrin (TAPP) and the 1230 and 1234 cm^{-1} bands of the Cu and Fe complexes were most likely attributed to the C_m-ph stretching. The band at 1076 cm^{-1} of the porphyrin ligand was most probably the vibration of the pyrrole C_β–H stretching, which appeared at the same wavenumber for the Cu complex and shifted to 1080 cm^{-1} for the Fe complex. The band at 997 cm^{-1} of the porphyrin ligand was most likely the vibration of pyrrole breathing and phenyl stretching, which shifted to 1001 cm^{-1} in the case of both the Cu and the Fe complexes as well. The band at 960 cm^{-1} of the porphyrin ligand was most probably the pyrrole breathing, but it could not be found in the complexes due to the exchange of the hydrogen atom with the metal ion in the N-H bonding. The band at 710 cm^{-1} was most likely the $\pi 3$, phenyl mode, which was not seen in the Cu complex but shifted to 714 cm^{-1} for the Fe complex. At 384 cm^{-1} for the Cu and at 385 cm^{-1} for the Fe complexes the bands are assigned to the M-N vibration, which cannot be seen in the porphyrin ligand [40]. For the porphyrin ligand, a weak Raman band can be seen at 330 cm^{-1}, which was most likely the in-plane translational motion of the pyrrole [41]. The peaks at around 1000 cm^{-1}, 1340 cm^{-1} and 1580 cm^{-1} indicate the presence of the aromatic hydrocarbons in which a hydrogen has been replaced by an amino group (benzene and pyrrole). These peaks appear in the spectra of the complexes as well as in that of the empty porphyrin ring but at a slightly different position (circled peaks in the spectra). For easier comparability, adifferent intensity scale was used in the case of the Cu-porphyrin complex since the intensities were too weak compared to the other two samples. Figure 4 shows the Raman spectra of the functionalized aerogels. Since the aerogels contain a little amount of the complexes, it was quite difficult to record acceptable quality spectra. Despite this hardship, the difference between the spectrum of the CuPA aerogel and that of the free complexes clearly indicate a change in the structure of the complex by which the formation of the covalent bond between the complex and silica framework is confirmed.

The results of the elemental analysis can be seen in Table 1. The non-zero carbon and nitrogen content of the blank aerogel sample indicates that after the supercritical drying adsorbed residues of the solvents and the catalyst ammonia may be present in the aerogels. Compared to that as a background, both the carbon and nitrogen contents of CuPA and FePA are higher in the obtained samples, indicating the successful incorporation of the porphyrin complexes.

Porosity of the aerogel samples was measured by nitrogen adsorption porosimetry at 77 K temperature, after degassing the samples at 100 °C for 24 h. Figure 5 shows the cumulative pore volumes and pore size distribution curves calculated by the BJH method. Nitrogen adsorption-desorption isotherms can be found in the Supplementary Materials (Figure S4). The specific surface area of the CuPA and FePA aerogels was 980 m^2/g, and 1019 m^2/g, respectively. The actual values are significantly higher than the 600–700 m^2/g values characteristic of pristine silica aerogels prepared by the same method. Bulk densities of the functionalized aerogels were in the range of 0.076–0.085 g/cm^3, which was a bit lower than 0.085–0.090 g/cm^3 obtained for the pristine silica aerogels. Measured and calculated parameters are given in Table 2. The values are characteristic of the silica-based aerogels in general, thus the incorporation of the porphyrin complexes did not alter the gel structure significantly. Most of the pores are in the 2–50 nm mesopore region, and only a negligible volume falls in the less than 2 nm diameter micropore region, as calculated by the t-plot method.

Scanning electron microscopy (SEM) pictures were in good agreement with the porosimetry results. The samples showed the homogeneous structure, and the visible pores were in the higher mesopore and lower macropore region (Figure 5). The size of the globules and the pores are also characteristic of the silica-based aerogels. Most importantly, no detectable agglomeration of the porphyrin complexes was observed. According to these results a uniform and molecular level distribution of the complexes was obtained in the silica matrix.

The metal content of the aerogels was determined by an inductively coupled plasma-optical emission spectrometry (ICP-OES) method, and the results are summarized in Table 3. The measured values are a bit higher than the values calculated from the chemical composition of the reaction mixtures, due to the leaching of short-chained partially hydrolized siloxanes in the ageing and solvent exchange steps. However, the colourless nature of the ageing solutions proved that no intense-coloured porphyrin complex was lost in the process, and their entire amount was incorporated.

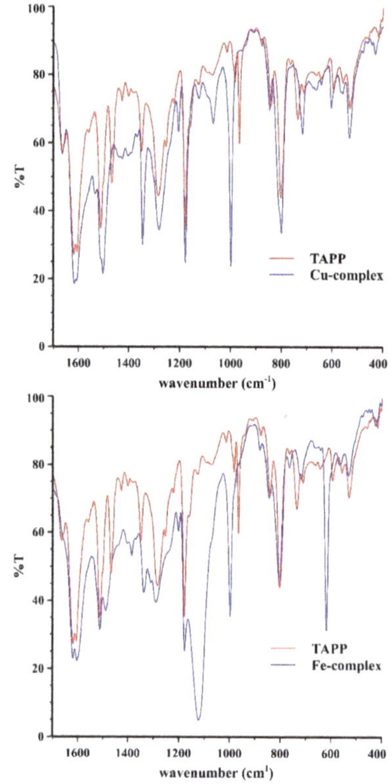

Figure 1. Fingerprint region of the FT-IR spectra of the free copper and iron complexes (blue curves) compared with that of the empty porphyrin ring (red curves). The full recorded spectra can be seen in the Supplementary Materials (Figure S3). Although accurate assignation of the peaks is not available, a detailed analysis of the IR spectra of tetraphenyl porphyrin complexes is available in the literature [42].

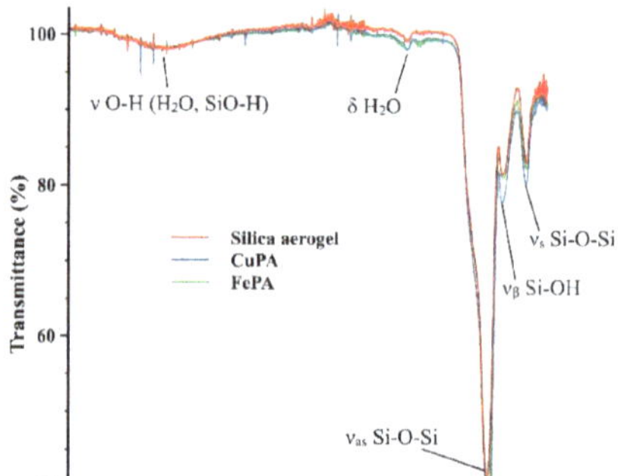

Figure 2. FT-IR spectra of the aerogels CuAP and FeAP functionalized with the copper and iron porphyrin complexes. They are in good agreement with the spectra published in the literature for silica aerogels [43] and silicas covalently coupled with porphyrins [44]. Due to the low concentration of the complexes in the silica aerogel matrix, the characteristic peaks of the complexes shown in Figure 1 are too weak to be observed.

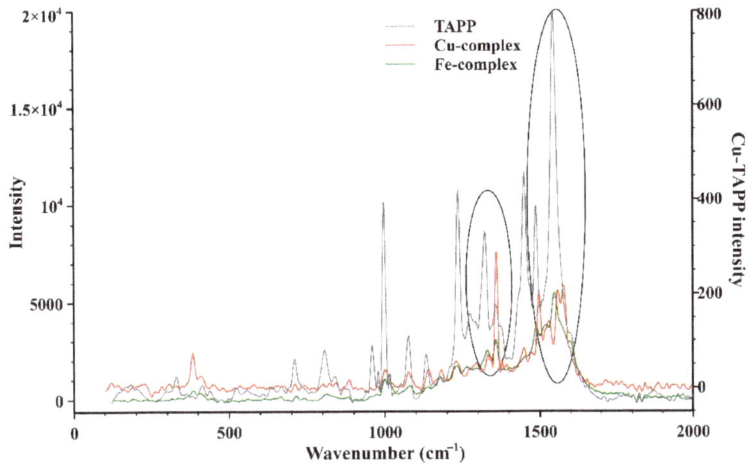

Figure 3. Raman-spectra of the empty 5,10,15,20-tetrakis(4-aminophenyl)porphyrin ring (TAPP) and its Cu- and Fe-complexes. In the case of Cu-complex, a different intensity scale was used for easier comparability of the spectra. The differences between the positions of the circled peaks—which can be assigned to the aromatic hydrocarbons, in which a hydrogen has been replaced by amino group—indicates change in the structure and the formation of the complexes.

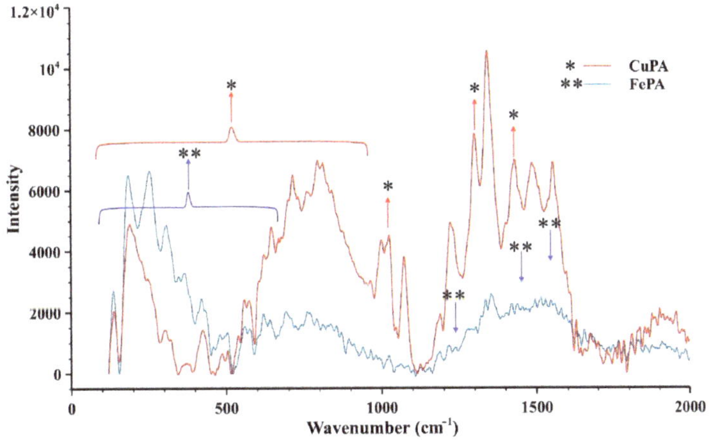

Figure 4. Raman-spectra of the aerogel materials CuPA and FePA. Due to the low concentration of the porphyrin complex, the spectrum of FePA is less informative than that of CuPA. The spectral differences and the direction of the intensity changes of the CuPA and FePA aerogels compared to the complexes (shown in Figure 3) clearly indicate the formation of covalent bonds between the functionalized porphyrin complexes and the silica aerogel matrix. (*) CuPA, (**) FePA.

Table 1. Elemental analysis of the hybrid aerogel samples. The increased carbon and nitrogen content indicate the presence of the porphyrin complexes in the silica matrix. The non-zero values of the blank silica sample are an indication of the presence of adsorbed organic residues and the catalyst ammonia after the supercritical CO_2 drying process.

Sample\Element	C (%)	H (%)	N (%)
"blank" silica aerogel	0.75	1.41	0.36
CuPA	1.93	1.70	0.51
FePA	3.21	1.80	0.69

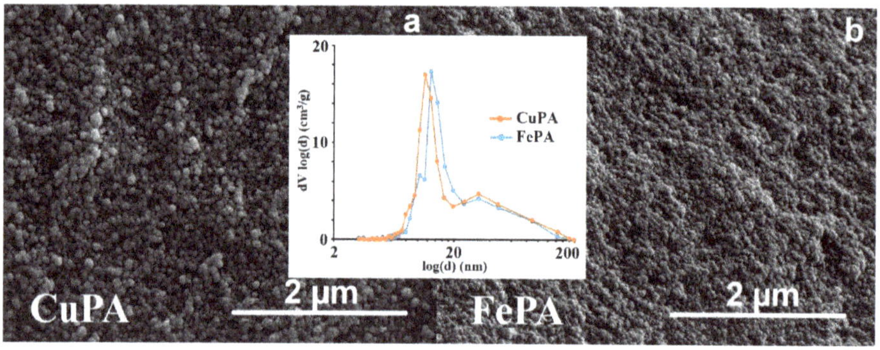

Figure 5. Scanning electron micrographs of the aerogel catalysts (**a**) CuPA, and (**b**) FePA. Insert: Pore size distribution curves calculated by the BJH method for CuPA (red line) and FePA (blue line). Both of the curves show the characteristic pore diameter of around 12 nm, which is typical for the silica aerogels.

Table 2. Summary of the nitrogen adsorption/desorption porosimetry results. Specific surface area (S_{BET}) was calculated by the multi-point BET method. Characteristic pore diameters (d), and total pore volumes (V_{total}) were calculated from the isotherms by the BJH method. Mesopore and macropore volumes (V_{meso}, V_{macro}) were calculated from the cumulative pore volume curves for the regions 2–50 nm and above 50 nm, respectively. Micropore contribution (V_{micr}) was calculated by the t-plot method.

	CuPA	FePA
S_{BET} (m^2/g)	980	1019
d (nm)	12.8	11.4
V_{total} (cm^3/g)	3.617	3.9185
V_{macro} (cm^3/g)	0.3702	0.417
V_{meso} (cm^3/g)	3.2218	3.4975
V_{micro} (cm^3/g)	0.025	0.004

Table 3. Metal ion content of the catalysts measured by the ICP-OES technique.

Sample	Theoretical (*w/w* %)	Measured (*w/w* %)
CuPA	0.068	0.072 ± 0.001
FePA	0.084	0.110 ± 0.001

2.3. Catalytic Activity

The catalytic activity of the samples was evaluated through the oxidation of phenol and chlorophenols with hydrogen peroxide in aqueous solutions at different temperatures. The pH of the reaction mixtures was left to change spontaneously in the course of the reactions in order to simulate the behaviour of real-life wastewaters. The main oxidation products of phenol were identified by HPLC measurements; further products and the proposed degradation pathways are published in a previous work [26]. In the case of chlorinated phenols, the oxidation products were undetectable by UV/HPLC, and mass spectrometry was used instead.

We made several attempts to determine the catalytic activities of the complexes in the homogeneous phase. Unfortunately, all of them failed, due to the insolubility of the porphyrin complexes in water, water–DMF and water–DMSO mixtures. The reaction mixture seemed to be homogeneous at the applied 90 °C, nevertheless when it started to cool down, solid particles appeared and the solution turned colourless (Figure S5). The reaction could have been tested in a homogeneous phase using DMF as the solvent, in which only the calculated volume of 30% (m/m) aqueous hydrogen peroxide was dissolved. However, none of the complexes showed any catalytic activity against the phenols. The reason for this can be either that a fast self-oxidation destroyed all the porphyrin complexes, or that the oxidation of the solvent DMF, which was present in large excess, consumed the oxidant.

2.3.1. Phenol Oxidation

Phenol oxidation and conversion was monitored by a reversed-phase HPLC technique, the concentrations were determined by using a five-point calibration curve and UV detection. Several catalyst-to-substrate, and catalyst-to-hydrogen peroxide molar ratios were tested in batch experiments. Conversion curves of phenol are shown in Figure 6. The main feature is that 80% of the phenol was converted within three hours even when the catalyst was applied only in a 0.33 mol%. Obviously, the free copper(II) ions in the homogeneous phase were much more effective. Nevertheless, in a homogeneous phase the separation of the catalyst Cu^{2+} would be difficult and it is not favorable in industrial use. As expected, the free iron(2+) ions were more effective catalysts in the homogeneous phase than the complex in the heterogeneous phase. However, when the catalyst was applied in 1 mol% quantity, more than 90% of the phenol was eliminated within one hour (Figure 6b). The FePA catalyst applied in a smaller amount did not show as high catalytic activity as the catalyst CuPA did. Both catalysts showed slightly S-shaped conversion curves, which may be the indication of autocatalytic reactions or the consequence of hindered diffusion and materials transport in the pores.

Depending on the applied metal complexes, a different intermediate profile was obtained (Figure 7). In the case of the copper complex, two main intermediates were detected: catechol and hydrochinon. Only catechol was detected as an intermediate when the iron complex was applied, supposedly due to the different reaction mechanisms. It was confirmed by mass spectrometry that the intermediates continued to transform into further products, and finally into short-chained carboxylic acids [45].

In order to compare the efficiency of the catalysts more expressively, the turnover frequencies (TOFs) were calculated. We selected a set of points from the initial linear section of the kinetic curves and applied linear regression (Figure 8). The TOF values and the regression coefficients of the fittings are shown in Table 4. As it can be seen, the rate of the phenol degradation is higher in the case of catalyst CuPA but both of the TOF values are comparable to that of the industrial catalysts, since for the most relevant industrial applications the TOF values are in the range of 10^{-2}–10^{2} s^{-1} [46].

Unfortunately, the catalysts lost their activity after the first cycle, most likely because the porphyrin ring suffered self-oxidation. Due to the static conditions applied during the reactions, the high excess of hydrogen peroxide could cause the oxidation of the porphyrin ring.

The reaction was carried out with lower H_2O_2 excess, the applied molar ratios were: catalyst:phenol:H_2O_2 = 1:100:1400. The kinetic curve can be seen in Figure 6. The conversion was higher than 90% after one hour in this case as well, but the colour of the catalyst changed this time too, which indicated the degradation of it. Therefore, we tried to optimize the temperature too. The reaction was carried out at different temperatures besides 90 °C: 30–70 °C using the catalyst:phenol:H_2O_2 = 1:100:1400 molar ratio. The final phenol conversion as a function of the temperature can be seen in Figure 9. We found that the highest conversion was achieved at 90 °C. There is a breakpoint between 50 °C and 60 °C, the conversion was 50% compared to 12% at 50 °C. At 60 °C the degradation of the catalyst was minimal, although the rate of the reaction was much lower than at 90 °C, the calculated turnover frequency was 4.06×10^{-4} s^{-1}.

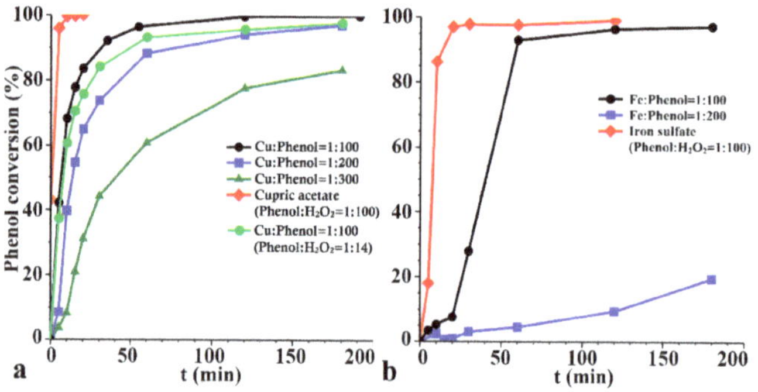

Figure 6. Kinetic curves of phenol conversion over time using the CuPA catalyst (**a**) and FePA catalyst (**b**), and the free metal ions using different catalyst-to-phenol (1:100, 1:200, 1:300), and phenol-to-H_2O_2 molar ratios (1:100 and 1:14).

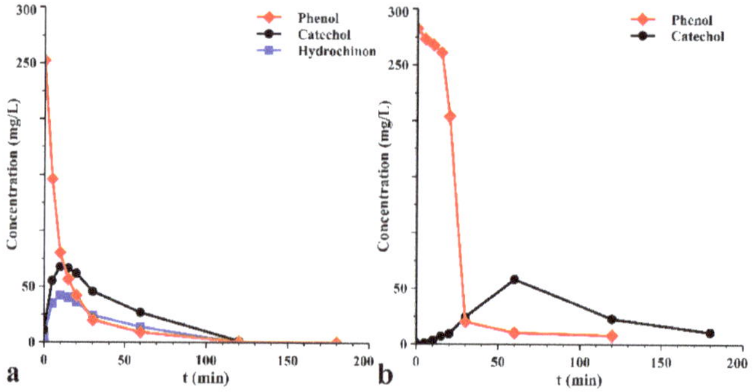

Figure 7. Intermediates profile of phenol oxidation by CuPA (**a**) and FePA (**b**) catalysts. In the case of the copper complex, catechol and hydrochinon were detected, and only catechol was observed when iron ions were applied.

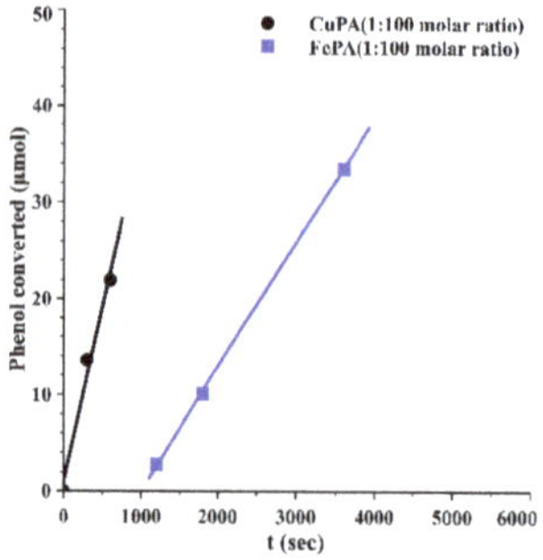

Figure 8. Linear trend lines fitted to points selected from the initial linear phase of the kinetic curves of phenol oxidation using CuPA (•) and FePA (■) catalysts.

Table 4. Turnover frequencies (TOFs) of the immobilized complexes applied for oxidation of phenol.

Sample	TOF (s^{-1})	Regression Coefficient
CuPA	1.144×10^{-1}	0.9816
FePA	4.05×10^{-2}	0.9998

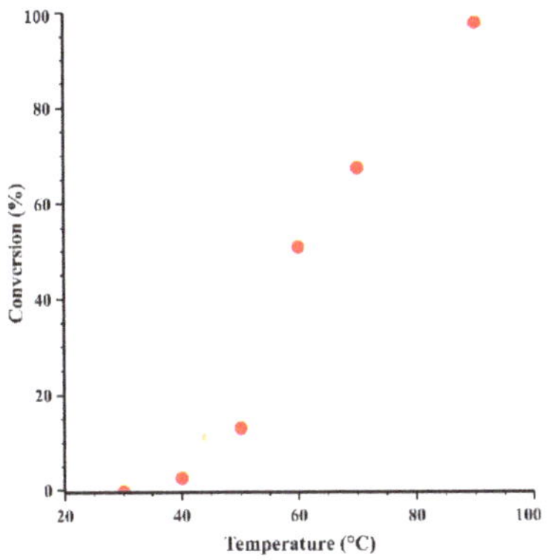

Figure 9. Phenol conversion after 180 min as a function of the reaction temperature. The highest conversion was achieved at 90 °C. Nevertheless, at 60 °C the degradation of the catalyst was minimal, so this value is found to be the most advantageous to carry out the reaction.

The quantity of the free metal ions in the reaction mixture after the decomposition of phenol was determined by an ICP-OES method. After three hours reaction time, 36% of the total copper ion, and 89% of the total iron ion was free. In a control experiment, when only the catalysts were suspended in distilled water and heated for three hours, only 3.7% of the total copper-ion was measured in the supernatant. This was in good agreement with the increasing solubility of the amorphous silica aerogel in water at elevated temperatures. The concentration of the free iron ions in the control experiment was under the limit of detection (LOD). Both experiments proved that the presence of the oxidizing agent hydrogen peroxide was the prerequisite for the degradation of the porphyrin rings, which led then to the release of metal ions. That is the reason why all of the regeneration attempts failed. The degradation of the porphyrin ring was discernible through the colour change of the catalysts. It gradually turned from red to brown in the case of CuPA, and the brown colour of the FePA catalyst slowly disappeared during the reaction.

Beyond the batch experiments, a custom made continuous-flow reactor (see in Supplementary Materials, Figure S6) was tested as well. Despite of the short contact time (9 min) we observed good catalytic activity, although the efficiency gradually decreased as the reaction proceeded (Figure 10). However, it might be approaching a constant value in longer times when the neighboring catalytic centers became so distant that they could not oxidize each other. The results proved that our aerogel-based materials can be used as catalyst beds in continuous flow processes, which is more advantageous and manageable for the industry.

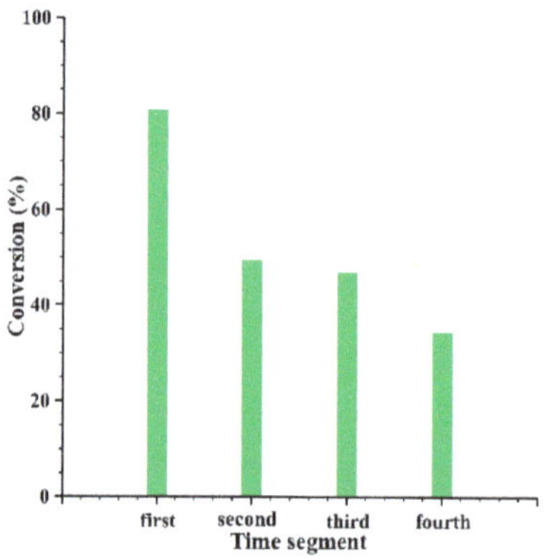

Figure 10. Conversion of phenol in a short-bed continuous-flow tubular reactor determined for 30 min time segments. (temperature 90 °C).

2.3.2. Oxidation of 3-Chloro- and 2,4-Dichlorophenol

The kinetic curves of the conversion of 3-chlorophenol (3-CP) can be seen in Figure 11. In contrast to the phenol oxidation, the catalytic activity of the CuPA catalyst was smaller than that of the FePA catalyst. Furthermore, the immobilized complex showed higher activity than the free iron ions even when it was applied in Fe:3-CP = 1:200 molar ratio. The decreased activity of the free iron ions is most likely due to the hydrolysis of Fe^{3+} ions forming catalytically inactive hydroxo-iron precipitates under the solution pH. The "S" shape of the curve in that case may also indicate either autocatalytic or diffusion controlled processes. The oxidation products were identified by high-resolution mass spectrometry (HRMS), both in the positive and in the negative ion mode. Based on the results, we suggested a reaction pathway, which is given in Scheme 2. The aromatic ring was hydroxylated first, then dechlorinated, split open and fragmented into short-chained carboxylic acids. In each case, fragmentation occurred through the loss of carbon dioxide. In the case of 2,4-dichlorophenol (2,4-DCP) (Figure 12) a significant decrease in the catalytic activity of the CuPA catalyst was observed compared to the 3-chlorophenol. Although the activity of the FePA decreased as well, it was almost as effective as in the case of the monochlorophenol. The free iron ions showed poor efficiency due to their hydrolysis under the reaction conditions, as mentioned above. Considering the complexity of the reaction, as well as the lack of appropriate analytical standards, the details of the mechanism has not been explored.

The turnover frequencies were also calculated to compare the efficiency of the catalysts (Figure 13), the results are summarized in Table 5. According to the results, the efficiency of the CuPA catalyst did not reach that of the catalysts applied in the industry if we used it for oxidation of a dichloro derivative of the phenol. In contrast to that, the TOF values of the FePA catalyst are still comparable with the industrial catalysts' TOF values.

The chlorophenols (mono- and dichlorinated) were identified by capillary electrophoresis coupled to a mass spectrometer (CE-MS). The advantage of using this technique over HPLC is the higher sensitivity of the MS detector, which enables detection of chlorophenols in a concentration range more relevant to environmental regulations [47]. A simple method has developed for the CE separation of the chlorophenols, which also made it possible to set the limit of quantitation (LOQ) below 1 ppm, which is the regulatory limit for phenol in wastewaters. Chlorophenols are ionized more readily in the negative ion mode, therefore, the negative ion mode was applied along with separation in basic background electrolyte (40 mM ammonium formate/ammonia, pH = 9.5).

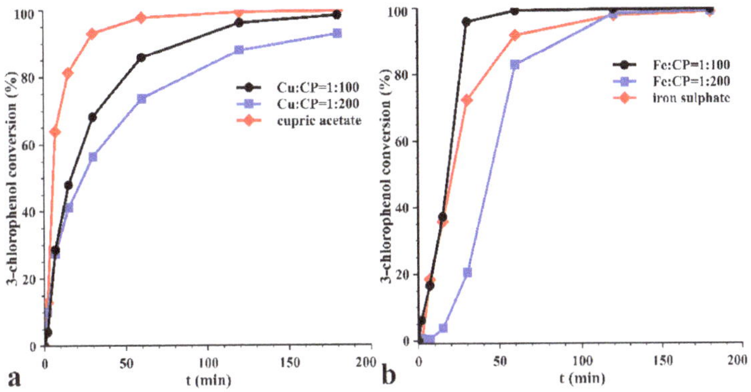

Figure 11. Kinetic curves of 3-chlorophenol conversion over time using the CuPA catalyst (**a**) and FePA catalyst (**b**) in different catalyst-to-phenol molar ratios (1:100 and 1:200), as well as free metal ions. The 3-chlorophenol-to-hydrogen peroxide ratio was at a constant value of 1:100.

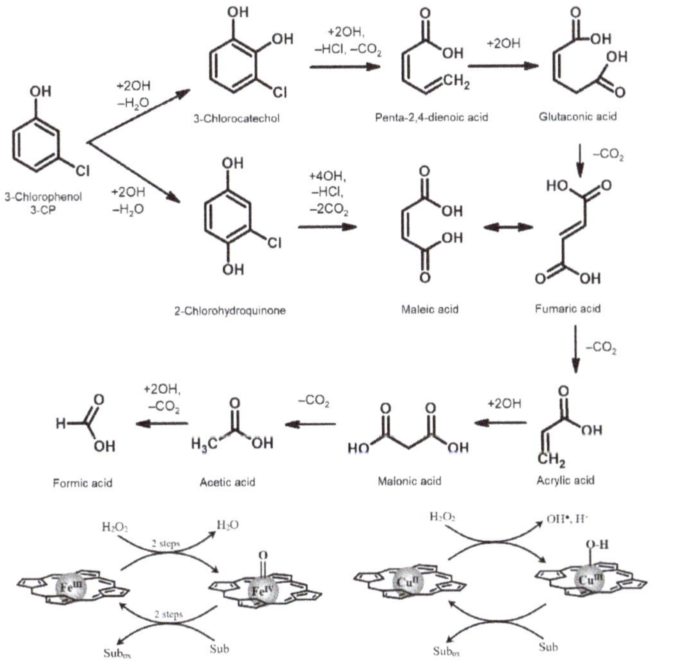

Scheme 2. Proposed reaction pathway, formal stoichiometry, and reaction mechanisms of Fenton-oxidation of 3-chlorophenol. Through catalytic hydroxylation and dechlorination, chlorophenol was transformed into non-toxic short-chained carboxylic acids. All of the intermediates shown were identified by high-resolution mass spectrometry (HRMS).

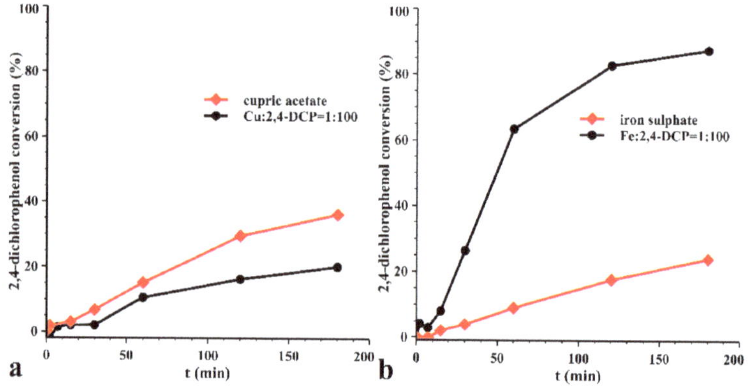

Figure 12. Kinetic curves of 2,4-dichlorophenol (2,4-DCP) conversion over time using the CuPA catalyst (**a**) and FePA catalyst (**b**), as well as free metal ions. The 2,4-dichlorophenol-to-hydrogen peroxide ratio was kept at a constant value of 1:100.

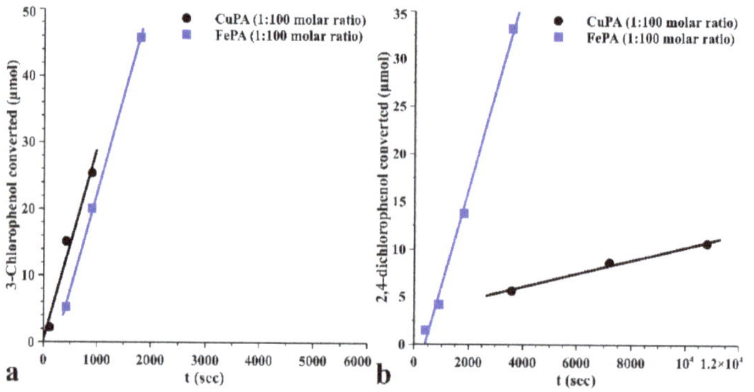

Figure 13. Linear trend lines fitted to points selected from the initial linear phase of the kinetic curves of 3-chlorophenol (**a**) and 2,4-dichlorophenol (**b**) oxidation using CuPA (•) and FePA (■) catalysts.

Table 5. Turnover frequencies (TOFs) of the immobilized complexes applied for oxidation of 3-chlorophenol and 2,4-dichlorophenol.

	3-Chlorophenol		2,4-Dichlorophenol	
	TOF (s^{-1})	Regression Coefficient	TOF (s^{-1})	Regression Coefficient
CuPA	5.0×10^{-2}	0.9604	1.5×10^{-3}	0.9884
FePA	5.0×10^{-2}	0.9996	2.2×10^{-2}	0.995

3. Conclusions

Copper and iron complexes of 4-aminophenylporphyrin have been immobilized successfully in silica aerogels using isocyanopropyl triehoxysilane as a bifunctional linker reagent. By functionalizing the porphyrin rings, followed by complexation with selected metal ions, we were able to bind the complexes to the silica aerogel matrix with strong covalent bonds. The as-obtained catalysts have a large specific surface area and an open mesoporous structure, which are important features of the heterogeneous catalysts.

The catalytic activity of the samples was tested through the oxidation of phenol, 3-chlorophenol and 2,4-dichlorophenol by hydrogen peroxide. The oxidation products were identified by high-pressure liquid chromatography in the case of phenol and by high-resolution mass spectrometry in the case of 3-chlorophenol. All intermediates of the degradation process were identified and a reaction scheme was proposed to describe the entire process. The FePA and CuPA catalysts showed different selectivity towards the substrates. In the case of phenol, the copper complex proved to be more efficient, while for the chlorinated derivatives the iron complex showed significantly higher activity. We have demonstrated that they can be used both in the batch and the continuous-flow modes.

Both the CuPA and FePA catalyst oxidized the dangerous phenolic pollutants in water by the safe agent hydrogen peroxide. Phenol, 3-chlorophenol and 2,4-dichlorophenol were converted to non-toxic short-chained carboxylic acids, and then finally mineralized into carbon dioxide, water and hydrochloric acid in the process. A special advantage of the process is that it can be used directly with contaminated natural waters, as the process does not require any buffering or use of additives.

Our study clearly showed that the covalent incorporation of the ab ovo water-insoluble porphyrin complexes in a hydrophilic silica aerogel matrix made them compatible with the aqueous medium and allowed their active use in such an environment. By immobilization, the extent and the rate of the catalysts' self-oxidation was reduced, although it was still present at a lower level. Since the direct interaction of the catalytically active centers of the complexes was not possible due to their immobilization, the degradation was most likely the consequence of the attack of active hydroxyl radicals generated in the catalytic cycles. For future application, the chemical stability of the complexes should be improved for example by changing the nature of the connecting pendant arms [48]. In order to minimize the degradation of the catalysts but keep the reaction fast enough, the temperature was optimized to 60 °C and the phenol:hydrogen peroxide molar ratio was cut back to 1:14, which is close to the theoretical limit of 1:12–14 molar ratio required for complete mineralization of chlorinated phenols. However, under such conditions, the turnover frequency dropped from the 10^{-1}–10^{-2} s^{-1} range to as low as 4.1×10^{-4} s^{-1}. Our results show that the as-obtained catalysts CuPA and FePA may be considered as potential alternatives for the mineralization of phenols with the environmentally safe oxidizing agent hydrogen peroxide.

4. Materials and Methods

4.1. Materials

Methanol (technical grade), acetone (technical grade), 30% hydrogen peroxide solution (analytical reagent grade), N,N-dimethylformamide (DMF) (analytical reagent grade) and 25 m/m% ammonia solution (analytical reagent grade) were purchased from Molar Chemicals Kft. (Hungary). Tetramethyl orthosilicate (TMOS) (purum), 3-chlorophenol (98 %) and 2,4-dichlorophenol (99 %) were obtained from Sigma-Aldrich Ltd. (St. Louis, MO, USA). 5,10,15,20-Tetrakis(4-aminophenyl)porphyrin (TAPP) and 3-isocyanatopropyltriethoxysilane (3-IPTES) were purchased from ABCR GmbH (Germany). Phenol (analytical reagent grade), copper acetate (reagent grade) and iron(II) sulfate (reagent grade) were acquired from Reanal Finomvegyszergyár Zrt. (Hungary). Carbon dioxide cylinder was purchased from Linde Gáz Magyarország Zrt. (Debrecen, Hungary). All the reagents were used without any further purification.

4.2. Synthesis of Porphyrin Complexes

The synthesis of the complexes was carried out in two consecutive steps. First, the porphyrin ring was functionalized with 3-isocyanatopropyltriethoxysilane (3-IPTES) in a two neck flask under anhydrous conditions. A total of 20 mg of 5,10,15,20-tetrakis(4-aminophenyl)porphyrin (2.97×10^{-2} mmol) was dissolved in 4.0 mL anhydrous DMF and kept under a dry argon atmosphere. 0.33 mL (1.191 mmol) of 3-IPTES was added to the solution also under an argon atmosphere. The mixture was stirred at 70 °C for five days. Next, the complexes were obtained by mixing 1.0 mL of 3-IPTES-TAPP with either copper or iron salts. The reaction mixtures were diluted to 4.0 mL final volume with DMF and stirred at 116 °C in sealed vessels. The metal ions were dosed in small portions until the red fluorescence of the free base 3-IPTES-TAPP disappeared, indicating the complete formation of the complexes. The reaction time depended on the metal-ions. The Cu^{2+}-TAPP-3-IPTES complex formed typically in 10 min, while the formation of the complex with iron ion took up to 72 h. The progress of the reactions was monitored by TLC.

4.3. Synthesis of Heterogeneous Catalysts

The catalysts (denoted as CuPA and FePA) were obtained by covalently binding the complexes to the silica precursors and then co-gelated with TMOS to develop the hybrid aerogel matrix. Two monoliths were prepared via the following general recipe. Two solutions (labelled "A" and "B") were prepared. Solution "A" contained 12.0 mL MeOH (297 mmol) and 3.0 mL TMOS (20.33 mmol). Solution "B" was made from 12.0 mL MeOH (297 mmol), 2.00 mL of the complex solution (3.705×10^{-3} mmol), 1.0 mL distilled water (5.54 mmol) and 1.00 mL of diluted (1:1) NH_3 solution (7.34 mmol). Solutions "A" and "B" were mixed together, then poured into cylindrical plastic molds (66×28 mm) and sealed with parafilm. The gels were kept in the molds overnight then they were transferred into perforated aluminum frames. The frames provided mechanical support and allowed for quick and efficient exchange of the solvents before supercritical drying. All gels were soaked in the following solvents in order to purify them and to remove the water: methanol-cc NH_3 (8:1), pure methanol; each for a day. Then methanol was gradually replaced by acetone, changed in 25% steps. There was no indication of leaching of the porphyrin complexes during the process. Finally, the gels were stored in a copious volume (2 L) of freshly distilled dry acetone for three days. After the change of the solvents, the aerogels were obtained by supercritical drying, which was carried out in a custom-made high-pressure reactor according to a general procedure published in a previous work [39].

4.4. Characterization of the Catalysts

Nitrogen gas porosimetry measurements were performed on a Quantachrome Nova 2200e surface area and porosity analyzer (Quantachrome Instruments, Boynton Beach, FL, USA). Pieces of the samples were ground in a mortar and outgassed under vacuum at 100 °C for 24 h before the measurements.

1H-NMR measurements were performed on a Bruker 360 spectrometer (Bruker Billerica, MA, USA).

FT-IR spectra were recorded on a Jasco FT/IR-4100 instrument (Easton, MD, USA).

Raman measurements were performed on a Renishaw InVia Raman microscope (Renishaw, Wotton-under-Edge, United Kingdom). It was used for the characterization of the aerogel samples in the range 100–5000 Raman shift/cm^{-1}. The laser used for the measurements was a 532 nm, 50 mW diode laser. All spectra were recorded at a 10 s exposure time each, utilizing 2400 L/mm grating. Beam centering and Raman spectra calibration were performed daily before spectral acquisition using the inbuilt Si standard.

Elemental analysis was performed in a varioMICRO element analyzer (Elementar Analysensysteme GmbH, Hanau, Germany).

Scanning electron micrographs (SEM) were recorded on a Hitachi S-4300 instrument (Hitachi Ltd., Tokyo, Japan) equipped with a Bruker energy dispersive X-ray spectroscope (Bruker Corporation, Billerica, MA, USA). The surfaces were covered by a sputtered gold conductive layer and a 5–15 kV accelerating voltage was used for taking high resolution pictures.

The analysis of the metal content of the aerogels was carried out by an Agilent ICP-OES 5100 SVDV device (Agilent Technologies, Santa Clara, CA, USA) after nitric acid and hydrogen peroxide microwave digestion.

4.5. Study of Catalytic Activity

The catalytic activity of the aerogels was tested through the decomposition of phenol and chlorophenols in an aqueous solution.

In the case of phenol, typically the phenol solution was placed into a flask along with the appropriate amount of catalyst; the molar ratios were catalyst to phenol 1:100, 1:200 or

1:300. The required amounts of the aerogels were calculated according to the ICP results. The mixture was thermally equilibrated for several minutes before hydrogen peroxide (molar ratios phenol to hydrogen peroxide 1:100 or 1:14) was added to start the reaction. The total volume was 15.0 mL. The initial concentration of phenol was 250 ppm. The reaction mixtures were stirred in closed vessels between 30–70 °C and at 90 °C for three hours. A total of nine samples were taken (1.0 mL each) and centrifuged. The supernatants were analyzed with HPLC, applying the following parameters: the column was a Phenomenex Phenyl Hexyl column (150 × 4.6 mm, particle size: 5 μm); the composition of the mobile phase was 50% H_2O, 50% MeOH; the flow rate was 1.0 mL/min and the analysis time was 5 min. The components were detected by a UV detector at 270 nm.

Beyond the batch experiments, the oxidation of phenol was carried out in a continuous-flow reactor as well. The applied catalyst to phenol and phenol to hydrogen peroxide molar ratios were both 1:100. The phenol concentration was 500 ppm. The reaction mixture—including the phenol solution, water and hydrogen peroxide solution—was filled in a syringe and was dosed by a syringe pump with 2 mL/h rate. The catalyst was filled in a U-shaped glass tube of 0.3 cm inner diameter. The bed length was 4.3 cm, the contact time was set to 9 min. The reactor was immersed in a 90 °C water bath.

In the case of the chlorophenols, the chlorophenol solution was placed into a flask along with the appropriate amount of catalyst, the molar ratios were catalyst to chlorophenol 1:100. The mixture was thermally equilibrated at 70 °C for several minutes before hydrogen peroxide (molar ratio: chlorophenol: hydrogen peroxide = 1:100) was added to start the reaction. The total volume was 15.0 mL. The initial concentration of the chlorophenols was 500 ppm. The reaction mixtures were stirred in closed vessels at 70 °C for three hours. In total, eight samples were taken (1.0 mL each) and centrifuged. The supernatants were analyzed with HPLC, applying the following parameters. The column was a Supelcosil LC-18 (250 mm × 4.6 mm, particle size: 5 μm) column; the composition of the mobile phase was mixture of 50% ammonium acetate (50 mM)—50% MeOH in the case of 3-chlorophenol; and ammonium acetate:methanol = 20:80 in the case of 2,4-DCP; the flow rate was 1.0 mL/min, the analysis time was 8 min in both cases. 3-CP and 2,4-DCP was detected by a UV detector at 280 nm and 287 nm, respectively.

Oxidation products of chlorophenols were detected by high resolution mass spectrometry (maXis II UHR ESI-QTOF MS instrument, Bruker, Karlsruhe, Germany), both in the positive and in the negative ion mode. For positive mode ESI, the following parameters were used: capillary voltage: 3.5 kV, nebulizer pressure: 0.5 bar, dry gas flow rate: 4.5 L/min, temperature: 200 °C. For negative mode ESI, the following parameters were used: capillary voltage: 2.5 kV, nebulizer pressure: 0.5 bar, dry gas flow rate: 4 L/min, temperature: 200 °C. MS tuning parameters were optimized in both cases to measure the relevant m/z range for chlorophenols (and their oxidation products), which was 50–600 m/z, to generally detect any possible products.

A more sensitive (CE-)MS method was developed to separate and detect chlorophenols in low concentrations (~1 ppm). The abovementioned MS instrument was coupled to a capillary electrophoresis (7100 CE System, Agilent, Waldbronn, Germany) instrument via a coaxial CE-ESI sprayer interface (G1607B, Agilent). Sheath liquid was transferred with a 1260 Infinity II isocratic pump (Agilent). CE instrument was operated by OpenLAB CDS Chemstation software.

Parameters for the capillary zone electrophoretic separation: capillary: 90 cm × 50 μm fused silica; background electrolyte: 40 mM ammonium formate/ammonia (pH 9.5); applied voltage: 20 kV; hydrodynamic injection: 500 mbar·s; sheath liquid: iso-propanol: water = 1:1 containing 5 mM ammonia; sheath liquid flow rate: 10 μL/min. After each injection, a small amount of background electrolyte was also injected from a distinct vial (150 mbar) to eliminate carry-over effects. During electrophoresis, 35 mbar pressure was applied to the inlet buffer reservoir to decrease migration times significantly. The MS method in the negative mode was tuned according to the desired mass range (80–250 m/z), a much narrower one compared to the general detection of products, to obtain better sensitivity, although parameters for the ESI source were the same. Applied spectra rate was 3 Hz. Electropherograms were recorded by otofControl version 4.1 (build: 3.5, Bruker). Spectral background correction and internal calibration were performed on each electropherogram and peaks were integrated by Compass DataAnalysis version 4.4 (build: 200.55.2969).

Leaching of metal ion from the catalysts was studied by the ICP-OES technique after the following procedure: the remaining reaction mixture was centrifuged and the supernatant was filtered through a PTFE membrane filter (pore size: 0.45 μm), after which the solution was filled up in a volumetric flask. The concentrations were determined against standard solutions.

Supplementary Materials: The following are available online at https://www.mdpi.com/article/10.3390/gels8040202/s1, Figure S1: ^{1}H NMR spectra of the 5,10,15,20-tetrakis(4-aminophenyl)porphyrin and the coupling agent: 3-isocyanatopropyltriethoxysilane, Figure S2: ^{1}H NMR spectra of the mixture of 5,10,15,20-tetrakis(4-aminophenyl)porphyrin and the coupling agent 3-isocyanatopropyltriethoxysilane. The spectra of the reaction mixture were recorded directly after mixing and after 96 hours reaction time. The changes—especially in the aromatic region—in the spectra indicate change in the chemical structure of the porphyrin ring, meaning that the functionalization was successful, Figure S3: FT-IR spectra of the empty porphyrin ring (a), and the complexes with Cu(II) (b), and Fe(II) ions (c), Figure S4: Nitrogen adsorption-desorption isotherms of the catalysts denoted as CuPA (left) and the iron-containing one, denoted as FePA (right). The shapes, as expected, are almost perfectly alike, Figure S5: Initially the reaction mixture seemed to be homogeneous (left), but after a while without heating solid particles of the porphyrin complex appeared (center), settled down and the solution became colorless (right), Figure S6: Drawing of the continuous-flow tubular reactor for phenol oxidation.

Author Contributions: Conceptualization, I.L.; methodology, E.G. and I.L.; formal analysis, E.G., Á.K. and I.L.; investigation, E.G., M.S., Á.K. and I.L.; resources, I.L. and I.F.; data evaluation, E.G. and Á.K.; writing—original draft preparation, E.G.; writing—review and editing, I.L. and I.F.; visualization, E.G.; supervision, I.L.; project administration, E.G., I.L. and I.F.; funding acquisition, I.F. and I.L. All authors have read and agreed to the published version of the manuscript.

Funding: The research was supported by the EU and co-financed by the European Regional Development Fund under the project GINOP-2.3.2-15-2016-00041, GINOP-2.2.1-15-2017-00068, GINOP-2.3.2-15-2016-00008, GINOP-2.3.3-15-2016-00004 and COST Action CA18125. The authors also acknowledge the financial support provided to this project by the Hungarian Science Foundation (OTKA: 17-124983) and the National Research, Development and Innovation Office, Hungary (K127931). The RAMAN measurements were financed by the GINOP-2.3.3-15-2016-00029 'HSLab' project.

Institutional Review Board Statement: Not applicable.

Informed Consent Statement: Not applicable.

Data Availability Statement: Not applicable.

Acknowledgments: The authors would like to express their thanks to Edina Baranyai, Sándor Harangi and Petra Herman for the ICP-OES measurements, Lajos Daróczi for the SEM measurements and László Tóth for the FT/IR measurements.

Conflicts of Interest: The authors declare no conflict of interest.

References

1. Ahlborg, U.G.; Thunberg, T.M.; Spencer, H.C. Chlorinated Phenols: Occurrence, Toxicity, Metabolism, And Environmental Impact. *CRC Crit. Rev. Toxicol.* **1980**, *7*, 1–35. [CrossRef] [PubMed]
2. Decision No 2455/2001/EC of the European Parliament and of the Council of 20 November 2001 Establishing the List of Priority Substances in the Field of Water Policy and Amending Directive 2000/60/EC. Available online: http://eur-lex.europa.eu/LexUriServ/LexUriServ.do?uri=OJ:L:2001:331:0001:0005:EN:PDF (accessed on 2 February 2022).
3. Goel, M.; Chovelon, J.-M.; Ferronato, C.; Bayard, R.; Sreekrishnan, T.R. The remediation of wastewater containing 4-chlorophenol using integrated photocatalytic and biological treatment. *J. Photochem. Photobiol. B Biol.* **2010**, *98*, 1–6. [CrossRef] [PubMed]
4. Sahinkaya, E.; Dilek, F.B. Biodegradation of 4-CP and 2,4-DCP mixture in a rotating biological contactor (RBC). *Biochem. Eng. J.* **2006**, *31*, 141–147. [CrossRef]

5. Field, J.A.; Sierra-Alvarez, R. Microbial degradation of chlorinated phenols. *Rev. Environ. Sci. Bio/Technol.* **2008**, *7*, 211–241. [CrossRef]
6. Barik, A.J.; Gogate, P.R. Hybrid treatment strategies for 2,4,6-trichlorophenol degradation based on combination of hydrodynamic cavitation and AOPs. *Ultrason. Sonochem.* **2018**, *40*, 383–394. [CrossRef]
7. Tai, C.; Jiang, G. Dechlorination and destruction of 2,4,6-trichlorophenol and pentachlorophenol using hydrogen peroxide as the oxidant catalyzed by molybdate ions under basic condition. *Chemosphere* **2005**, *59*, 321–326. [CrossRef]
8. Xiong, J.; Hang, C.; Gao, J.; Guo, Y.; Gu, C. A novel biomimetic catalyst templated by montmorillonite clay for degradation of 2,4,6-trichlorophenol. *Chem. Eng. J.* **2014**, *254*, 276–282. [CrossRef]
9. Wang, L.; Kong, D.; Ji, Y.; Lu, J.; Yin, X.; Zhou, Q. Formation of halogenated disinfection byproducts during the degradation of chlorophenols by peroxymonosulfate oxidation in the presence of bromide. *Chem. Eng. J.* **2018**, *343*, 235–243. [CrossRef]
10. Li, Y.; Liu, L.-D.; Liu, L.; Liu, Y.; Zhang, H.-W.; Han, X. Efficient oxidation of phenol by persulfate using manganite as a catalyst. *J. Mol. Catal. A Chem.* **2016**, *411*, 264–271. [CrossRef]
11. Bossmann, S.H.; Oliveros, E.; Göb, S.; Siegwart, S.; Dahlen, E.P.; Payawan, J.L.; Straub, M.; Wörner, A.M.; Braun, A.M. New Evidence against Hydroxyl Radicals as Reactive Intermediates in the Thermal and Photochemically Enhanced Fenton Reactions. *J. Phys. Chem. A* **1998**, *102*, 5542–5550. [CrossRef]
12. Barbusinski, K. Fenton Reaction—Controversy Concerning The Chemistry. *Ecol. Chem. Eng. S* **2009**, *16*, 347–358.
13. Ross, J.R.H. *Heterogeneous Catalysis: Fundamentals and Applications*; Elsevier: Amsterdam, The Netherlands, 2012; ISBN 978-0-444-53363-0.
14. Davis, M.E.; Davis, R.J. *Fundamentals of Chemical Reaction Engineering*, International ed.; McGraw-Hill Chemical Engineering Series; McGraw-Hill: Boston, MA, USA, 2003; ISBN 978-0-07-245007-1.
15. Jiang, S.; Zhang, H.; Yan, Y. Catalytic wet peroxide oxidation of phenol wastewater over a novel Cu–ZSM-5 membrane catalyst. *Catal. Commun.* **2015**, *71*, 28–31. [CrossRef]
16. Shukla, P.; Wang, S.; Singh, K.; Ang, H.; Tade, M. Cobalt exchanged zeolites for heterogeneous catalytic oxidation of phenol in the presence of peroxymonosulphate. *Appl. Catal. B Environ.* **2010**, *99*, 163–169. [CrossRef]
17. Yan, Y.; Jiang, S.; Zhang, H.; Zhang, X. Preparation of novel Fe-ZSM-5 zeolite membrane catalysts for catalytic wet peroxide oxidation of phenol in a membrane reactor. *Chem. Eng. J.* **2015**, *259*, 243–251. [CrossRef]
18. El-Remaily, M.A.E.A.A.A.; Elhady, O. Cobalt(III)–porphyrin complex (CoTCPP) as an efficient and recyclable homogeneous catalyst for the synthesis of tryptanthrin in aqueous media. *Tetrahedron Lett.* **2016**, *57*, 435–437. [CrossRef]
19. Miyamoto, T.; Zhu, Q.; Igrashi, M.; Kodama, R.; Maeno, S.; Fukushima, M. Catalytic oxidation of tetrabromobisphenol A by iron(III)-tetrakis(p-sulfonatephenyl)porphyrin catalyst supported on cyclodextrin polymers with potassium monopersulfate. *J. Mol. Catal. B Enzym.* **2015**, *119*, 64–70. [CrossRef]
20. Najafian, A.; Rabbani, M.; Rahimi, R.; Deilamkamar, M.; Maleki, A. Synthesis and characterization of copper porphyrin into SBA-16 through "ship in a bottle" method: A catalyst for photo oxidation reaction under visible light. *Solid State Sci.* **2015**, *46*, 7–13. [CrossRef]
21. Fukushima, M.; Tatsumi, K. Complex formation of water-soluble iron(III)-porphyrin with humic acids and their effects on the catalytic oxidation of pentachlorophenol. *J. Mol. Catal. A Chem.* **2006**, *245*, 178–184. [CrossRef]
22. Nappa, M.J.; Tolman, C.A. Steric and electronic control of iron porphyrin catalyzed hydrocarbon oxidations. *Inorg. Chem.* **1985**, *24*, 4711–4719. [CrossRef]
23. Rubin, M.; Lampert, C.M. Transparent silica aerogels for window insulation. *Sol. Energy Mater.* **1983**, *7*, 393–400. [CrossRef]
24. Tabata, M.; Adachi, I.; Hatakeyama, Y.; Kawai, H.; Morita, T.; Sumiyoshi, T. Large-area silica aerogel for use as Cherenkov radiators with high refractive index, developed by supercritical carbon dioxide drying. *J. Supercrit. Fluids* **2016**, *110*, 183–192. [CrossRef]
25. Maleki, H.; Durães, L.; García-González, C.A.; del Gaudio, P.; Portugal, A.; Mahmoudi, M. Synthesis and biomedical applications of aerogels: Possibilities and challenges. *Adv. Colloid Interface Sci.* **2016**, *236*, 1–27. [CrossRef]
26. Bereczki, H.F.; Daróczi, L.; Fábián, I.; Lázár, I. Sol-gel synthesis, characterization and catalytic activity of silica aerogels functionalized with copper(II) complexes of cyclen and cyclam. *Microporous Mesoporous Mater.* **2016**, *234*, 392–400. [CrossRef]
27. Lázár, I.; Kalmár, J.; Peter, A.; Szilágyi, A.; Győri, E.; Ditrói, T.; Fábián, I. Photocatalytic performance of highly amorphous titania–silica aerogels with mesopores: The adverse effect of the in situ adsorption of some organic substrates during photodegradation. *Appl. Surf. Sci.* **2015**, *356*, 521–531. [CrossRef]
28. Yoo, J.; Bang, Y.; Han, S.J.; Park, S.; Song, J.H.; Song, I.K. Hydrogen production by tri-reforming of methane over nickel–alumina aerogel catalyst. *J. Mol. Catal. A Chem.* **2015**, *410*, 74–80. [CrossRef]
29. Amiri, T.Y.; Moghaddas, J. Cogeled copper-silica aerogel as a catalyst in hydrogen production from methanol steam reforming. *Int. J. Hydrog. Energy* **2015**, *40*, 1472–1480. [CrossRef]
30. Hu, E.; Wu, X.; Shang, S.; Tao, X.; Jiang, S.; Gan, L. Catalytic ozonation of simulated textile dyeing wastewater using mesoporous carbon aerogel supported copper oxide catalyst. *J. Clean. Prod.* **2015**, *112*, 4710–4718. [CrossRef]
31. Zhao, L.; Wang, Z.-B.; Li, J.-L.; Zhang, J.-J.; Sui, X.; Zhang, L.-M. Hybrid of carbon-supported Pt nanoparticles and three dimensional graphene aerogel as high stable electrocatalyst for methanol electrooxidation. *Electrochim. Acta* **2016**, *189*, 175–183. [CrossRef]

32. Liu, M.; Peng, C.; Yang, W.; Guo, J.; Zheng, Y.; Chen, P.; Huang, T.; Xu, J. Pd nanoparticles supported on three-dimensional graphene aerogels as highly efficient catalysts for methanol electrooxidation. *Electrochim. Acta* **2015**, *178*, 838–846. [CrossRef]
33. Balantseva, E.; Miletto, I.; Coluccia, S.; Berlier, G. Immobilisation of Zinc porphyrins on mesoporous SBA-15: Effect of bulky substituents on the surface interaction. *Microporous Mesoporous Mater.* **2014**, *193*, 103–110. [CrossRef]
34. Silva, M.; Azenha, M.; Pereira, M.; Burrows, H.; Sarakha, M.; Forano, C.; Ribeiro, M.F.; Fernandes, A. Immobilization of halogenated porphyrins and their copper complexes in MCM-41: Environmentally friendly photocatalysts for the degradation of pesticides. *Appl. Catal. B Environ.* **2010**, *100*, 1–9. [CrossRef]
35. McDonald, A.R.; Franssen, N.; van Klink, G.P.; van Koten, G. 'Click' silica immobilisation of metallo-porphyrin complexes and their application in epoxidation catalysis. *J. Organomet. Chem.* **2009**, *694*, 2153–2162. [CrossRef]
36. Ribeiro, S.; Serra, A.; Gonsalves, A.R. Covalently immobilized porphyrins as photooxidation catalysts. *Tetrahedron* **2007**, *63*, 7885–7891. [CrossRef]
37. García-Sánchez, M.; de la Luz, V.; Estrada-Rico, M.; Murillo-Martínez, M.; Coahuila-Hernández, M.; Sosa-Fonseca, R.; Tello-Solís, S.; Rojas, F.; Campero, A. Fluorescent porphyrins covalently bound to silica xerogel matrices. *J. Non-Cryst. Solids* **2009**, *355*, 120–125. [CrossRef]
38. Lente, G.; Fábián, I. Kinetics and mechanism of the oxidation of water soluble porphyrin FeIIITPPS with hydrogen peroxide and the peroxomonosulfate ion. *Dalton Trans.* **2007**, 4268–4275. [CrossRef]
39. Lázár, I.; Fábián, I. A Continuous Extraction and Pumpless Supercritical CO_2 Drying System for Laboratory-Scale Aerogel Production. *Gels* **2016**, *2*, 26. [CrossRef]
40. Lian, W.; Sun, Y.; Wang, B.; Shan, N.; Shi, T. Synthesis and properties of 5,10,15,20-tetra[4-(3,5-dioctoxybenzamidephenyl] porphyrin and its metal complexes. *J. Serb. Chem. Soc.* **2012**, *77*, 335–348. [CrossRef]
41. Kozlowski, P.M.; Jarzęcki, A.A.; Pulay, P.; Li, X.-Y.; Zgierski, M.Z. Vibrational Assignment and Definite Harmonic Force Field for Porphine. 2. Comparison with Nonresonance Raman Data. *J. Phys. Chem.* **1996**, *100*, 13985–13992. [CrossRef]
42. Pop, S.-F.; Ion, R.-M.; Corobea, M.C.; Raditoiu, V. Spectral and Thermal Investigations of Porphyrin and Phthalocyanine Nanomaterials. *J. Optoelectron. Adv. Mater.* **2011**, *13*, 906–911.
43. Al-Oweini, R.; El-Rassy, H. Synthesis and characterization by FTIR spectroscopy of silica aerogels prepared using several $Si(OR)_4$ and $R''Si(OR')_3$ precursors. *J. Mol. Struct.* **2009**, *919*, 140–145. [CrossRef]
44. Ribeiro, S.M.; Serra, A.C.; Gonsalves, A.R. Covalently immobilized porphyrins on silica modified structures as photooxidation catalysts. *J. Mol. Catal. A Chem.* **2010**, *326*, 121–127. [CrossRef]
45. Munoz, M.; de Pedro, Z.M.; Pliego, G.; Casas, J.A.; Rodriguez, J.J.; Garcia, M.M. Chlorinated Byproducts from the Fenton-like Oxidation of Polychlorinated Phenols. *Ind. Eng. Chem. Res.* **2012**, *51*, 13092–13099. [CrossRef]
46. Hagen, J. *Industrial Catalysis: A Practical Approach*, 3rd Completely Revised and Enlarged ed.; Wiley-VCH: Weinheim, Germany, 2015; ISBN 978-3-527-33165-9.
47. Toxicological Profile for Chlorophenols Draft for Public Comment. Available online: https://www.atsdr.cdc.gov/toxprofiles/tp107.pdf (accessed on 4 February 2022).
48. Zhao, X.; Liu, X.; Yu, M.; Wang, C.; Li, J. The highly efficient and stable Cu, Co, Zn-porphyrin–TiO_2 photocatalysts with heterojunction by using fashioned one-step method. *Dye. Pigment.* **2016**, *136*, 648–656. [CrossRef]

gels

Article

Features of Luminescent Properties of Alginate Aerogels with Rare Earth Elements as Photoactive Cross-Linking Agents

Vladislav Kaplin [1,*], Aleksandr Kopylov [1,2], Anastasiia Koryakovtseva [1], Nikita Minaev [3], Evgenii Epifanov [3], Aleksandr Gulin [1], Nadejda Aksenova [1,4], Peter Timashev [1,4,5], Anastasiia Kuryanova [1], Ilya Shershnev [1] and Anna Solovieva [1]

1 N.N. Semenov Federal Research Center for Chemical Physics, Russian Academy of Sciences, 119991 Moscow, Russia
2 Institute of Fine Chemical Technologies, Russian Technological University, 119571 Moscow, Russia
3 Federal Research Centre "Crystallography and Photonics", Institute of Photonic Technologies, Russian Academy of Sciences, Troitsk, 108840 Moscow, Russia
4 Institute for Regenerative Medicine, I.M. Sechenov First Moscow State Medical University, 119991 Moscow, Russia
5 Chemistry Department, Lomonosov Moscow State University, 119991 Moscow, Russia
* Correspondence: vladislav.s.kaplin@gmail.com

Abstract: Luminescent aerogels based on sodium alginate cross-linked with ions of rare earth elements (Eu^{3+}, Tb^{3+}, Sm^{3+}) and containing phenanthroline, thenoyltrifluoroacetone, dibenzoylmethane, and acetylacetone as ligands introduced into the matrix during the impregnation of alginate aerogels (AEG), were obtained for the first time in a supercritical carbon dioxide medium. The impregnation method used made it possible to introduce organically soluble sensitizing ligands into polysaccharide matrices over the entire thickness of the sample while maintaining the porous structure of the aerogel. It is shown that the pore size and their specific area are 150 nm and 270 m^2/g, respectively. Moreover, metal ions with content of about 23 wt.%, acting as cross-linking agents, are uniformly distributed over the thickness of the sample. In addition, the effect of sensitizing ligands on the luminescence intensity of cross-linked aerogel matrices is considered. The interaction in the resulting metal/ligand systems is unique for each pair, which is confirmed by the detection of broad bands with individual positions in the luminescence excitation spectra of photoactive aerogels.

Keywords: aerogels; lanthanide luminescence; supercritical carbon dioxide; sodium alginate; luminescent sensor

check for updates

Citation: Kaplin, V.; Kopylov, A.; Koryakovtseva, A.; Minaev, N.; Epifanov, E.; Gulin, A.; Aksenova, N.; Timashev, P.; Kuryanova, A.; Shershnev, I.; et al. Features of Luminescent Properties of Alginate Aerogels with Rare Earth Elements as Photoactive Cross-Linking Agents. *Gels* **2022**, *8*, 617. https://doi.org/10.3390/gels8100617

Academic Editors: István Lázár and Melita Menelaou

Received: 17 August 2022
Accepted: 21 September 2022
Published: 27 September 2022

1. Introduction

The prospect of using rare earth elements (REE) in the creation of luminophores for analytical purposes, in particular for sensors, is usually associated not only with high quantum yields, long luminescence lifetimes, and a wide spectral range (from UV to IR), in which narrow-band luminescence of REE compounds is observed [1], but is determined by the possibility of regulating the functional characteristics of such systems when organic ligands of different natures are introduced (usually, the introduction of several ligands out of 4–6 possible options is conformationally acceptable) with the formation of specific photoactive centers "REE ion–ligands" [2–4]. Moreover, ligands also act as "antennas" [5–7] for radiation-initiating luminescence. This is even more important because of the low intensity of the luminescence of rare earth ions, caused by the forbidden electronic parity transitions [8]. The presence of third-party molecules or ions in the medium can affect the response of "antennas". Thus, cellulose aerogels cross-linked with terbium and europium ions and exhibiting luminescence sensitive to K^+, Ni^{2+}, Co^{2+}, Cu^{2+}, and Fe^{2+} ions are described in [9]. Zhang et al. demonstrate the quenching of the luminescence of Eu^{3+} ions in a complex with YVO_4 introduced into an alginate aerogel under the vapors of various organic solvents, including acetone, benzene, toluene, etc. [10]. Hai et al. show the binding of terbium ions with cellulose macromolecules using bridge ligands: 4-aminopyridine-2,6-dicarboxylic acid and 2-(2-aminobenzamido)benzoic acid. The resulting material exhibits reversible quenching/buildup of the luminescence of terbium ions in the presence of ClO^- and SCN^- ions, respectively [11].

The next range of problems that arise when creating analytical, in particular sensor, systems using luminescent complexes based on REE elements is associated with the formation of these complexes in polymer aerogels with a superporous cross-linked structure, bearing in mind the conformational possibilities for placement in such matrices of ions of organic ligands in the vicinity of REE. It is known [12] that for such purposes, the sodium salt of alginic acid can be chosen as the polymer base. Indeed, lanthanide ions, similar to calcium ions, are capable of binding with carboxyl groups of alginate polyanion macromolecules [13] to form three-dimensionally cross-linked gels, which acquire the structure of aerogels after treatment in a supercritical carbon dioxide (SC-CO_2) medium [14]. The formation of three-dimensionally cross-linked structures prevents the extraction of ionically bound luminophore centers from the matrix during SC-CO_2 drying [15], which occurs, for example, in lanthanide oxide aerogels [16]. At the same time, it should be noted that, although the methods for obtaining alginate aerogels lack such stages as hydrolysis, sol–gel formation, and sintering, which are characteristic of the synthesis of lanthanide-containing silicon and aluminum oxide aerogels [17], a certain problem associated with the search for common solvent for photoactive dopants and polymer matrix arises in the preparation of such aerogels. This problem is solved in this work based on a result previously obtained by the authors of the article. Thus, it is shown that organic sensitizing ligands, in particular phenanthroline, are easily introduced into cross-linked polymer matrices in a SC-CO_2 medium [18]. By this method, for the first time, this work demonstrates the sensitization of the luminescence of rare earth ions in the composition of alginate aerogel matrices, performed in a supercritical CO_2 medium. No similar researches were found in the literature. This made it possible to obtain luminescent aerogels based on sodium alginate cross-linked with lanthanide ions (Eu^{3+}, Tb^{3+}, Sm^{3+}), and to establish the effects of the exposure of the formed systems of introduced sensitizing ligands (phenanthroline, dibenzoylmethane, thenoyltrifluoroacetone, acetylacetonate) on the luminescence intensity, with a possible perspective of using such systems as optical sensors for volatile organic substances.

2. Results and Discussion

2.1. *Some Physicochemical Characteristics of Aerogel Matrices*

2.1.1. Specific Surface Area

Superporous aerogel structures were obtained by drying in a supercritical CO_2 medium (Figure 1). The average pore diameter is 149 nm $\pm$ 61 nm.

The data presented in Table 1 were obtained by the method of low-temperature adsorption of argon. The specific surface area (SSA) values of AEGs cross-linked by REE ions are comparable with similar values for some inorganic aerogels [19–22]. This makes it possible to use the obtained luminescent alginate aerogels as matrices in the development of sensors for the identification of gases and volatile substances.

Table 1. Specific surface area of alginate aerogel films cross-linked with REE ions.

Aerogel Matrix	SSA Average Value, m^2/g
Eu AEG	255 ± 22
Tb AEG	290 ± 43
Sm AEG	270 ± 9
Inorganic AEG	$\approx$100–2300

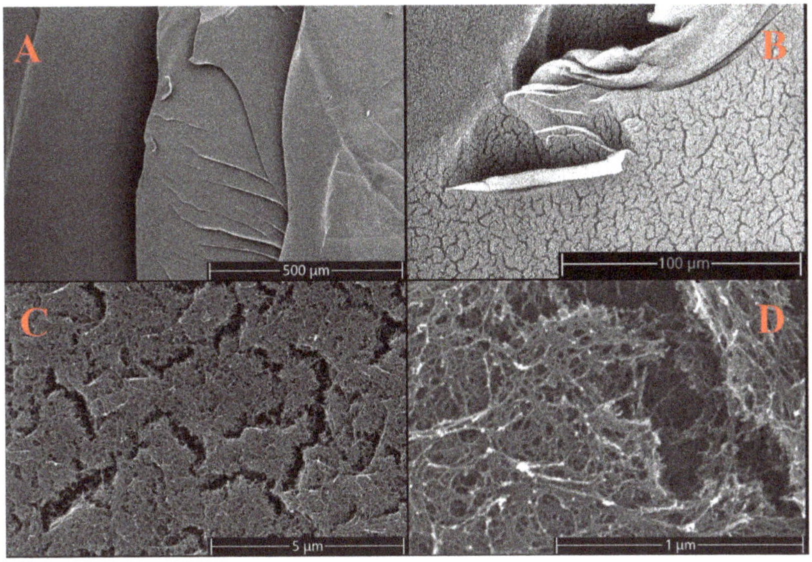

Figure 1. SEM images of the surface of an aerogel film cross-linked with Eu^{3+} ions: (**A**) magnification 100×, (**B**) 810×, (**C**) 15,000×, (**D**) 65,000×.

2.1.2. The Content of Rare Earth Metals in Cross-Linked Aerogel Matrices

Even after prolonged washing of the hydrogels, the metal content in the obtained aerogels remains constant, which indicates the fixation of all REE ions in the cross-linking sites of the alginate matrix. Therefore, the content of rare earth metals can provide information on the degree of cross-linking of the three-dimensional structure of the alginate aerogel. The elemental analysis data presented in Figure 2 as six data rows, correspond to six radial straight lines emerging from the center of the cylinder (200 µm) to its edges (6400 µm). The analysis was performed along each straight line at seven points. It can be seen that the mass content of europium ions is approximately the same over the entire cross-section of the Eu AEG sample (cylinder with a diameter of 13 mm), and is about 26%, which is close to the theoretical maximum content of europium (22.4 wt.%), corresponding to one trivalent europium ion per three carboxyl units. However, local measurements at the cut points lead to a high measurement error, about 23%. Therefore, the thermogravimetric method was used to determine the exact metal content in cross-linked aerogels.

The metal content was estimated by the thermogravimetric method, based on the fact that after thermal oxidation in air (at 1000 °C), the residue contains only metal oxide. The calculated maximum possible metal content (theoretical) and the content estimated using the thermogravimetric method (experimental) are presented in Table 2, and is also close to the EDS analysis data. The values that are lower than the theoretical ones are due to the incomplete reaction of the substitution of sodium ions by REE ions (about 85%), caused by steric hindrances in the formation of a cross-linked structure.

Table 2. Experimental and theoretical mass content of metal in alginate aerogel films cross-linked with Eu^{3+}, Tb^{3+}, and Sm^{3+} ions.

Sample	Metal Content (Experimental), wt.%	Metal Content (Theoretical), wt.%
Eu AEG	19.4 ± 0.3	22.4
Tb AEG	20.9 ± 0.2	23.2
Sm AEG	17.9 ± 0.2	22.2

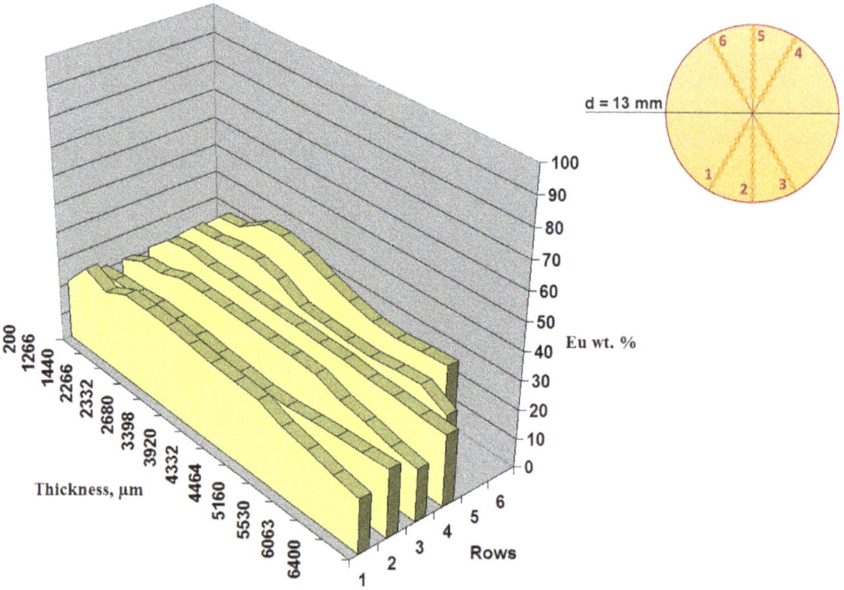

Figure 2. Mass content of Eu in a cross-linked matrix of sodium alginate.

For additional characterization of the complexes formed upon SC impregnation of cross-linked aerogels with organic ligands, the materials were analyzed using FTIR. The FTIR spectra of the initial films and ligands, as well as the resulting systems, are shown in Figure 3a–c. As can be seen from the spectra, during the cross-linking of sodium alginate, the bands at 1403 cm^{-1} and 1591 cm^{-1} (characteristic bands of symmetric and antisymmetric C=O vibrations for salts of carboxylic acids) shift towards each other up to 1415 cm^{-1} and 1585 cm^{-1}, respectively, when the Na^+ ion is replaced by the lanthanide ion. The 1024 cm^{-1} band (C-O hydroxyl groups) also shifts to 1032 cm^{-1} under the influence of a more electronegative ion. The IR spectra of the impregnated films are a superposition of the most intense bands of the ligand on the spectrum of cross-linked alginate, with the exception of the Acac ligand, whose bands are not detected due to the extremely low concentration (Figure S1). According to other publications, the absorption bands of Dbm carbonyl groups do not undergo significant shifts relative to the absorption of the free ligand upon coordination with lanthanides [23]. For the films SC -impregnated with Dbm, shifts of the C=O vibration bands from 1525 cm^{-1} to 1517 cm^{-1} and C-H from 1461 cm^{-1} to 1479 cm^{-1} are observed (Figure 3b). In this case, a band at 516 cm^{-1} appears, which is related to the new Ln–O bond, and is absent in the initial ligand and film. Due to the low concentration of the ligand, this band is be detected in SC-impregnated films with the Tta ligand. However, it is known that the position of the C=O band of the Tta ligand shift down by about 40 cm^{-1} when combined with a rare earth ion [24,25]. For Ln AEG + Tta films, a shift of this band from 1638 cm^{-1} to 1596 cm^{-1} is observed, which confirms the coordination. On the spectra of films SC-impregnated with Phen, only a few of the most intense bands belonging to phenanthroline are detected. However, their position is also shifted relative to the bands of free phenanthroline, which is typical for Phen complexes with metal: from 623 cm^{-1} to 635 cm^{-1}, from 737 cm^{-1} to 730 cm^{-1}, from 852 cm^{-1} to 841 cm^{-1}, and from 1504 cm^{-1} to 1518 cm^{-1} (Figure 3c).

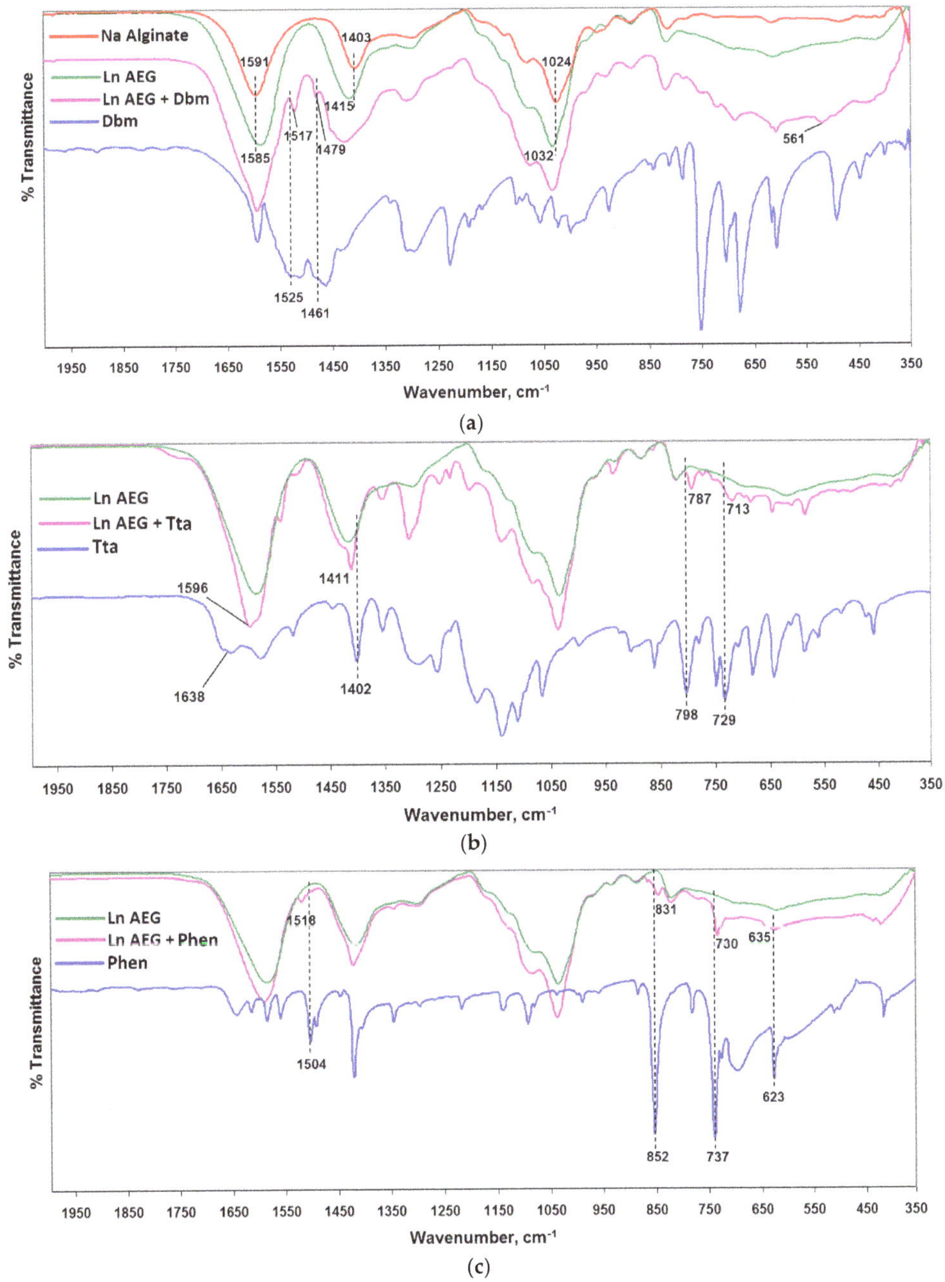

Figure 3. (**a**) FTIR spectra of sodium alginate, dibenzoylmethane, initial cross-linked films Ln AEG, and SC-impregnated films Ln AEG + Dbm. (**b**) FTIR spectra of thenoyltrifluoroacetone, initial cross-linked films Ln AEG, and SC-impregnated films Ln AEG + Tta. (**c**) FTIR spectra of phenanthroline, initial cross-linked films Ln AEG, and SC-impregnated films Ln AEG + Phen.

2.2. Effect of Organic Sensitizing Ligands on the Luminescent Properties of Aerogel Polysaccharide Matrices Cross-Linked with REE Ions

2.2.1. Luminescence of Aerogel Films

Well-studied ligands with known triplet energy levels close to the radiative levels of a given series of metals, as well as well-soluble in SC-CO_2 medium, were used for sensitization of luminescence: acetylacetone, phenanthroline, dibenzoylmethane, and thenoyltrifluoroacetone. Based on the difference between the energies of the triplet level of the ligand and the radiative level of the metal, which should be in the range of 1000 cm^{-1}–5500 cm^{-1} [26], one can predict in advance the efficiency of energy transfer from the ligand to the metal [27–30]. Table 3 lists the energies of the triplet levels of the abovementioned ligands and the radiative levels of Eu^{3+}, Tb^{3+}, and Sm^{3+} ions, as well as the difference between these energies (ΔE) for each metal/ligand pair. Green color indicates the pairs for which an effective sensitization process is expected and, as a result, an increase in the intensity of the luminescence of rare earth ions. Highlighted in red are ΔE, at which the energy transfer from the ligand to the metal either does not occur (ΔE > 5500 cm^{-1}), or at which the reverse energy transfer dominates (ΔE < 1500 cm^{-1}).

Table 3. The energies of the triplet levels of the Tta, Phen, Acac, and Dbm ligands, the energies of the excited radiative levels of the Eu^{3+}, Tb^{3+}, and Sm^{3+} ions, and their difference ΔE for each metal/ligand pair.

	Tta: 20,500 cm^{-1}	**Phen: 22,075 cm^{-1}**	**Acac: 25,310 cm^{-1}**	**Dbm: 20,300 cm^{-1}**
Eu ($^5D_0$) (17,267 cm^{-1})	3233 cm^{-1}	4808 cm^{-1}	8043 cm^{-1}	3033 cm^{-1}
Tb ($^5D_4$) (20,394 cm^{-1})	106 cm^{-1}	1681 cm^{-1}	4916 cm^{-1}	−94 cm^{-1}
Sm ($^4G_{5/2}$) (17,825 cm^{-1})	2675 cm^{-1}	4250 cm^{-1}	7485 cm^{-1}	2475 cm^{-1}

Thus, the effective REE–ligand interaction should be observed for Eu and Sm systems with Tta, Phen, and Dbm, and for Tb with Phen and Acac. The efficiency of the sensitization process after the SC introduction of ligands was evaluated by the increase in the luminescence intensity of aerogel matrices cross-linked with rare earth ions, and the occurrence of metal–ligand interaction by changes in the luminescence excitation spectra. Indeed, the change in the luminescence intensity for all AEGs occurs in accordance with the expected results, with the exception of the Eu AEG + Dbm pair. Dbm molecules show a weaker sensitizing ability compared to other ligands, and no sensitization is observed in the Eu AEG matrix. This can be explained by the features of the keto–enol equilibrium of the Dbm tautomers in the nonpolar SC-CO_2 medium [31,32]. In the luminescence spectra of aerogel matrices cross-linked with Eu^{3+} and Tb^{3+} ions, characteristic narrow bands of low intensity metal-centered luminescence are observed. Moreover, the luminescence excitation spectra are also represented by a set of narrow bands, the positions of which are given in Table 4. At the same time, the characteristic luminescence (bands at 563 nm, 598 nm, and 644 nm) are not detected in the samples cross-linked with Sm^{3+} ions. The excitation and luminescence spectra of cross-linked aerogels are presented in Figures S2 and S3. Before SC impregnation, the samples are transparent white or slightly yellow. After the introduction of ligands in the SC-CO_2 medium, the transparency is preserved, and the films acquire a pink (for Phen and Dbm) or yellow (for Acac and Tta) tint. Figure 4 shows the films of the original Eu AEG (1) and of the Eu AEG SC-impregnated with Phen (2) and Tta (3) ligands. In the first row (A), the films are placed on a light monitor. Transparency is also confirmed by the absorption spectra of the original Eu AEG film and of the Eu AEG impregnated with Phen ligands (Figure S4). The second row (B) shows the films in daylight, the third row (C) shows the films exposed to 365 nm UV light. The remaining samples have a similar appearance, except that their glow under the ultraviolet light is not apparent to the naked eye.

Table 4. The position of the maxima of the original alginate aerogels cross-linked with REE ions and aerogels after SC impregnation with organic ligands, as well as changes in the luminescence intensity. The letters "S", "M", and "W" denote strong, medium, and weak luminescence bands, respectively.

Sample	Position of Band Maxima in the Luminescence Excitation Spectra	Changes in the Intensity of Characteristic Luminescence Peaks of REE Ions
Initial alginate aerogels cross-linked with REE ions		
Eu AEG	393 nm; 464 nm	577, 590, 615 (S)
Tb AEG	317 nm; 340 nm; 351 nm; 368 nm; 377 nm	488, 543, 584, 620 (M)
Sm AEG	–	563, 598, 644 (Not detected)
Matrices impregnated with thenoyltrifluoroacetone (Tta)		
Eu AEG +Tta	363 nm	↑ (S)
Tb AEG +Tta	290 nm; 356 nm; 410 nm	↓ (W)
Sm AEG +Tta	368 nm	Luminescence (M)
Matrices impregnated with phenanthroline (Phen)		
Eu AEG +Phen	350 nm	↑ (S)
Tb AEG +Phen	347 nm	↑ (S)
Sm AEG +Phen	368 nm	↑ (M)
Matrices impregnated with acetylacetone (Acac)		
Eu AEG +Acac	338 nm	↓ (M)
Tb AEG +Acac	304 nm	↑ (S)
Sm AEG +Acac	–	No luminescence
Matrices impregnated with dibenzoylmethane (Dbm)		
Eu AEG +Dbm	386 nm	↓ (W)
Tb AEG +Dbm	295 nm	↓ (W)
Sm AEG +Dbm	395 nm	Luminescence (W)

In all cases, after SC impregnation of the matrices, the excitation spectra are represented by broad bands (from 140 to 240 nm wide) that are characteristic for organic molecules (Table 4). Correspondingly, ligands dissolved in SC fluid are adsorbed on the surface and in the volume of aerogels, and coordinate near REE ions, forming luminescent systems with them. All recorded spectra of the SC-impregnated aerogels are shown in Figures S5–S15.

In Table 4, green indicates an increase in luminescence intensity after impregnation with organic ligands, while red indicates a decrease in intensity.

It is known that the standard procedure for obtaining luminescent organic REE complexes includes mixing solutions of metal salts and ligands, adjusting pH to a certain value, and isolating and purifying the precipitated product [33]. The obtained complexes are no longer able to act as cross-linking agents for water-soluble polyanions, since they become insoluble in aqueous media. On the other hand, the introduction of a ready-made luminescent REE complex into cross-linked aerogels (for example, by impregnation in SC-CO_2) is limited by solubility in SC-CO_2, concentration quenching, and aggregation. For alginate aerogels cross-linked with REE, each luminescent center is located in the cross-link site and is shielded from neighboring centers by fragments of polymer molecules. Therefore, the sensitization of such distributed luminescent centers in alginate matrices can be considered as a way to bypass the problem of concentration quenching. However, luminescence quenching is observed in aerogels with an ion content of about 20%. Thus, the maximum luminescence intensity is achieved at a REE content of about 10 wt.%. Also, preliminary tests of the luminescence sensitivity of some aerogel matrices to the presence of organic and inorganic vapors (acetone, ammonia) were carried out. The original non-impregnated aerogels cross-linked with REE are not sensitive to the tested volatile compounds. It is interesting to note that only one matrix (AEG Eu, containing Tta) shows the reaction to acetone after SC impregnation: the intensity of characteristic luminescence bands in the presence of acetone vapor increase by 32% (Figure S16). Luminescence quenching by ammonia vapor is observed in AEG Eu and AEG Tb aerogels SC-impregnated with Phen. The luminescence intensity of the matrices drops by 18% and 60%, respectively (Figures S17 and S18). Thus, not only the structure of the organic ligand, but also the metal, has a significant effect on the nature of the response of luminescent aerogels.

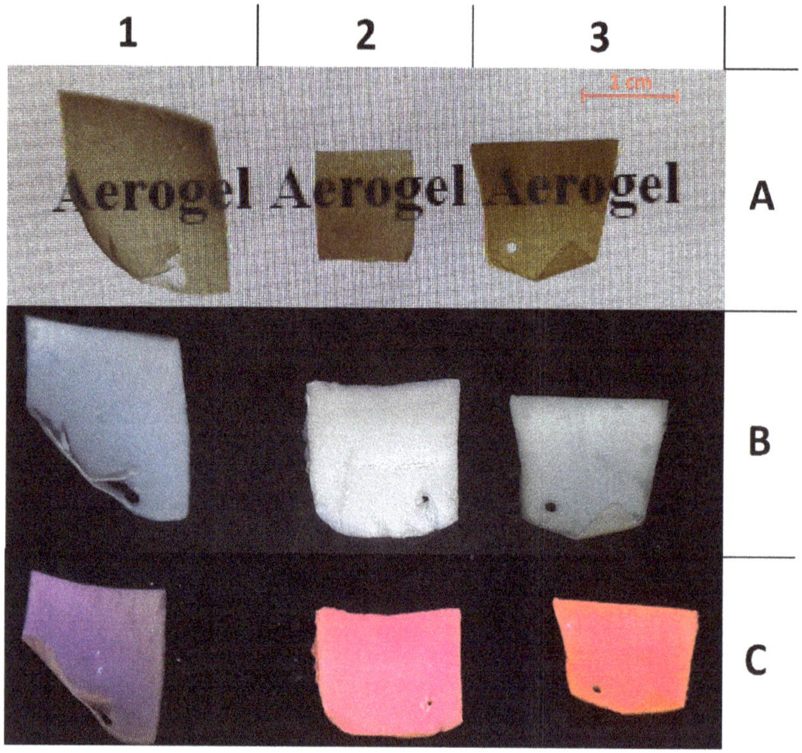

Figure 4. Eu AEG (**1**), Eu AEG + Phen (**2**), and Eu AEG + Tta (**3**) aerogels in transmitted light (row **A**), daylight (row **B**), and under 365 nm UV light (row **C**).

2.2.2. Features of the Distribution of Impregnated Ligands in the Volume of Aerogels

It is important to note that when mentioning the most characteristic properties of aerogel materials, such as porosity, mechanical properties, and refractive index, three-dimensional structure properties are implied. This also applies to luminescent properties: up to a certain thickness, aerogels are optically transparent for the UV–NIR range; accordingly, radiation should occur not only from rare earth ions localized on the surface of the matrix, but also in its volume. Therefore, it is necessary to make sure that the interaction of luminescent ions with the introduced sensitizing ligands occurs through the volume of the matrix. For example, Zhang et al. demonstrate the penetration of a sensitizing ligand into a sample layer no thicker than 130 μm, after impregnating an aerogel matrix in solution [10].

This work shows that the impregnation of alginate aerogels in the SC-CO_2 medium ensures the impregnation of the matrix to a depth of at least 3.3 mm (cylinder with a radius of 6.6 mm) (Figure 5). At the same time, this value is still limited only by the difficulties in obtaining thicker aerogel blocks, but not by the capabilities of the SC-fluid. Figure 5 shows the intensity distribution of the luminescence signal at 613 nm over the thickness of the Eu AEG (curve 1) and Eu AEG–Phen (after SC impregnation) matrices (curve 2). It can be seen that the increase in signal intensity occurs throughout the entire volume of the matrix, which indicates the penetration of the ligand dissolved in the SC-medium to a given depth. The increased values from the edges of the matrix (0–100 μm, 6000–6500 μm) are associated with a more intense diffusion of the solution into the near-surface layers.

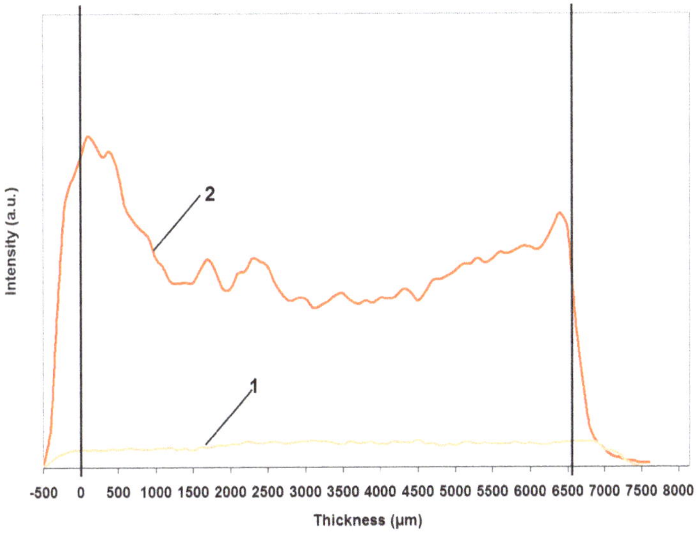

Figure 5. Distribution of luminescence intensity at 613 nm over a section of the Eu AEG–Phen sample: (1) luminescence intensity in the initial matrix, (2) intensity after impregnation with phenanthroline.

3. Conclusions

The luminescent aerogels based on sodium alginate, cross-linked with ions of rare earth elements (Eu^{3+}, Tb^{3+}, Sm^{3+}) and containing phenanthroline, thenoyltrifluoroacetone, dibenzoylmethane, and acetylacetonate as ligands, introduced into the matrix during SC impregnation of alginate aerogels, were obtained in a supercritical carbon dioxide medium for the first time. It is shown that the intensity of the luminescence bands change after impregnation. Moreover, the nature of the influence of organic additives (ligands) on the luminescent properties of REE ions depends on the nature of both the ion and the ligand. It is demonstrated that upon SC impregnation, ligands can penetrate and act as luminescence sensitizers of rare earth ions throughout the entire thickness of aerogels.

4. Materials and Methods

4.1. Preparation of Alginate Aerogels Cross-Linked with REE Ions

The following substances were used without additional preparation and purification: REE chloride hexahydrates: XCl_3x6H_2O, where X is Eu, Tb, Sm (Aldrich, St. Louis, MO, USA, 99.9%); gadolinium (III) acetylacetonate hydrate ($Gd(Acac)_3xH_2O$) (Aldrich, 99.9%); europium (III) theonyltrifluoroacetonate trihydrate ($Eu(Tta)_3x3H_2O$) (Acros Organics, Geel, Belgium 95%); sodium alginate (Rushim, Moscow, Russia); sensitizing ligands: 1,10-phenanthroline (Acros Organics, 99+%); thenoyltrifluoroacetone (Aldrich, 99+%); dibenzoylmethane (Aldrich, 99+%); isopropanol (HIMFARM, Moscow, Russia, TU 2632-181-44493179-2014) (hereinafter, coordination water is not indicated for REE compounds).

To create supercritical conditions for impregnation and drying, dry carbon dioxide, with the volume content of water vapor not exceeding 0.001%, according to the quality certificate, was utilized (OOO "NII KM" 99.8% All-Union State Standard 8050-85).

Alginate aerogels in the form of films and cylinders were obtained by the following method. First, hydrogel films were obtained by pouring 40 mL of an aqueous solution of REE chloride (5 wt.%) into 30 mL of an aqueous solution of sodium alginate (2 wt.%) in a plastic Petri dish (d = 85 mm). The thickness of the formed film varied from 1 mm along the edges to 3 mm in the center. Hydrogel cylinders of 2 cm in diameter were obtained by squeezing 10 mL of a 2% aqueous solution of sodium alginate into a 10-fold excess of a 5% aqueous solution of REE chloride from a 10 mL syringe. The hydrogels were kept in distilled water for 72 h, changing the water three times to remove unreacted REE chloride. Then, the water in the hydrogels was replaced with isopropanol: the hydrogels were kept in a mixture of isopropanol/water (25/75) for 24 h, and then the proportion of isopropanol was increased by 25% once a day, bringing it to 100%.

The cross-linked alginate hydrogels were dried in a high-pressure flow reactor in supercritical carbon dioxide at a temperature of 40 °C and a pressure of 115 bar. The diagram of the process is shown in Figure 6.

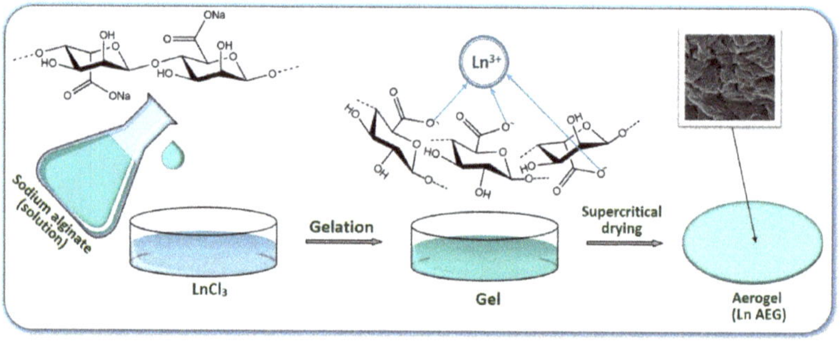

Figure 6. Diagram for obtaining aerogel films cross-linked with REE ions.

4.2. Impregnation of Aerogels Cross-Linked with REE Ions by the Organic Ligands

Aerogels were impregnated with the organic ligands in SC-CO_2 medium. The concentration of ligands in the supercritical solution was 0.25 mg/mL. The impregnation was carried out for 1 h at a pressure of 180 bar and a temperature of 90 °C. Previously, in our work it is shown that, under these conditions, it is possible to achieve a uniform distribution of impregnated compounds in various polymer matrices in a SC-CO_2 medium [34]. The reactor was then cooled to room temperature and depressurized to atmospheric pressure for 30 min (Figure 7).

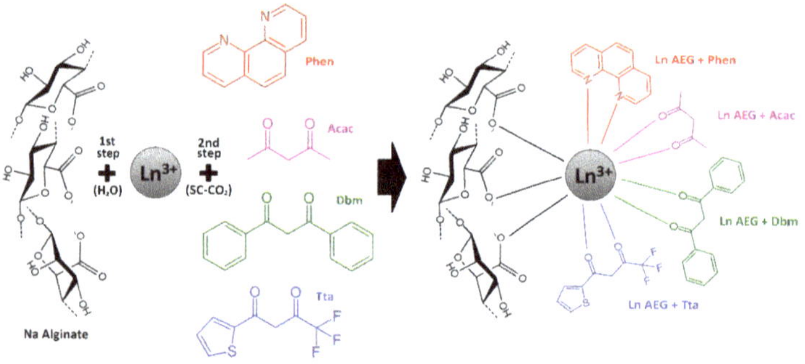

Figure 7. Diagram of complex formation of new systems of REE-containing alginate and sensitizing ligands Tta, Phen, Acac, and Dbm.

4.3. Determination of Luminescent and Physicochemical Characteristics of Cross-Linked Aerogel Matrices

The luminescence and luminescence excitation spectra of the aerogel films were recorded using a Horiba Fluoromax Plus (Horiba-Jobin-Yvon, Palaiseau, France) spectrofluorometer at room temperature. The distribution of the luminescence intensity over the thickness of the aerogel cylinders was determined using a flexible optical fiber with a diameter of 0.8 mm directed at the cross-section of the sample, and a QE Pro 65000 spectrometer (Ocean Insight, Orlando, FL, USA). The displacement was provided by a movable stage with a positioning accuracy of 10 ± 1 μm. The values were recorded from the surface of the cross-section of the cylinder along a straight line from the periphery to the center with a step of 100 μm.

The specific surface area (SSA) of polysaccharide aerogels was determined by the low-temperature argon adsorption method (BET method). The analysis was carried out at the V.V. Voevodsky Laboratory of Kinetics of Mechanochemical and Free-Radical Processes (N.N. Semenov Federal Research Center for Chemical Physics, RAS, Moscow, Russia).

SEM images of the porous structure of aerogels were obtained using a scanning electron microscope Prisma E (Thermo Fisher Scientific, Scheepsbouwersweg, The Netherlands) after deposition of a layer of gold (10 nm). Data on the metal content in aerogel matrices were obtained from the surface of a cross-section of a cylindrical sample using a Phenom ProX scanning electron microscope (Thermo Fisher Scientific, Scheepsbouwersweg, The Netherlands) equipped with an energy-dispersive spectroscopy (EDS) silicon drift detector, which allows the performance of elemental analysis. Also, the metal content in the matrices was determined by the gravimetric method, based on the residue after burning the samples in a Saturn 1 high-temperature furnace at a temperature of 1000 °C.

FTIR analysis of the initial components and the synthesized system was carried out using a spectrum two FT-IR spectrometer (PerkinElmer, Waltham, MA, USA) in attenuated total reflectance (ATR) mode. The spectrometer features were as follows: high-performance, room-temperature $LiTaO_3$ MIR detector, standard optical system with KBr windows for data collection over a spectral range of 4000–350 cm^{-1} at a resolution of 0.5 cm^{-1}. All spectra were initially collected in ATR mode and converted into IR transmittance mode.

Supplementary Materials: The following supporting information can be downloaded at: https://www.mdpi.com/article/10.3390/gels8100617/s1, Figure S1: FTIR spectra of initial cross-linked films Ln AEG and SC-impregnated films Ln AEG + Acac; Figures S2 and S3: Luminescence spectra (orange curve) and luminescence excitation spectra (green curve) of alginate aerogels cross-linked with Eu^{3+} and Tb^{3+} ions, respectively; Figure S4: Absorption spectra of Eu AEG (curve 1) and Eu AEG film after SC impregnation of Phen (curve 2). Figures S5–S15: Luminescence (orange curve) and luminescence excitation (green curve) spectra of alginate aerogels cross-linked with Eu^{3+}, Tb^{3+}, and Sm^{3+} ions, SC-impregnated with Tta, Phen, Acac, and Dbm ligands; Figure S16: Luminescence spectra: 1.AEG Eu SC-impregnated with Tta before and 2. after exposure to acetone vapor; Figure S17: Luminescence spectra: 1. AEG Eu SC-impregnated with Phen before and 2. after exposure to ammonia vapor, Figure S18: Luminescence spectra: 1. AEG Tb SC-impregnated with Phen before and 2. after exposure to ammonia vapor.

Author Contributions: Writing—original draft preparation, V.K.; conceptualization and supervision, A.K. (Aleksandr Kopylov); methodology and investigation, V.K., A.K. (Anastasiia Koryakovtseva), E.E. and N.A.; resources, N.M. and A.G.; project administration, P.T.; visualization, A.K. (Anastasiia Kuryanova); data curation, I.S.; writing—review and editing and funding acquisition, A.S. All authors have read and agreed to the published version of the manuscript.

Funding: This study was conducted in the framework of the Russian Government assignment № 122040400099-5. The work was supported by the Ministry of Science and Higher Education as part of the work under the state task of the Federal Research Center "Crystallography and Photonics" of the Russian Academy of Sciences in terms of using the equipment of the Center for Collective Use "Structural Diagnostics of Materials" when characterizing samples using the energy-dispersive spectroscopy (EDX) method and measuring local luminescence spectra of aerogel samples. SEM images and luminescence spectra were obtained with the equipment of the FRCCP RAS shared research facilities (No. 506694).

Data Availability Statement: Data are contained within the article or supplementary material.

Conflicts of Interest: The authors declare no conflict of interest.

References

1. Huang, C.-H. *Rare Earth Coordination Chemistry: Fundamentals and Applications*; John Wiley& Sons: Singapore; Hoboken, NJ, USA, 2010.
2. Liu, J.; Liang, Q.-B.; Wu, H.-B. Synthesis, photophysics, electrochemistry, thermal stability and electroluminescent performances of a new europium complex with bis(β-diketone) ligand containing carbazole group: Luminescent Performances of a New Europium Complex. *Luminescence* **2016**, *32*, 460–465. [CrossRef] [PubMed]
3. Turchetti, D.A.; Domingues, R.A.; Zanlorenzi, C.; Nowacki, B.; Atvars, T.D.Z.; Akcelrud, L.C. A Photophysical Interpretation of the Thermochromism of a Polyfluorene Derivative–Europium Complex. *J. Phys. Chem. C* **2014**, *118*, 30079–30086. [CrossRef]
4. Turchetti, D.A.; Nolasco, M.M.; Szczerbowski, D.; Carlos, L.D.; Akcelrud, L.C. Light emission of a polyfluorene derivative containing complexed europium ions. *Phys. Chem. Chem. Phys.* **2015**, *17*, 26238–26248. [CrossRef] [PubMed]
5. George, M.R.; Critchley, P.E.; Whitehead, G.F.; Bailey, A.J.; Cuda, F.; Murdin, B.N.; Grossel, M.C.; Curry, R.J. Modified pyridine-2,6-dicarboxylate acid ligands for sensitization of near-infrared luminescence from lanthanide ions (Ln3+ = Pr3+, Nd3+, Gd3+, Dy3+, Er3+). *J. Lumin* **2021**, *230*, 117715. [CrossRef]
6. Wang, S.; Chu, X.; Xiang, X.; Cao, Y. Highly selective antenna effect of graphene quantum dots (GQDs): A new fluorescent sensitizer for rare earth element terbium in aqueous media. *Talanta* **2019**, *209*, 120504. [CrossRef] [PubMed]
7. Sun, N.-N.; Yan, B. Near-infrared emission sensitization of lanthanide cation based on Ag+ functionalized metal-organic frameworks. *J. Alloy. Compd.* **2018**, *765*, 63–68. [CrossRef]
8. Sabbatini, N.; Guardigli, M.; Lehn, J.-M. Luminescent lanthanide complexes as photochemical supramolecular devices. *Co-ord. Chem. Rev.* **1993**, *123*, 201–228. [CrossRef]
9. Fan, W.; Du, J.; Kou, J.; Zhang, Z.; Liu, F. Hierarchical porous cellulose/lanthanide hybrid materials as luminescent sensor. *J. Rare Earths* **2018**, *36*, 1036–1043. [CrossRef]
10. Zhang, Z.-Y.; Zhu, H.; Xu, Q.-Q.; Liu, F.-Y.; Zhu, A.-X.; Kou, J.-F. Hybrid luminescent alginate hydrogels containing lanthanide with potential for acetone sensing. *New J. Chem.* **2019**, *43*, 13205–13211. [CrossRef]
11. Hai, J.; Li, T.; Su, J.; Liu, W.; Ju, Y.; Wang, B.; Hou, Y.; Liu, W. Reversible Response of Luminescent Terbium(III)-Nanocellulose Hydrogels to Anions for Latent Fingerprint Detection and Encryption. *Angew. Chem.* **2018**, *130*, 6902–6906. [CrossRef]
12. Zhang, Z.; Liu, F.; Xu, Q.; Zhu, H.; Zhu, A.; Kou, J. Covalent Grafting Terbium Complex to Alginate Hydrogels and Their Application in Fe^{3+} and pH Sensing. *Glob. Challenges* **2019**, *3*, 1800067. [CrossRef]
13. Liu, F.; Carlos, L.D.; Ferreira, R.A.S.; Rocha, J.; Gaudino, M.C.; Robitzer, M.; Quignard, F. Photoluminescent Porous Alginate Hybrid Materials Containing Lanthanide Ions. *Biomacromolecules* **2008**, *9*, 1945–1950. [CrossRef]
14. Robitzer, M.; David, L.; Rochas, C.; Di Renzo, F.; Quignard, F. Nanostructure of Calcium Alginate Aerogels Obtained from Multistep Solvent Exchange Route. *Langmuir* **2008**, *24*, 12547–12552. [CrossRef]
15. Sorensen, L.; Strouse, G.F.; Stiegman, A.E. Fabrication of Stable Low-Density Silica Aerogels Containing Luminescent ZnS Capped CdSe Quantum Dots. *Adv. Mater.* **2006**, *18*, 1965–1967. [CrossRef]
16. Tillotson, T.M.; Sunderland, W.E.; Thomas, I.M.; Hrubesh, L.W. Synthesis of lanthanide and lanthanide-silicate aerogels. *J. Sol.-Gel Sci. Technol.* **1994**, *1*, 241–249. [CrossRef]
17. Małecka, M.A.; Kępiński, L. Solid state reactions in highly dispersed single and mixed lanthanide oxide–SiO_2 systems. *Catal. Today* **2012**, *180*, 117–123. [CrossRef]
18. Kaplin, V.S.; Kopylov, A.S.; Zarhina, T.S.; Timashev, P.S.; Solov'Eva, A.B. Luminescent Properties of Mixed-Ligand Neodymium β-Diketonates Obtained in Supercritical Carbon Dioxide in Polymer Matrices of Various Nature. *Opt. Spectrosc.* **2020**, *128*, 869–876. [CrossRef]
19. Alwin, S.; Ramasubbu, V.; Shajan, X.S. TiO_2 aerogel–metal organic framework nanocomposite: A new class of photoanode material for dye-sensitized solar cell applications. *Bull. Mater. Sci.* **2018**, *41*, 27. [CrossRef]
20. Nguyen, B.N.; Meador, M.A.B.; Scheiman, D.; McCorkle, L. Polyimide Aerogels Using Triisocyanate as Cross-linker. *ACS Appl. Mater. Interfaces* **2017**, *9*, 27313–27321. [CrossRef]
21. Maleki, H.; Durães, L.; Portugal, A. An overview on silica aerogels synthesis and different mechanical reinforcing strategies. *J. Non-Crystalline Solids* **2014**, *385*, 55–74. [CrossRef]
22. Li, H.; Li, J.; Thomas, A.; Liao, Y. Ultra-High Surface Area Nitrogen-Doped Carbon Aerogels Derived From a Schiff-Base Porous Organic Polymer Aerogel for CO_2 Storage and Supercapacitors. *Adv. Funct. Mater.* **2019**, *29*, 1904785. [CrossRef]
23. Kanimozhi, A.J.; Alexander, V. Synthesis and photophysical and magnetic studies of ternary lanthanide(iii) complexes of naphthyl chromophore functionalized imidazo[4,5-f][1,10]phenanthroline and dibenzoylmethane. *Dalton Trans.* **2017**, *46*, 8562–8571. [CrossRef]
24. Wang, Y.-P.; Luo, Y.; Wang, R.-M.; Yuan, L. Synthesis and fluorescence properties of the mixed complexes of Eu(III) with polymer ligand and thenoyl trifluoroacetone. *J. Appl. Polym. Sci.* **1997**, *66*, 755–760. [CrossRef]

25. Wang, L.-H.; Wang, W.; Zhang, W.-G.; Kang, E.-T.; Huang, W. Synthesis and Luminescence Properties of Novel Eu-Containing Copolymers Consisting of Eu(III)−Acrylate−β-Diketonate Complex Monomers and Methyl Methacrylate. *Chem. Mater.* **2000**, *12*, 2212–2218. [CrossRef]
26. Xu, H.; Sun, Q.; An, Z.; Wei, Y.; Liu, X. Electroluminescence from europium(III) complexes. *Coord. Chem. Rev.* **2015**, *293–294*, 228–249. [CrossRef]
27. Freidzon, A.Y.; Kurbatov, I.A.; Vovna, V.I. *Ab initio* calculation of energy levels of trivalent lanthanide ions. *Phys. Chem. Chem. Phys.* **2018**, *20*, 14564–14577. [CrossRef] [PubMed]
28. Smirnova, T.D.; Shtykov, S.N.; Kochubei, V.I.; Khryachkova, E.S. Excitation energy transfer in europium chelate with doxycycline in the presence of a second ligand in micellar solutions of nonionic surfactants. *Opt. Spectrosc.* **2011**, *110*, 60–66. [CrossRef]
29. Jinghe, Y.; Xuezhen, R.; Huabin, Z.; Ruiping, S. Enhanced luminescence of the europium(III)-terbium(III)-dibenzoylmethane-ammonia-acetone system and its application to the determination of europium. *Analyst* **1990**, *115*, 1505–1508. [CrossRef]
30. Gusev, A.N.; Hasegawa, M.; Shimizu, T.; Fukawa, T.; Sakurai, S.; Nishchymenko, G.A.; Shul'Gin, V.F.; Meshkova, S.B.; Linert, W. Synthesis, structure and luminescence studies of Eu(III), Tb(III), Sm(III), Dy(III) cationic complexes with acetylacetone and bis(5-(pyridine-2-yl)-1,2,4-triazol-3-yl)propane. *Inorganica Chim. Acta* **2013**, *406*, 279–284. [CrossRef] [PubMed]
31. Kojić, M.; Lyskov, I.; Milovanović, B.; Marian, C.M.; Etinski, M. The UVA response of enolic dibenzoylmethane: Beyond the static approach. *Photochem. Photobiol. Sci.* **2019**, *18*, 1324–1332. [CrossRef] [PubMed]
32. AL-Hilfi, J.A. A Structural Study of 2-Thenoyltrifluoroacetone Schiff Bases and Their Thione Derivatives: Synthesis, NMR and IR. AIP Conference Proceedings. In Proceedings of the 8th International Conference on Applied Science and Technology (ICAST 2020), Karbala, Iraq, 4 December 2020; Volume 2290, p. 030030. [CrossRef]
33. Ugale, A.; Kalyani, T.N.; Dhoble, S.J. Potential of Europium and Samarium β-Diketonates as Red Light Emitters in Organic Light-Emitting Diodes. In *Lanthanide-Based Multifunctional Materials*; Elsevier: Amsterdam, The Netherlands, 2018; pp. 59–97. [CrossRef]
34. Kopylov, A.S.; Yusupov, V.I.; Cherkasova, A.V.; Shershnev, I.V.; Timashev, P.S.; Solovieva, A.B. The Distribution Features of Photoactive Fillers in Different-Nature Polymer Matrices upon Their Impregnation in a Supercritical Carbon Dioxide Medium. *Russ. J. Phys. Chem. B* **2018**, *12*, 1298–1305. [CrossRef]

gels

Article

Efficient Removal of Polyvalent Metal Ions (Eu(III) and Th(IV)) from Aqueous Solutions by Polyurea-Crosslinked Alginate Aerogels

Efthalia Georgiou [1], Ioannis Pashalidis [1,*], Grigorios Raptopoulos [2] and Patrina Paraskevopoulou [2,*]

1 Laboratory of Radioanalytical and Environmental Chemistry, Department of Chemistry, University of Cyprus, P.O. Box 20537, Cy-1678 Nicosia, Cyprus; georgiou.efthalia@ucy.ac.cy
2 Inorganic Chemistry Laboratory, Department of Chemistry, National and Kapodistrian University of Athens, Panepistimiopolis Zografou, 15771 Athens, Greece; grigorisrap@chem.uoa.gr
* Correspondence: pspasch@ucy.ac.cy (I.P.); paraskevopoulou@chem.uoa.gr (P.P.)

Abstract: The removal of polyvalent metal ions Eu(III) and Th(IV) from aqueous solutions using polyurea-crosslinked calcium alginate (X-alginate) aerogels has been investigated by batch-type experiments under ambient conditions and pH 3. The material presents relatively high sorption capacity for Eu(III) (550 g kg^{-1}) and Th(IV) (211 g kg^{-1}). The lower sorption capacity for Th(IV) compared to Eu(III) is attributed to the net charge of the dominant species in solution under the given experimental conditions, which is Eu^{3+} for Eu(III), and $Th(OH)_2{}^{2+}$ and $Th(OH)_3{}^{+}$ for Th(IV). Generally, the sorption is an endothermic and entropy-driven process, and it follows the Langmuir isotherm model. According to the FTIR spectra, sorption occurs via formation of inner-sphere complexes between the surface functional groups and the *f*-metal cationic species. The presence of europium and thorium in the adsorbent material was confirmed and quantified with EDS analysis. To the best of our knowledge, this is the first report of an aerogel material used as an adsorbent for Eu(III). Compared to other materials used for the sorption of the specific ions, which are mostly carbon-based, X-alginate aerogels show by far the highest sorption capacity. Regarding Th(IV) uptake, X-alginate aerogels show the highest capacity per volume (27.9 g L^{-1}) among the aerogels reported in the literature. Both Eu(III) and Th(IV) could be recovered from the beads by 65% and 70%, respectively. Furthermore, Th(VI) could also be quantitatively removed from wastewater, while Eu(III) could be removed by 20%. The above, along with their stability in aqueous environments, make X-alginate aerogels attractive candidates for water treatment and metal recovery applications.

Keywords: alginate aerogels; polymer-crosslinked aerogels; polyurea-crosslinked alginate aerogels; Th(IV) sorption; Eu(III) sorption; thermodynamic; environmental remediation; water decontamination

Citation: Georgiou, E.; Pashalidis, I.; Raptopoulos, G.; Paraskevopoulou, P. Efficient Removal of Polyvalent Metal Ions (Eu(III) and Th(IV)) from Aqueous Solutions by Polyurea-Crosslinked Alginate Aerogels. *Gels* **2022**, *8*, 478. https://doi.org/10.3390/gels8080478

Academic Editors: Melita Menelaou and Istvan Lazar

Received: 27 May 2022
Accepted: 24 July 2022
Published: 29 July 2022

1. Introduction

Europium is a member of the lanthanide series and one of the rarest members of the rare earth elements. Similar to other lanthanide elements, europium in aqueous solutions exists basically in the trivalent oxidation state and, under ambient conditions, hydrolysis and carbonate complexation determine its chemical behavior. Although a heavy metal, europium is relatively non-toxic, because it has no significant role in biological processes. Europium compounds have found industrial applications because of their optical/luminescent properties. In addition, Eu(III) is used as an analogue for trivalent lanthanides and actinides, such as Am(III) and Cm(III), because it presents similar aquatic chemistry to its radioactive counterparts, is non-(radio)toxic and possesses useful fluorescent properties, which make spectroscopic studies on its chemical behavior and speciation possible [1].

Efficient collection of lanthanides (including europium) from process waters and industrial wastewaters prior to their discharge into environmental receivers is nowadays mandatory in order to protect the environment and secure the future availability of lanthanide resources. Lanthanides play a main role in the fabrication of different industrial products (e.g., automotive catalysts, magnets, optical devices, ceramics, etc.), as well as in green and sustainable energy production [2,3].

Thorium belongs to the actinide series of elements. Monazite contains relatively high amounts of thorium together with other rare earth elements, and it is extracted from the ore by liquid−liquid extraction using tributyl phosphate [4]. The naturally occurring thorium isotope (Th-232) is a weakly radioactive nuclide, and not fissile, but it can be transmuted via neutron absorption into the fissile uranium isotope U-233. Over the next decades thorium could play a major role as a nuclear fuel in liquid fluoride thorium reactors (LFTR), which, compared to uranium-fueled light water reactors, are significantly more energy effective, and produce far less nuclear waste [5]; and they are also characterized of higher safety, which is associated with considerably lower chances of getting into a nuclear accident [6]. In addition, thorium is more abundant than uranium, which results to lower energy production prices [6].

In the past, thorium has been widely used in ceramic glazes, lantern mantles, and welding rods. In addition, thorium in the form of colloidal thorium dioxide was used until the 1950s as a contrast agent in medical radiology. Studies of people and patients who were exposed to thorium have shown increased risk of liver tumors, pancreatic, lung, and bone cancer. The latter is mainly because thorium is stored in bones [7]. It is expected that increased levels of thorium in the biosphere could have negative effects on the environment and human health [8]. Therefore, removal of thorium from industrial process waters and wastewaters prior to their discharge into environmental receivers is of particular importance regarding the protection of the associated ecosystems. In addition, possible recovery of thorium from industrial wastewaters is of particular interest because of its use in nuclear fuel production.

Metal removal technologies based on adsorption have attracted a significant interest because of their ease of operation, possible reuse of absorbents, and efficient recovery of precious elements [9]. Aerogel materials as adsorbents are very attractive because of their diverse chemical composition, high porosity, and a wide range of pore dimensions, including micro-, meso-, and macropores, resulting in effective and low-cost processes, and often also in selectivity towards the desirable adsorbates [10,11].

One example of aerogel materials studied as adsorbents are the polyurea-crosslinked alginate (X-alginate) aerogels, which are a new class of biopolymer-based aerogels with high mechanical strength [12–14]. These aerogels were prepared according to the polymer-cross-linked (X-aerogel) technology [15–20], by the reaction of aliphatic or aromatic triisocyanates with pre-formed M-alginate wet-gels (M refers to divalent or trivalent metal cation), which results in the formation of a nano-thin layer of polyurea over the M-alginate skeleton, leaving the primary M-alginate structure practically undisturbed. That polyurea layer renders the materials hydrophobic, in contrast to the corresponding M-alginate aerogels that are extremely hydrophilic. The nature of polyurea (aliphatic or aromatic) is also significant, as the aliphatic polyurea layer is more compact compared to the aromatic polyurea layer, which is a more randomly oriented polymer structure, leaving more exposed the coordination sites [21].

Based on the above, X-alginate aerogels (Figure 1) derived from pre-formed Ca-alginate (Figure 1) gels and an aromatic triisocyanate (Desmodur RE; Figure 1) have been found to be excellent candidate materials for environmental remediation. These materials combine hydrophobicity, high mechanical strength, and high stability (no swelling, shrinking, or disintegration) in all aqueous environments, including seawater. Therefore they have been used for decontamination of seawater from Pb(II), organic solvents, and oil [22], and also for U(VI) uptake from various water environments [23]. In the case of U(VI), the adsorption process was especially fast, and X-alginate aerogels showed an extremely high adsorption capacity uptake—twice the mass of the material, which compared to that of other aerogel materials is one of the highest sorption capacities per weight and the highest per volume. In addition to this, uranium could be recovered almost quantitatively.

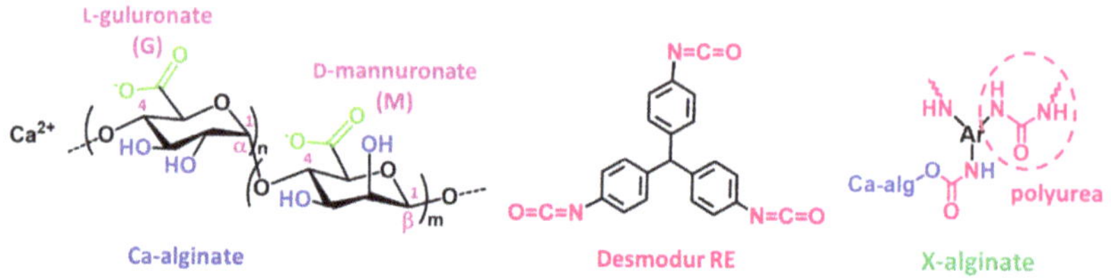

Figure 1. The structures of Ca-alginate, triphenylmethane-4,4′,4″-triisocyanate (TIPM; Desmodur RE) and the corresponding X-alginate aerogels. Ca-alginate is a block copolymer of β-(1→4)-linked D-mannuronate (M) and α-(1→4)-linked L-guluronate (G).

In this study, X-alginate aerogels described above have been studied in terms of their ability to sorb Eu(III) and Th(IV) from aqueous solutions. In addition to batch-type experiments, which were carried out to investigate the effect of various parameters on the sorption efficiency of the aerogels towards Eu(III) and Th(IV), FTIR and EDS data were used to confirm the post-adsorption presence of Eu(III) and Th(IV) on the adsorbent and to evaluate the sorption mechanism.

2. Results and Discussion

2.1. Eu(III) Sorption

The adsorption capacity of the X-alginate aerogels for Eu(III) has been studied with batch-type experiments, under ambient conditions and at pH 3.0. The experiments were carried out at pH 3.0 to avoid the formation of hydrolysis products and enable the use of increased metal ion concentrations without considering complex polynucleation and surface precipitation reactions. The dominant species in solution under the given experimental conditions is Eu^{3+}. The initial Eu(III) concentration varied between 10^{-5} and 0.1 mol L^{-1}, and the isothermal data were fitted with the Langmuir model (Equation (1)), where q_e is the Eu(III) uptake (in mol kg^{-1}) at equilibrium, C_e is the Eu(III) concentration in the solution at equilibrium (in mol L^{-1}), q_{max} is the adsorption capacity (in mol kg^{-1}), and K_L the *Langmuir* equilibrium constant (in L mol^{-1}). Equation (1) is as follows:

$$q_e = \frac{q_{max}\, K_L\, C_e}{1+K_L\, C_e} \quad (1)$$

The isothermal data are shown in Figure 2; these data are best fitted with the Langmuir isotherm model, which assumes that specific sites are available for Eu(III) coordination. The plateau of this curve is associated with saturation of these sites, which occurs at increased Eu(III) concentrations. Evaluation of the experimental data by the Langmuir isotherm model resulted in an adsorption capacity (q_{max}) of 3.62 mol Eu(III) per kg of X-alginate aerogel (550 g kg^{-1}), which is a supreme value when considering corresponding literature values (90 g kg^{-1} < q_{max} < 360 g kg^{-1}) for carbon-based adsorbents, particularly under acidic conditions (pH 3) [24,25].

The FTIR spectra in Figure 3, which correspond to X-alginate aerogel beads (black line) and dried X-alginate beads after sorption of different amounts of Eu(III) (colored lines), clearly show that the shape and the relative intensity of the bands at 1622 and 1406 cm^{-1} (attributed to $-COO^-$ coordinated to Ca(II)) [26], 1506 cm^{-1} (attributed to –NH groups [14]) and 1016 cm^{-1} (attributed to –C–O–C– groups of the sugar ring [14]) change upon increasing Eu(III) adsorption. This is an indication that there is a direct interaction between Eu(III) and the respective groups ($-COO^-$ and –NH) that leads to the formation of inner-sphere complexes. Analogous observations were made previously for the adsorption of U(VI) on X-alginate beads [23].

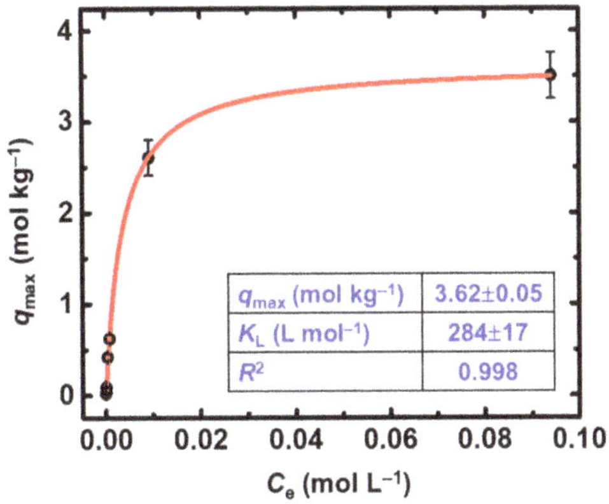

Figure 2. Experimental data and Langmuir sorption isotherm of Eu(III) on X-alginate aerogels at 283 K and varying initial Eu(III) concentrations (10^{-5}–0.1 mol L^{-1}). Experimental conditions are as follows: contact time 24 h, pH 3.0, adsorbent dosage 0.4 g L^{-1}, agitation rate 125 rpm.

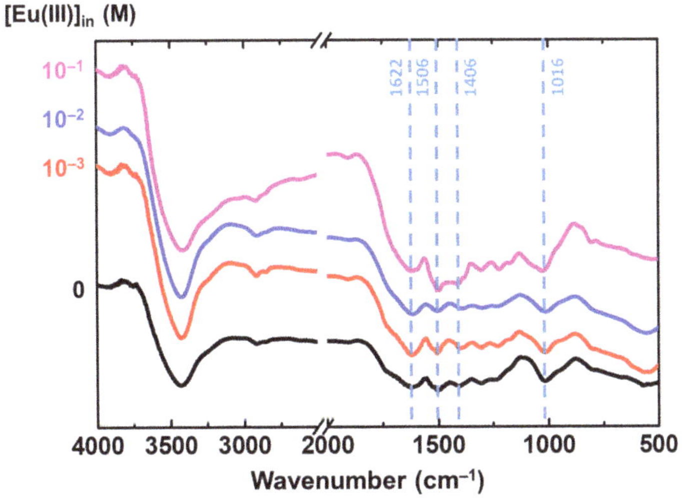

Figure 3. FTIR spectra of X-alginate aerogel beads (black line) and dried X-alginate beads after sorption of different amounts of Eu(III) (colored lines) from solutions with different initial Eu(III) concentrations (10^{-3}–0.1 mol L^{-1}, as indicated).

An EDS analysis was performed on selected samples, i.e., on X-alginate beads that were used for adsorption from solutions with initial Eu(III) concentrations equal to 10^{-5} and 10^{-1} mol L^{-1} (Figure 4). The EDS analysis revealed that Ca is present in both samples, and that the atomic ratio Eu/Ca changes from 0.55 to 55, respectively. These results are consistent with a cation exchange process between Eu(III) and Ca(II). However, the moles of Eu(III) adsorbed are up to seven times more than the moles of Ca(II) present in the same mass of the X-alginate beads, which clearly shows that Ca(II)-coordination sites are not the only sites available for Eu(III) coordination; indeed, -COO−, -NH, and −OH groups on the framework of the aerogel can also act as coordination sites. Analogous observations were previously made for the adsorption of U(VI) on X-alginate beads [23].

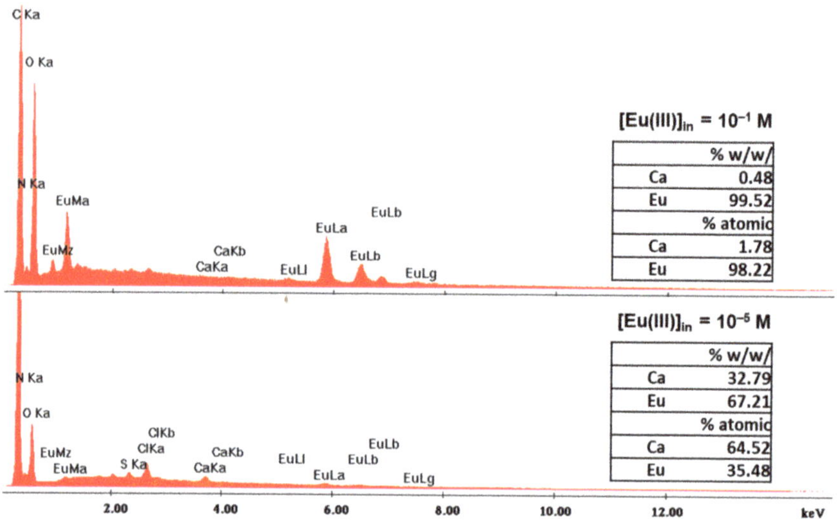

Figure 4. Energy-dispersive X-ray spectra (EDS) of dried X-alginate beads after sorption of different amounts of Eu(III) from solutions with different initial Eu(III) concentrations, as indicated.

The data obtained from the temperature effect experiments are graphically presented in Figure 5. The calculated thermodynamic parameters (ΔH^o = 155.3 kJ mol^{-1}; ΔS^o= 113.7 J K^{-1} mol^{-1}) clearly indicate that Eu(III) sorption is an endothermic and entropy-driven process, similar to the thermochemical behavior previously observed for U(VI) [23], suggesting a similar sorption mechanism based on inner-sphere complex formation between the surface moieties and Eu(III) cations.

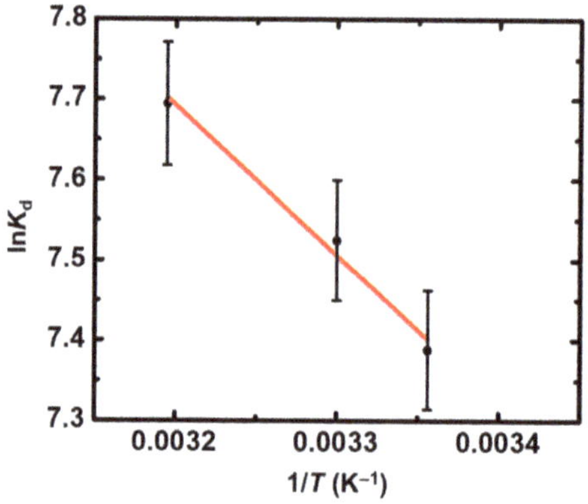

Figure 5. lnK_d as a function of $1/T$ for the sorption of Eu(III) by X-alginate aerogels at an initial Eu(III) concentration equal to 5×10^{-4} mol L^{-1}. Experimental conditions are as follows: contact time 24 h, pH 3.0, adsorbent dosage 0.4 g L^{-1}, agitation rate 125 rpm.

2.2. Th(IV) Sorption

Thorium sorption experiments were also performed at pH 3 in order to obtain comparable results and to avoid extensive hydrolysis, polynucleation, and surface precipitation reactions. The concentrations used in this study ([Th(IV)] < 0.001 M) are below saturation but, at pH 3, Th(IV) exists in solution mainly in the form of the $Th(OH)_2^{2+}$ and $Th(OH)_3^+$ species [27]. The affinity of these Th(IV) species to form complexes is expected to be significantly reduced due to their significantly lower charge density compared to the non-hydrolyzed aquo cation (Th^{4+}).

The maximum sorption capacity calculated using the Langmuir model (Figure 6) is q_{max} = 0.91 mol Th(IV) per kg of X-alginate aerogel (211 g kg^{-1}). This value is significantly lower than the corresponding value for Eu(III) and U(VI) [23], in agreement with the fact that the effective charge plays a central role regarding the stability of the surface species. Nevertheless, this value is higher than the corresponding values obtained for modified biochars (20 g kg^{-1} < q_{max} < 176 g kg^{-1}) [28–33]. Compared to other aerogel materials that have been reported in the literature for Th(IV) sorption (all graphene-based; Table 1) [34–36], X-alginate aerogels do not show the highest sorption capacity per weight but, interestingly, they show the highest sorption capacity per volume (27.9 g L^{-1}; 6–12 times higher than the other aerogels). Thus, for example, the volume of X-alginate aerogel needed to adsorb a certain amount of Th(IV) is 6–12 times the volume of other aerogels.

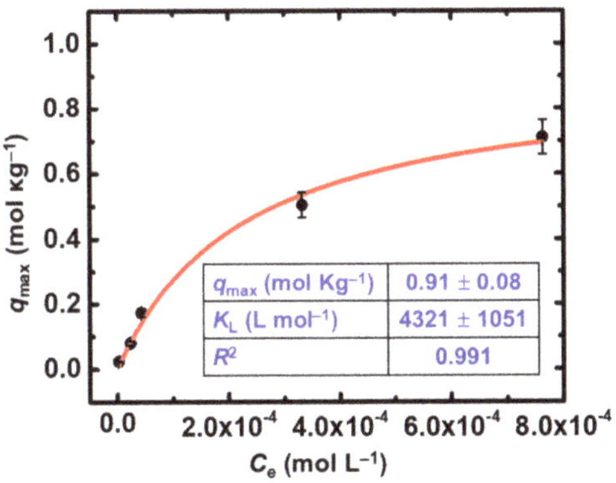

Figure 6. Experimental data and Langmuir sorption isotherm of Th(IV) on X-alginate aerogels at 283 K and varying initial Th(IV) concentrations (10^{-5}–10^{-3} mol L^{-1}). Experimental conditions are as follows: contact time 24 h, pH 3.0, adsorbent dosage 0.4 g L^{-1}, agitation rate 125 rpm.

Table 1. Th(IV) sorption capacity from laboratory solutions and selected material properties of different aerogel adsorbents.

Material	T (K)	pH	Max. Sorption Capacity (Langmuir) q_{max} (mg g^{-1})	Langmuir Constant K_L (L mg^{-1})	BET Surf. Area σ (m^2 g^{-1})	Bulk Density ρ_b (mg cm^{-3})	Max. Sorption Capacity q_{max} (mg cm^{-3})	Ref.
X-alginate aerogel	298	3	211.12	0.019	459 [a]	150 [a]	27.9	this work
graphene nanoribbons aerogel	298	3	380.4	0.020	597.4	6.2	2.36	[34]
poly(TRIM/VPA)-functionalized graphene oxide nanoribbons aerogel [b]	298	3	457.9	0.045	433.2	10.6	4.85	[35]
PEI-functionalized graphene aerogel [c]	298	2	38.17	0.0088	36.06			[36]

[a] Data taken from ref. [22]. [b] Abbreviations are defined as follows: TRIM, trimethylolpropane trimethacrylate; VPA, vinylphosphonic acid. [c] The abbreviation PEI is defined as polyethylenimine.

From the FTIR spectra shown in Figure 7, which correspond to dried X-alginate beads after sorption of different amounts of Th(IV), similar observations can be made to those described above for Eu(III), as in Section 2.1. The changes are not as intense as in the case of Eu(III) or U(VI), in agreement with smaller sorption capacity of Th(IV) compared to Eu(III) and U(VI), but they indicate again a direct interaction between Th(IV) species and the respective groups (–COO– and –NH) of X-alginate, resulting in the formation of inner-sphere complexes [23].

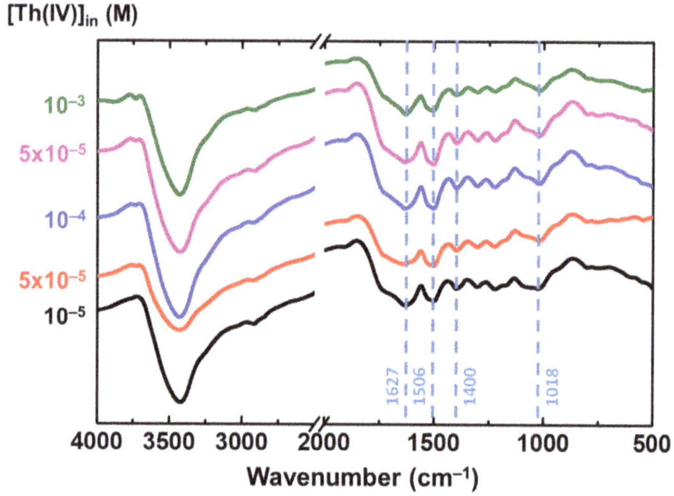

Figure 7. FTIR spectra of dried X-alginate beads after sorption of different amounts of Th(IV) from solutions with different initial Th(IV) concentrations (10^{-5}–10^{-3} mol L^{-1}, as indicated).

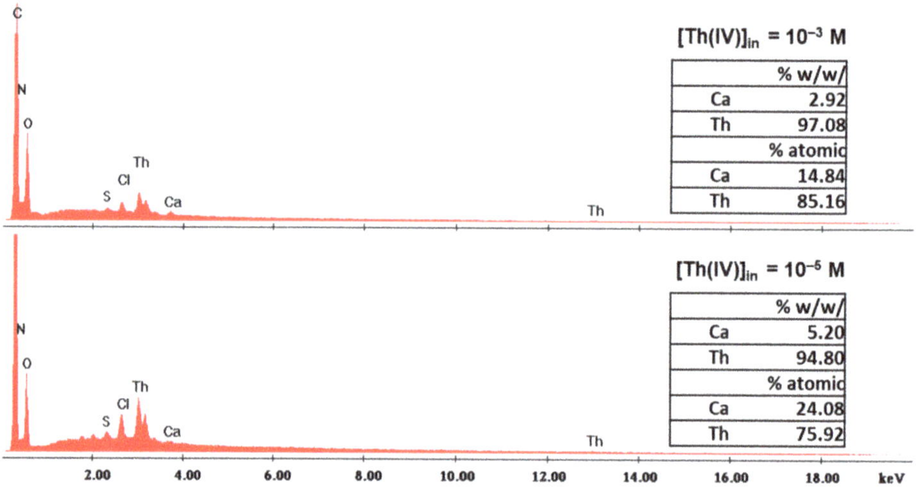

Figure 8. Energy-dispersive X-ray spectra (EDS) of dried X-alginate beads after sorption of different amounts of Th(IV) from solutions with different initial Th(IV) concentrations, as indicated.

An EDS analysis was performed on selected samples, i.e., on X-alginate beads that were used for adsorption from solutions with initial Th(IV) concentrations equal to 10^{-5} and 10^{-3} mol L^{-1} (Figure 8). The EDS results showed that the atomic ratio Th/Ca in the beads changes from 3.2 to 5.7, respectively. These results not only indicate cation exchange between Th(IV) species and Ca(II), but that Ca(II) are being replaced to a great extent even at low concentrations (i.e., 10^{-5} mol L^{-1}), showing a great affinity of Th(IV) species to X-alginate. Again, Ca(II)-occupied sites are not the only coordination sites available for Th(IV) coordination, as the moles of Th(IV) adsorbed are up to 1.4 times more than the moles of Ca(II) present in the same mass of the aerogel. As has been previously discussed for Eu(III) (see Section 2.1) and U(VI) [23], –COO–, –NH, and –OH groups on the framework of the aerogel provide numerous sites available for coordination with Th(IV).

The thermodynamic parameters (ΔH^o = 140.6 kJ mol^{-1}; ΔS^o = 109.2 J K^{-1} mol^{-1}) obtained from the evaluation of the experimental data corresponding to the temperature effect (Figure 9) suggest that sorption is an endothermic and entropy-driven process. This is similar to the thermochemical behavior observed for U(VI) [23] and Eu(III) and indicates a similar sorption mechanism, which is based on the direct interaction between the surface moieties and the Th(IV) species.

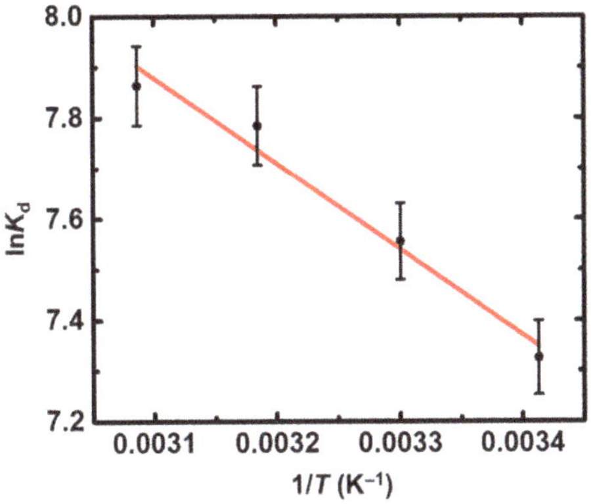

Figure 9. lnK_d as a function of $1/T$ for the sorption of Th(IV) by X-alginate aerogels at an initial Th(IV) concentration equal to 5×10^{-4} mol L^{-1}. Experimental conditions are as follows: contact time 24 h, pH 3.0, adsorbent dosage 0.4 g L^{-1}, agitation rate 125 rpm.

2.3. Desorption Studies and Application to Wastewater Solutions

Eu(III) and Th(IV) desorption has been studied with batch-type recovery experiments, i.e., via extraction with an aqueous solution of EDTA (pH 10). The recovery was calculated using Equation (3), and it was evaluated to 65 ± 2% and 70 ± 3% for Eu(III) and Th(IV), respectively. After desorption, the beads remained apparently intact and could be reused for further sorption experiments.

Application of X-alginate aerogel beads to Th(IV) and Eu(III) recovery from wastewater solutions under similar conditions to de-ionized water test solutions have shown that, in contrast to Eu(III), which is only partially (~20%) adsorbed, Th(IV) is quantitatively (100%) removed from the solution. The latter can be attributed to the following facts: (a) Th(IV) under the given conditions (pH 8.1) exists in the form of neutral $Th(OH)_4$ which presents very low solubility and high sorption tendency, and (b) the use of an extremely low Th(IV) concentration ([Th] = 0.1 nmol L^{-1}) to avoid colloid formation and/or solid phase precipitation.

3. Conclusions

X-alginate aerogel beads derived from Ca-alginate gels and the aromatic triisocyanate Desmodur RE show high adsorption capacities for Eu(III) and Th(IV). The concentration range studied covers a wide range of Eu(III) and Th(IV) concentrations expected in acidic process waters and wastewaters. The experimental data were fitted to the Langmuir isotherm model and the maximum adsorption capacity, q_{max}, was found to be equal to 3.6 mol (550 g) of Eu(III) per kg of aerogel and 0.9 mol (211 g) of Th(IV) per kg of aerogel. The lower sorption capacity for Th(IV) compared to Eu(III) is attributed to the net charge of the dominant species in solution under the given experimental conditions (pH = 3), which is Eu^{3+} for Eu(III), and $Th(OH)_2^{2+}$ and $Th(OH)_3^{+}$ for Th(IV). Evaluation of the thermodynamic data indicated that in both cases sorption is an endothermic, entropy-driven process. Furthermore, FTIR spectroscopy indicated the formation of inner-sphere complexes between the surface functional groups of X-alginate beads and Eu(III) or Th(IV) species, and EDS analysis confirmed the post-adsorption presence of europium and thorium in the adsorbent. The sorption of these *f*-metal species is very similar to the one observed previously for U(VI).

To the best of our knowledge, this is the first aerogel material used for the adsorption of Eu(III). Compared to other materials used for the sorption of Eu(III), which are mostly carbon-based, X-alginate aerogels show by far the highest sorption capacity. Regarding Th(IV) species, X-alginate aerogels show the highest capacity per volume (27.9 g L^{-1}) among the aerogels reported in the literature.

Both Eu(III) and Th(IV) could be recovered from the beads by 65% and 70%, respectively. Th(VI) could also be quantitatively removed from wastewater, while Eu(III) could be removed by 20%.

The above, along with their stability in aqueous environments, make X-alginate aerogels attractive candidates for water treatment and metal recovery applications. Future studies will focus on the application of X-alginate aerogels in packed columns and pilot scale water treatment units, which are of particular interest for the potential commercialization of the material.

4. Experimental Section

4.1. Materials and Methods

Sodium alginate PROTANAL LF 240 D (G/M = 0.43–0.54) was used as starting material. Then, $Eu(NO_3)_3{\cdot}6H_2O$ (analytical grade), $CaCl_2$ and Arsenazo-III were purchased from Sigma-Aldrich (Saint Louis, MO, USA), and $Th(NO_3)_4{\cdot}5H_2O$ was purchased from Merck (Darmstadt, Germany). Desmodur RE (27% *w*/*w* triphenylmethane-4,4′,4″-triisocyanate (TIPM) solution in ethyl acetate) was generously provided by Covestro AG (Leverkusen, Germany). MeCN (HPLC grade) was purchased from Fisher Scientific (Waltham, MA, USA), and acetone (P.A., ISO reagent) was purchased from Lach-Ner (Neratovice, Czechia). All solvents were used as received.

All UV–Vis spectrophotometry measurements were carried out using a UV-2401 PC Shimadzu (Kyoto, Japan) spectrophotometer. Alpha-spectroscopy was performed using a high-resolution alpha-spectrometer (Alpha Analyst Integrated Alpha Spectrometer, Canberra, Meriden, CT, USA) equipped with semiconductor detectors. All pH measurements were performed using a commercial glass electrode (Sentek, Braintree, UK), which was calibrated prior to and after each experiment using a series of buffer solutions (pH 2, 4, 7 and 10, Scharlau, Barcelona, Spain). The FTIR spectroscopic measurements were carried out using an FTIR spectrometer 8900 (Shimadzu, Kyoto, Japan). Samples were prepared as pellets with KBr.

Energy-dispersive X-ray spectra (EDS) analysis was performed using an FEI Quanta Inspect (FEI Company, Hillsboro, OR, USA) scanning electron microscope (SEM) equipped with an EDAX EDS system.

4.2. Preparation and Characterization of X-Alginate Aerogels

X-alginate aerogels were prepared and characterized as previously described [14]. The concentration of the initial aqueous solution of sodium alginate was 3% *w*/*w*.

4.3. Adsorption Experiments

The adsorption experiments of Eu(III) and Th(IV) on X-alginate aerogels were conducted as described elsewhere [37–39]. Briefly, aqueous solutions (25 mL) containing 0.01 g (0.4 g L^{-1}) of X-alginate aerogel beads and varying metal ion concentrations at pH 3.0 were prepared, and the adsorption of Eu(III) or Th(IV) (M(z)) was investigated under ambient conditions. The initial metal ion concentration was varied between 10^{-5} mol L^{-1} and 0.1 mol L^{-1} for Eu(III) and 10^{-5} mol L^{-1} and 0.001 mol L^{-1} for Th(IV). The contact time to reach equilibrium was set at 24 h and the determination of metal ions in solution was carried out using UV–Vis spectrophotometry directly at higher Eu(III) concentrations and using Arsenazo-III at lower Eu(III) and Th(IV) concentrations [40]. Thorium analysis was also performed by alpha-spectroscopy. The effect of temperature was investigated at 298 K, 303 K, 314 K, and 324 K at a metal ion concentration of 5×10^{-4} mol L^{-1}. The spectrophotometric method was calibrated prior to and after each experiment using reference solutions, which were prepared by dissolution of analytical grade nitrate salts of the respective metal ions (i.e., $Eu(NO_3)_3 \cdot 5H_2O$) and $Th(NO_3)_4 \cdot 5H_2O$) in de-ionized water under similar conditions. The amount of M(z) adsorbed, q_e (mol kg^{-1}), was calculated using Equation (2), where C_o (mol L^{-1}) is the initial M(z) concentration in the solution, C_e (mol L^{-1}) is the final M(z) concentration in the solution at equilibrium, V (L) is the solution volume, and m (g) is the weight of X-alginate aerogel beads. Equation (2) is as follows:

$$q_e = \frac{(C_o - C_e)V}{m} \quad (2)$$

After adsorption, X-alginate beads were dried overnight in a vacuum oven at 80 °C and were characterized by FTIR (KBr) spectroscopy and EDS analysis.

The applicability of X-alginate aerogel beads to remove Eu(III) and Th(IV) from "real" samples was investigated by contacting wastewater obtained from a wastewater treatment plant in Nicosia contaminated with Eu(III) ([Eu] = 1×10^{-4} mol L^{-1}) and Th(IV) ([Th-230] = 0.1 nmol L^{-1}). Sampling was carried out according to the Standard Methods. Water sample analysis was performed using the Standard Methods for the examination of water and wastewater [41]. The uptake experiments were performed using 0.4 g of X-alginate aerogel beads per liter of solution.

4.4. Desorption Experiments

In this study, Eu(III) and Th(IV) batch-type recovery experiments were performed using an aqueous EDTA solution 0.1 M (pH 10). After the adsorption experiments, the beads were separated from the suspension and 30 mL of the EDTA solution were added. Subsequently, the new suspension was shaken for 7 h in a thermostatic orbital shaker (100 rpm, 23 ± 2 °C). Afterwards, the beads were separated and the europium or thorium (M(z)) concentration in solution was determined using UV–Vis spectrophotometry. The desorption efficiency (% Desorption) was determined using Equation (3) ($[M(z)]_{ads}$ and $[M(z)]_{des}$ are the concentrations of ions adsorbed and desorbed, respectively).

$$\%\ Desorption = \frac{[M(z)]_{des}}{[M(z)]_{ads}} \times 100 \quad (3)$$

Author Contributions: Conceptualization, I.P. and P.P.; formal analysis, E.G., I.P., G.R. and P.P.; resources, I.P. and P.P.; writing—original draft preparation, E.G. and I.P.; writing—review and editing, E.G., I.P., G.R. and P.P.; funding acquisition, I.P. and P.P. All authors have read and agreed to the published version of the manuscript.

Funding: This publication is based upon work from COST Action "Advanced Engineering of aeroGels for Environment and Life Sciences" (AERoGELS, ref. CA18125), supported by COST (European Cooperation in Science and Technology). The General Secretariat for Research and Innovation, Greece, and the Special Account of Research Grants of the National and Kapodistrian University of Athens are also acknowledged.

Institutional Review Board Statement: Not applicable.

Informed Consent Statement: Not applicable.

Data Availability Statement: The data presented in this study are available on request from the corresponding authors.

Acknowledgments: We thank Nikos Boukos and Elias Sakellis (Electron Microscopy and Nanomaterials group, Institute of Nanoscience and Nanotechnology, NCSR "Demokritos", Athens, Greece) for the EDS spectra. We are also grateful to Covestro AG for their kind supply of Desmodur RE.

Conflicts of Interest: The authors declare no conflict of interest.

References

1. Choppin, G.R. Lanthanide Complexation in Aqueous Solutions. *J. Less Common Met.* **1984**, *100*, 141–151. [CrossRef]
2. Anastopoulos, I.; Bhatnagar, A.; Lima, E.C. Adsorption of Rare Earth Metals: A Review of Recent Literature. *J. Mol. Liq.* **2016**, *221*, 954–962. [CrossRef]
3. U.S. Environmental Protection Agency. *Rare Earth Elements: A Review of Production, Processing, Recycling, and Associated Environmental Issues*; U.S. Environmental Protection Agency: Washington, DC, USA, 2012.
4. Orabi, A.H.; Mohamed, B.T.; Ismaiel, D.A.; Elyan, S.S. Sequential Separation and Selective Extraction of Uranium and Thorium from Monazite Sulfate Leach Liquor Using Dipropylamine Extractant. *Miner. Eng.* **2021**, *172*, 107151. [CrossRef]
5. I.A.E. Agency. *Thorium Fuel Cycle: Potential Benefits and Challenges*; International Atomic Energy Agency: Vienna, Austria, 2005; ISBN 978-92-0-103405-2.
6. Cooper, N.; Minakata, D.; Begovic, M.; Crittenden, J. Should We Consider Using Liquid Fluoride Thorium Reactors for Power Generation? *Environ. Sci. Technol.* **2011**, *45*, 6237–6238. [CrossRef] [PubMed]
7. U.S. Environmental Protection Agency. *Radiation Protection: Radionuclide Basics: Thorium*; U.S. Environmental Protection Agency: Washington, DC, USA, 2015.
8. Keith, S.; Wohlers, D.; Ingerman, L. *Toxicological Profile for Thorium*; US Department of Health and Human Service: North Syracuse, NY, USA, 2019.
9. Burakov, A.E.; Galunin, E.V.; Burakova, I.V.; Kucherova, A.E.; Agarwal, S.; Tkachev, A.G.; Gupta, V.K. Adsorption of Heavy Metals on Conventional and Nanostructured Materials for Wastewater Treatment Purposes: A Review. *Ecotoxicol. Environ. Saf.* **2018**, *148*, 702–712. [CrossRef] [PubMed]
10. Singh, B.; Dhiman, M. Chapter 12—Carbon Aerogels for Environmental Remediation. In *Advances in Aerogel Composites for Environmental Remediation*; Khan, A.A.P., Ansari, M.O., Khan, A., Asiri, A.M., Eds.; Elsevier: Amsterdam, The Netherlands, 2021; pp. 217–243. ISBN 978-0-12-820732-1.
11. Ihsanullah, I.; Sajid, M.; Khan, S.; Bilal, M. Aerogel-Based Adsorbents as Emerging Materials for the Removal of Heavy Metals from Water: Progress, Challenges, and Prospects. *Sep. Purif. Technol.* **2022**, *291*, 120923. [CrossRef]
12. Paraskevopoulou, P.; Smirnova, I.; Athamneh, T.; Papastergiou, M.; Chriti, D.; Mali, G.; Čendak, T.; Chatzichristidi, M.; Raptopoulos, G.; Gurikov, P. Mechanically Strong Polyurea/Polyurethane-Cross-Linked Alginate Aerogels. *ACS Appl. Polym. Mater.* **2020**, *2*, 1974–1988. [CrossRef]
13. Paraskevopoulou, P.; Smirnova, I.; Athamneh, T.; Papastergiou, M.; Chriti, D.; Mali, G.; Čendak, T.; Raptopoulos, G.; Gurikov, P. Polyurea-Crosslinked Biopolymer Aerogel Beads. *RSC Adv.* **2020**, *10*, 40843. [CrossRef]
14. Raptopoulos, G.; Papastergiou, M.; Chriti, D.; Effraimopoulou, E.; Čendak, T.; Samartzis, N.; Mali, G.; Ioannides, T.; Gurikov, P.; Smirnova, I.; et al. Metal-Doped Carbons from Polyurea-Crosslinked Alginate Aerogel Beads. *Mater. Adv.* **2021**, *2*, 2684–2699. [CrossRef]
15. Leventis, N. Three-Dimensional Core-Shell Superstructures: Mechanically Strong Aerogels. *Acc. Chem. Res.* **2007**, *40*, 874–884. [CrossRef]
16. Leventis, N.; Sotiriou-Leventis, C.; Zhang, G.; Rawashdeh, A.-M.M. Nanoengineering Strong Silica Aerogels. *Nano Lett.* **2002**, *2*, 957–960. [CrossRef]
17. Mandal, C.; Donthula, S.; Far, H.M.; Saeed, A.M.; Sotiriou-Leventis, C.; Leventis, N. Transparent, Mechanically Strong, Thermally Insulating Cross-Linked Silica Aerogels for Energy-Efficient Windows. *J. Sol-Gel Sci. Technol.* **2019**, *92*, 84–100. [CrossRef]
18. Mohite, D.P.; Mahadik-Khanolkar, S.; Luo, H.; Lu, H.; Sotiriou-Leventis, C.; Leventis, N. Polydicyclopentadiene Aerogels Grafted with PMMA: I. Molecular and Interparticle Crosslinking. *Soft Matter* **2013**, *9*, 1516–1530. [CrossRef]
19. Mohite, D.P.; Mahadik-Khanolkar, S.; Luo, H.; Lu, H.; Sotiriou-Leventis, C.; Leventis, N. Polydicyclopentadiene Aerogels Grafted with PMMA: II. Nanoscopic Characterization and Origin of Macroscopic Deformation. *Soft Matter* **2013**, *9*, 1531–1539. [CrossRef]
20. Mulik, S.; Sotiriou-Leventis, C.; Leventis, N. Macroporous Electrically Conducting Carbon Networks by Pyrolysis of Isocyanate-Cross-Linked Resorcinol-Formaldehyde Aerogels. *Chem. Mater.* **2008**, *20*, 6985–6997. [CrossRef]
21. Paraskevopoulou, P.; Raptopoulos, G.; Len, A.; Dudás, Z.; Fábián, I.; Kalmár, J. Fundamental Skeletal Nanostructure of Nanoporous Polymer-Cross-Linked Alginate Aerogels and Its Relevance to Environmental Remediation. *ACS Appl. Nano Mater.* **2021**, *4*, 10575–10583. [CrossRef]
22. Paraskevopoulou, P.; Raptopoulos, G.; Leontaridou, F.; Papastergiou, M.; Sakellari, A.; Karavoltsos, S. Evaluation of Polyurea-Crosslinked Alginate Aerogels for Seawater Decontamination. *Gels* **2021**, *7*, 27. [CrossRef]
23. Georgiou, E.; Raptopoulos, G.; Papastergiou, M.; Paraskevopoulou, P.; Pashalidis, I. Extremely Efficient Uranium Removal from Aqueous Environments with Polyurea-Cross-Linked Alginate Aerogel Beads. *ACS Appl. Polym. Mater.* **2022**, *4*, 920–928. [CrossRef]
24. Hadjittofi, L.; Charalambous, S.; Pashalidis, I. Removal of Trivalent Samarium from Aqueous Solutions by Activated Biochar Derived from Cactus Fibres. *J. Rare Earths* **2016**, *34*, 99–104. [CrossRef]
25. Liatsou, I.; Pashalidis, I.; Oezaslan, M.; Dosche, C. Surface Characterization of Oxidized Biochar Fibers Derived from Luffa Cylindrica and Lanthanide Binding. *J. Environ. Chem. Eng.* **2017**, *5*, 4069–4074. [CrossRef]
26. Paraskevopoulou, P.; Gurikov, P.; Raptopoulos, G.; Chriti, D.; Papastergiou, M.; Kypritidou, Z.; Skounakis, V.; Argyraki, A. Strategies toward Catalytic Biopolymers: Incorporation of Tungsten in Alginate Aerogels. *Polyhedron* **2018**, *154*, 209–216. [CrossRef]

27. Moulin, C.; Amekraz, B.; Hubert, S.; Moulin, V. Study of Thorium Hydrolysis Species by Electrospray-Ionization Mass Spectrometry. *Anal. Chim. Acta* **2001**, *441*, 269–279. [CrossRef]
28. Kütahyalı, C.; Eral, M. Sorption Studies of Uranium and Thorium on Activated Carbon Prepared from Olive Stones: Kinetic and Thermodynamic Aspects. *J. Nucl. Mater.* **2010**, *396*, 251–256. [CrossRef]
29. Dong, L.; Chang, K.; Wang, L.; Linghu, W.; Zhao, D.; Asiri, A.M.; Alamry, K.A.; Alsaedi, A.; Hayat, T.; Li, X.; et al. Application of Biochar Derived from Rice Straw for the Removal of Th(IV) from Aqueous Solution. *Sep. Sci. Technol.* **2018**, *53*, 1511–1521. [CrossRef]
30. Salem, N.A.; Yakoot, E.; Sobhy, M. Adsorption Kinetic and Mechanism Studies of Thorium on Nitric Acid Oxidized Activated Carbon. *Desalination Water Treat.* **2016**, *57*, 28313–28322. [CrossRef]
31. Liatsou, I.; Christodoulou, E.; Pashalidis, I. Thorium Adsorption by Oxidized Biochar Fibres Derived from Luffa Cylindrica Sponges. *J. Radioanal. Nucl. Chem.* **2018**, *317*, 1065–1070. [CrossRef]
32. Hadjittofi, L.; Pashalidis, I. Thorium Removal from Acidic Aqueous Solutions by Activated Biochar Derived from Cactus Fibers. *Desalination Water Treat.* **2016**, *57*, 27864–27868. [CrossRef]
33. Wang, W.D.; Cui, Y.X.; Zhang, L.K.; Li, Y.M.; Sun, P.; Han, J.H. Synthesis of a Novel ZnFe2O4/Porous Biochar Magnetic Composite for Th(IV) Adsorption in Aqueous Solutions. *Int. J. Environ. Sci. Technol.* **2021**, *18*, 2733–2746. [CrossRef]
34. Li, Y.; He, H.; Liu, Z.; Lai, Z.; Wang, Y. A Facile Method for Preparing Three-Dimensional Graphene Nanoribbons Aerogel for Uranium(VI) and Thorium(IV) Adsorption. *J. Radioanal. Nucl. Chem.* **2021**, *328*, 289–298. [CrossRef]
35. Wang, Y.; Chen, X.; Hu, X.; Wu, P.; Lan, T.; Li, Y.; Tu, H.; Liu, Y.; Yuan, D.; Wu, Z.; et al. Synthesis and Characterization of Poly(TRIM/VPA) Functionalized Graphene Oxide Nanoribbons Aerogel for Highly Efficient Capture of Thorium(IV) from Aqueous Solutions. *Appl. Surf. Sci.* **2021**, *536*, 147829. [CrossRef]
36. Bai, R.; Yang, F.; Meng, L.; Zhao, Z.; Guo, W.; Cai, C.; Zhang, Y. Polyethylenimine Functionalized and Scaffolded Graphene Aerogel and the Application in the Highly Selective Separation of Thorium from Rare Earth. *Mater. Des.* **2021**, *197*, 109195. [CrossRef]
37. Hadjittofi, L.; Pashalidis, I. Uranium Sorption from Aqueous Solutions by Activated Biochar Fibres Investigated by FTIR Spectroscopy and Batch Experiments. *J. Radioanal. Nucl. Chem.* **2015**, *304*, 897–904. [CrossRef]
38. Liatsou, I.; Michail, G.; Demetriou, M.; Pashalidis, I. Uranium Binding by Biochar Fibres Derived from Luffa Cylindrica after Controlled Surface Oxidation. *J. Radioanal. Nucl. Chem.* **2017**, *311*, 871–875. [CrossRef]
39. Philippou, K.; Savva, I.; Pashalidis, I. Uranium(VI) Binding by Pine Needles Prior and after Chemical Modification. *J. Radioanal. Nucl. Chem.* **2018**, *318*, 2205–2211. [CrossRef]
40. Savvin, S.B. Analytical Use of Arsenazo III: Determination of Thorium, Zirconium, Uranium and Rare Earth Elements. *Talanta* **1961**, *8*, 673–685. [CrossRef]
41. APHA; AWWA; WEF. *Standard Methods for the Examination of Water and Wastewater*, 21st ed.; Water Environment Federation: Alexandria, VA, USA, 2005.

gels

Article

Photocatalytic Hydrogen Production Using Porous 3D Graphene-Based Aerogels Supporting Pt/TiO$_2$ Nanoparticles

Márta Kubovics [1,*], Cláudia G. Silva [2,3], Ana M. López-Periago [1], Joaquim L. Faria [2,3] and Concepción Domingo [1,*]

1 Instituto de Ciencia de Materiales de Barcelona, CSIC, Campus UAB s/n, 8193 Bellaterra, Spain
2 LSRE-LCM-Laboratory of Separation and Reaction Engineering–Laboratory of Catalysis and Materials, Faculty of Engineering, University of Porto, Rua Dr. Roberto Frias, 4200-465 Porto, Portugal
3 ALiCE-Associate Laboratory in Chemical Engineering, Faculty of Engineering, University of Porto, Rua Dr. Roberto Frias, 4200-465 Porto, Portugal
* Correspondence: mkubovics@icmab.es (M.K.); conchi@icmab.es (C.D.)

Abstract: Composites involving reduced graphene oxide (rGO) aerogels supporting Pt/TiO$_2$ nanoparticles were fabricated using a one-pot supercritical CO$_2$ gelling and drying method, followed by mild reduction under a N$_2$ atmosphere. Electron microscopy images and N$_2$ adsorption/desorption isotherms indicate the formation of 3D monolithic aerogels with a meso/macroporous morphology. A comprehensive evaluation of the synthesized photocatalyst was carried out with a focus on the target application: the photocatalytic production of H$_2$ from methanol in aqueous media. The reaction conditions (water/methanol ratio, catalyst concentration), together with the aerogel composition (Pt/TiO$_2$/rGO ratio) and architecture (size of the aerogel pieces), were the factors that varied in optimizing the process. These experimental parameters influenced the diffusion of the reactants/products inside the aerogel, the permeability of the porous structure, and the light-harvesting properties, all determined in this study towards maximizing H$_2$ production. Using methanol as the sacrificial agent, the measured H$_2$ production rate for the optimized system (18,800 $\mu mol_{H2} h^{-1} g_{NPs}{}^{-1}$) was remarkably higher than the values found in the literature for similar Pt/TiO$_2$/rGO catalysts and reaction media (2000–10,000 $\mu mol_{H2} h^{-1} g_{NPs}{}^{-1}$).

Keywords: aerogel; graphene oxide; supercritical CO$_2$; photocatalysis; H$_2$ production

check for updates

Citation: Kubovics, M.; Silva, C.G.; López-Periago, A.M.; Faria, J.L.; Domingo, C. Photocatalytic Hydrogen Production Using Porous 3D Graphene-Based Aerogels Supporting Pt/TiO$_2$ Nanoparticles. *Gels* **2022**, *8*, 719. https://doi.org/10.3390/gels8110719

Academic Editor: Miguel Sanchez-Soto

Received: 30 September 2022
Accepted: 2 November 2022
Published: 7 November 2022

1. Introduction

Hydrogen (H$_2$) is one of the most promising carbon-neutral alternatives as a renewable energy source, mainly due to its high calorific value and attainable purity [1]. Large-scale H$_2$ production via photocatalytic water splitting is a simple and cheap method, although the low reached conversion values (ca. 1%) cause the procedure to still be inefficient and economically unviable [2]. Consequently, the study of new catalytic systems that enhance the conversion is important to improve the efficiency and sustainability of the process [3]. Heterogeneous photocatalysis, working via water splitting or by the photoreforming of organic waste, is an attractive solution for H$_2$ production, since the utilization of solar energy moderates urgent environmental and energy issues [4].

In the photocatalytic H$_2$ production process, a light beam with sufficient energy irradiates a semiconductor material. The thus attained excited electrons (e^-) and holes (h^+) migrate onto the surface of the catalyst and act as reducing and oxidizing agents, respectively. Hence, the reduction and oxidation potentials of the reactant (e.g., water, alcohol, glycerol) must be within the band gap of the photocatalyst [5]. Semiconductors, such as titanium dioxide (TiO$_2$), cadmium sulfide (CdS), and carbon nitride (C$_3$N$_4$), fulfill this condition and are often used as photocatalytic systems [6–8]. TiO$_2$ is traditionally one of the most utilized semiconductors owing to its highly negative conduction band potential, and thus strong reduction ability. This oxide is chemically stable, cheap, and abundant. However, its use also experienced some drawbacks. First, due to its wide band gap, UV light is necessary for activation, and this light is present in a percentage lower than 5% in sunlight [9]. Second, TiO_2 experienced an extremely rapid recombination rate of the photogenerated h^+ and e^-, in the order of $10^{-12} - 10^{-11}$ s, while a value in the interval of $10^{-9} - 10^{-7}$ s is required for capturing the generated species in a successful redox reaction [7]. To improve the photocatalytic efficiency, the semiconductor must be combined with agents that can scavenge the photogenerated e^-, e.g., Pt [1,10]. It is worth mentioning that Pt has been recently included in the list of "critical raw materials -with economic importance, but high supply risk-" by the European Commission [11]. The target objective must be to decrease our dependence on these critical raw materials by minimizing their percentage of use in the designed product [12].

A newly designed and feasible strategy to moderate the detrimental effects of the large band gap and high e^-/h^+ recombination rate of TiO_2 is to incorporate graphene into the catalyst [13]. Undoubtedly, 2D graphene sheets are becoming a top choice as catalyst compartment/supports due to their unique physicochemical properties, related to their large surface area, high thermal and electrical conductivity, and ability to tailor the band gap energy level of the semiconductor [14–16]. Furthermore, due to its high work function, e^- from the conduction band of the semiconductor can be accepted and transferred by the graphene [17]. In parallel, the 2D sheets ensure an appropriate surface for extensively anchoring the semiconductor in the form of nanoparticles (NPs), which can be deposited on both sides of exfoliated graphene flakes, displaying the end composite with a high concentration of active sites for the catalytic reaction [18,19]. The main drawback of using this support is the large tendency of 2D graphene flakes to aggregate, which results in the poor accessibility of the reagents and the light to the catalytically active NPs trapped between the flakes, e^- transport hindering, and the poor diffusion of the gaseous product [7,20,21]. The penetration depth of the UV light in TiO_2 is limited to ca. 100 nm; thus, the formation of large aggregates of NPs would result in an increased amount of semiconductors not affected by the radiation [22]. Several works can be found in the literature using composites of (Pt)TiO_2 NPs and reduced graphene oxide (rGO) [22–28]. These systems are composed of either multiple stacked layers of rGO covered with NP agglomerates [26,27] or photoactive NPs covered with an rGO layer [28]. In other works, the aggregation of the Pt/TiO_2 NPs has been directed to build 3D porous composites with rGO added as an additive [29,30].

In this study, a different approach is presented to build a 3D structure, in which an rGO aerogel matrix provides support for the photoactive NPs: Pt/TiO_2@rGO. The precursor is a 3D aerogel structure of graphene oxide (GO): Pt/TiO_2@GO. GO is a highly oxygenated precursor, with mainly hydroxyl, epoxy, and carboxylic functionalities, easily exfoliated in polar liquids, and capable of establishing strong metal–support interactions to ensure NP dispersion and to avoid NP leaching during catalytic reaction [31–33]. As with other common aerogels [34,35], those of GO have a low-density network with a meso/macroporous structure. The Pt/TiO_2@GO intermediate composite was synthesized in the form of a monolith using a previously described one-pot supercritical CO_2 ($scCO_2$) methodology [36]. After aerogel synthesis, the number of oxygenated groups on the support can be modulated by thermal treatment to prepare the desired Pt/TiO_2@rGO end product [37]. The macroscopic size and variable shape of the synthesized 3D aerogel macrostructures bring advantages of operability and recoverability. The obtained aerogel composites were structural and texturally characterized. Moreover, the new catalytic system was evaluated for its photocatalytic H_2 production in aqueous methanol solutions. The process was adjusted for effective H_2 production regarding the reaction conditions (catalyst concentration, composition of reaction mixture), catalyst composition (Pt:TiO_2:rGO ratios), and architecture (one-piece monolith or smashed aerogel). Optimizing the Pt/TiO_2@rGO composite leads to H_2 production rates in an aqueous methanol solution of ca. 2–10 times higher than the values reported for similar systems in the literature [25,29].

2. Results and Discussion

2.1. Aerogel Synthesis

The xPt/TiO_2@rGO composite aerogels were synthesized through the intermediate xPt/TiO_2@GO aerogel (Figure 1), involving a non-reduced GO matrix containing a high amount of oxygenated functional groups, mainly hydroxyl, epoxy, and carboxylic, which are located on the basal plane and at the edges of the 2D platelets. The oxygenated functionalities facilitate the dispersion and exfoliation of GO in aqueous and polar solutions via simple sonication. Moreover, preserving the oxygenated functional groups in GO during $scCO_2$ aerogel synthesis was essential to ensure the presence of many anchoring points on the substrate for the NPs, which guarantees the establishment of strong interactions with a net or composed hydrophilic TiO_2 involving hydroxyl groups on the surface [32,38,39]. Three different compositions for the NPs in the intermediate were tested, namely, 1Pt/TiO_2@GO, 0.5Pt/TiO_2@GO, and 0.1Pt/TiO_2@GO, corresponding to an initially mixed amount of Pt with the TiO_2 NPs of 1, 0.5 and 0.1 wt%, respectively. To activate the aerogel for the catalytic process, the xPt/TiO_2@GO intermediate samples were exposed to a temperature of 300 °C in a N_2 atmosphere. This treatment eliminates most of the oxygenated groups in GO. The reduction step is crucial to further achieve an efficient photocatalytic reaction, since important graphene-like characteristics, such as high e^- mobility, are partially restored by removing some of the oxygenated groups. Hence, the rGO matrix can act as an efficient sink, where the photogenerated e^- are stored and transferred [13,40]. During reduction, ca. 30 wt% is eliminated from the sample, corresponding mostly to oxygenated functionalities. Taking this into account, the estimated ratios of the NPs:rGO phase in the xPt/TiO_2@rGO samples were calculated as 3:1 and 9:1, corresponding to the intermediates with NPs:GO ratios of 2:1 and 6:1, respectively. The Pt content in the reduced composites was measured by ICP-MS, giving values close to the expected quantity, 0.9, 0.5, and 0.1 wt%, percentages calculated concerning TiO_2 weight. Thus, practically, no noble metal loss occurs during the preparation procedure. The obtained samples were named as 0.9Pt/TiO_2@rGO, 0.5Pt/TiO_2@rGO, and 0.1Pt/TiO_2@rGO.

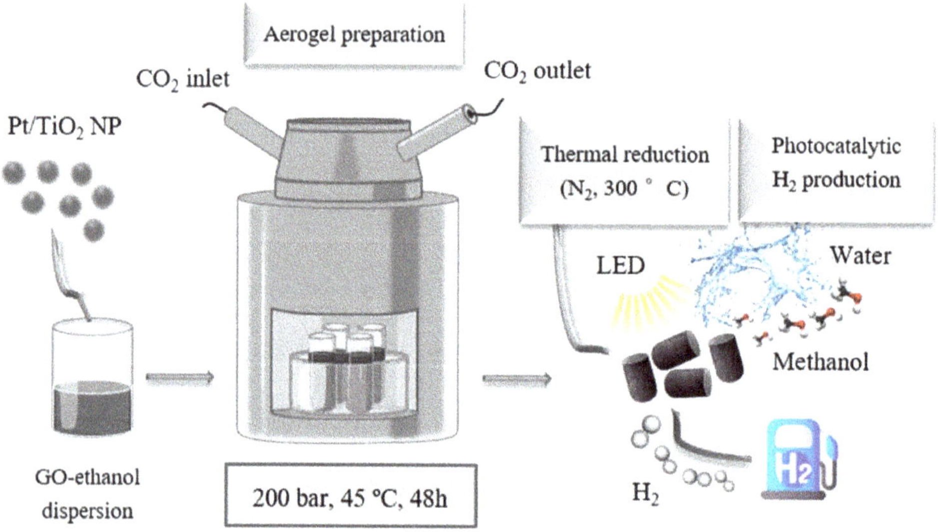

Figure 1. Scheme of the $scCO_2$-assisted synthesis for Pt/TiO_2@rGO aerogels.

2.2. Aerogels Structure

The composite components, as well as the intermediate xPt/TiO_2@GO and reduced xPt/TiO_2@rGO aerogels, were structurally analyzed by PXRD. Figure 2a shows the main signals in the patterns obtained in the 2θ interval of 20 to 40°, with the lines corresponding to anatase (2θ = 25.4, 37.0, 37.9, and 38.7°) and rutile (2θ = 27.4 and 36.2°), which were identical in the bare TiO_2 P25 and composed Pt/TiO_2 patterns. The signal of GO is described to appear at low angles, ca. 11° [41]. This signal could be observed for the intermediate non-reduced sample as a minor peak at this 2θ (Figure S1), while it disappears from the pattern of the reduced composite. The broadening of the diffraction lines was used to estimate NPs diameter by using the Scherrer equation. For bare TiO_2 and binary xPt/TiO_2 NPs, a size of ca. 21–22 nm was estimated, similar to that observed with TEM microscopy (Figure S2a,b). The NPs size in the reduced composites was similar, ca. 19–20 nm. The estimated particle size was, in all cases, in the range of the mean value given for the commercial TiO_2 P25 (ca. 20 nm). Consequently, no significant alteration of the crystalline structure or in the particle size of the TiO_2 took place throughout the deposition of Pt on its surface or the during aerogel formation and reduction.

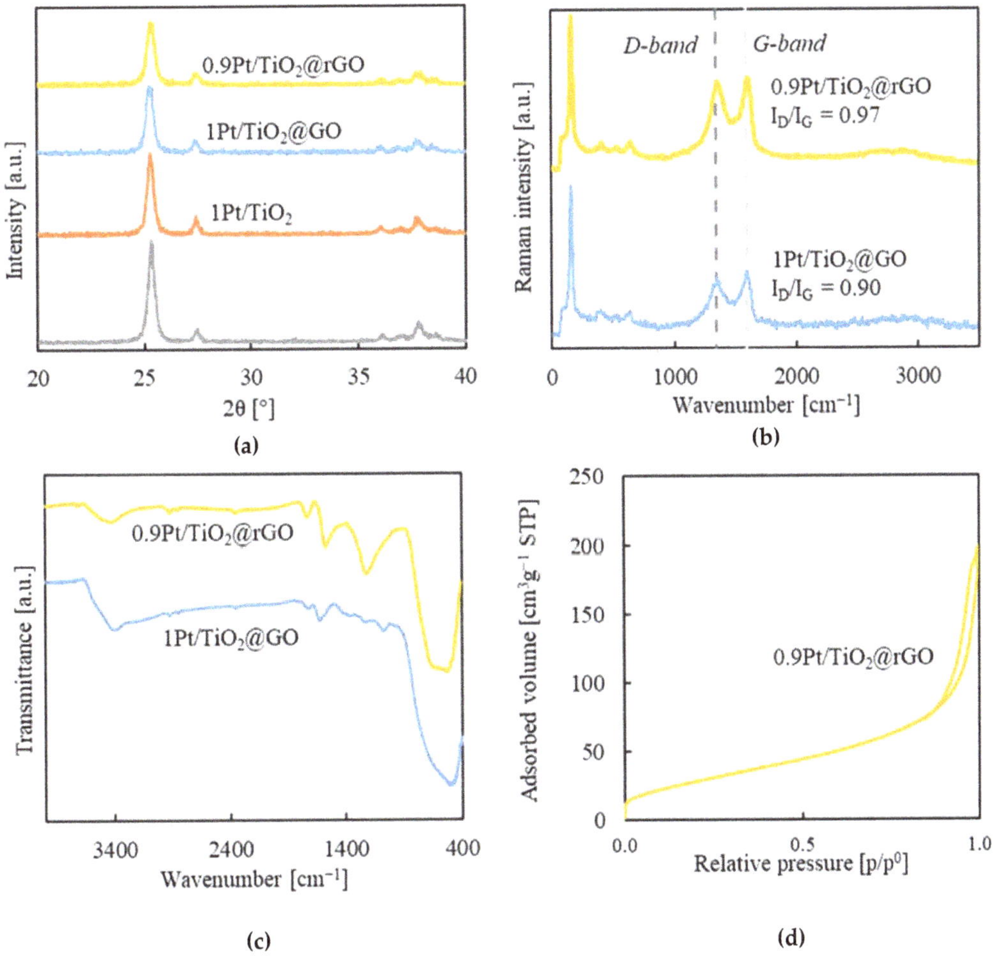

Figure 2. Structural characterization of bare NPs and as-synthesized or reduced aerogel composite for the $1Pt/TiO_2@GO$ (2:1) and $0.9Pt/TiO_2@rGO$ (3:1) samples: (**a**) PXRD patterns, (**b**) Raman spectra, in which D and G bands are indicated with dashed lines, (**c**) FTIR spectra, and (**d**) N_2 adsorption/desorption analysis of the reduced aerogel.

Raman spectroscopy was used to investigate the structural changes occurring in the GO functional groups during the reduction process of the aerogel. The as-synthesized and reduced aerogels recorded spectra were characterized by the presence of the typical G and D bands at 1590 and 1345 cm^{-1}, respectively (Figure 2b). The G band is generated by the in-plane vibrations of sp^2-bonded carbon atoms (C–C stretching), while the D-band represents the out-of-plane sp^3 vibrations corresponding to the defects in the graphitic structure. The ratios of the intensities of the D and G bands (I_D/I_G) were 0.90 and 0.97 for the GO and rGO composite aerogels, respectively. This result indicates that both synthesized and reduced samples have an elevated degree of sp^3 defects in the graphitic structure. For the xPt/TiO_2@GO samples, the defects represent the highly oxygenated character of GO. For the xPt/TiO_2@rGO composites, defects not only originated from the residual oxygenated groups but there are also structural defects (holes, vacancies, dislocations, etc.) caused by the thermal treatment applied for sample reduction. It has been described that reduction under relatively mild conditions, such as the ones used in this work, triggers the formation of these defects in rGO [42]. A considerable number of new sp^2 graphitic domains are formed, but of small size [43]. Moreover, a broadened band observed in the 2700–3000 cm^{-1} region of the Raman spectra, the usual position of the 2D peak in graphene, is indicative of randomly oriented multilayer graphene composing the aerogels [44]. Finally, the presence of the NPs in the composite was noticed by the detection of the TiO_2 bands at 151, 394, 515, and 630 cm^{-1}. The oxygenated character of GO in the intermediate composite and reduced aerogel was further analyzed by FTIR spectroscopy. The spectra of the intermediate samples (Figure 2c) show the GO functional groups by displaying the vibrational modes of C=O at 1719 cm^{-1}, C-OH at 1221 cm^{-1}, C-O at 1060 cm^{-1}, C-O-C at 1370 cm^{-1}, and OH at 1618 and 3400 cm^{-1}, the latter indicating also adsorbed water. The reduced sample displayed an intense C=C band at 1550 cm^{-1}, indicating the partial restoration of the graphitic structure. However, the bands of most of the oxygen groups in GO were somehow preserved in the rGO samples, only the peaks corresponding to epoxy vanished totally, while the peak at 3400 cm^{-1} corresponding to the hydroxyl groups diminished. The presence of the NPs in the composites is shown by the intense and broad stretching band appearing at 500–800 cm^{-1}. Raman and FTIR characterization indicate that, under the used experimental conditions for the reduction, 300 °C and a N_2 atmosphere, the GO phase was partially reduced, but a significant amount of oxygenated functionalities was conserved in the structure of rGO.

2.3. Textural Properties and Morphology

The textural properties of the synthesized aerogels were analyzed by N_2 adsorption/desorption at low temperatures. Figure 2d shows the isotherm recorded for the sample 0.9Pt/TiO_2@rGO (3:1), representative of all the studied systems, while the isotherms of the precursors can be found in the SI (Figure S3). The isotherm of the 0.9Pt/TiO_2@rGO is described as type IV at low and medium relative pressure and type II at high relative pressure, which is characteristic of nanoporous structures with both meso- and macropores and a negligible contribution of microporosity. A similar shape was found for the non-reduced GO precursor, while the 0.9Pt/TiO_2 NPs constitute a mesoporous system originated by particle aggregation. The S_a value for the 0.9Pt/TiO_2@rGO sample was in the order of 110 m^2g^{-1}. This value is inferior to that found in the non-reduced GO sample (*ca.* 150 m^2g^{-1}) but is superior to pristine NPs (ca. 50 m^2g^{-1}). Drying gels with $scCO_2$ is known to produce relatively denser aerogels than drying at the critical point of the alcohol due to moderate shrinkage occurring upon gelation and drying. In this work, a diameter of ca. 0.8 cm was measured for the cylindrical aerogel intermediates xPt/TiO_2@GO, indicating that they suffered some contraction in the axial direction since they were synthesized in a vial of 1 cm diameter. Further, some extra tightening occurs during reduction, leading to monoliths of ca. 0.7 cm diameter for the end products xPt/TiO_2@rGO. Aerogels with mesoporosity homogeneously distributed along all the mesopore range were obtained (Figure S4), with a BJH V_p of 0.30 cm^3g^{-1} and an average mesopore size of 10 nm.

All the different xPt/TiO_2@GO synthesized intermediate aerogels have similar morphology; they are highly porous with a sponge-like macrostructure, as shown in the SEM image of Figure 3a. The SEM images of the reduced monoliths displayed, similarly to the intermediates, a 3D structure with interconnected meso- and macropores (Figure 3b).

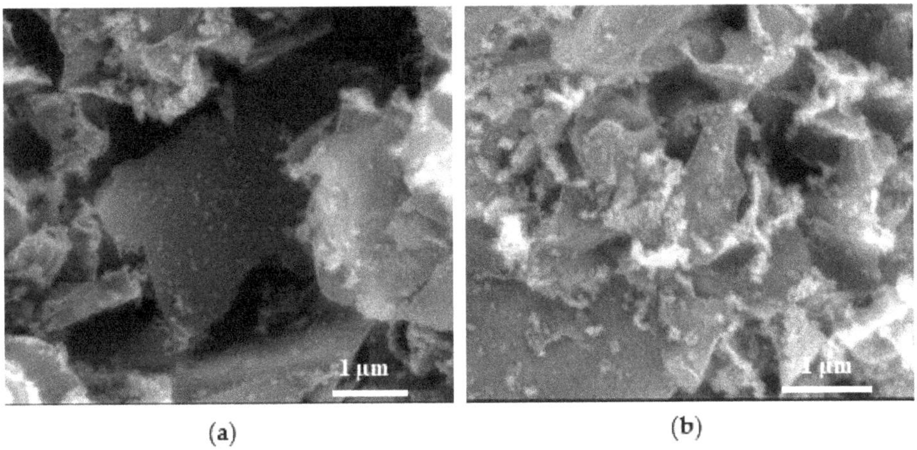

Figure 3. SEM images of samples: (**a**) 1Pt/TiO_2@GO (2:1), and (**b**) 0.9Pt/TiO_2@rGO (3:1), representative of the morphology of as-synthesized and reduced aerogels.

In most of the as-synthesized and reduced samples, the NPs can be discerned as a highly dispersed phase deposited on the surface of the GO or rGO plates (Figure 4a). However, the formation of large aggregates was also detected for the samples with the highest NPs:rGO (9:1) ratio (Figure 4b).

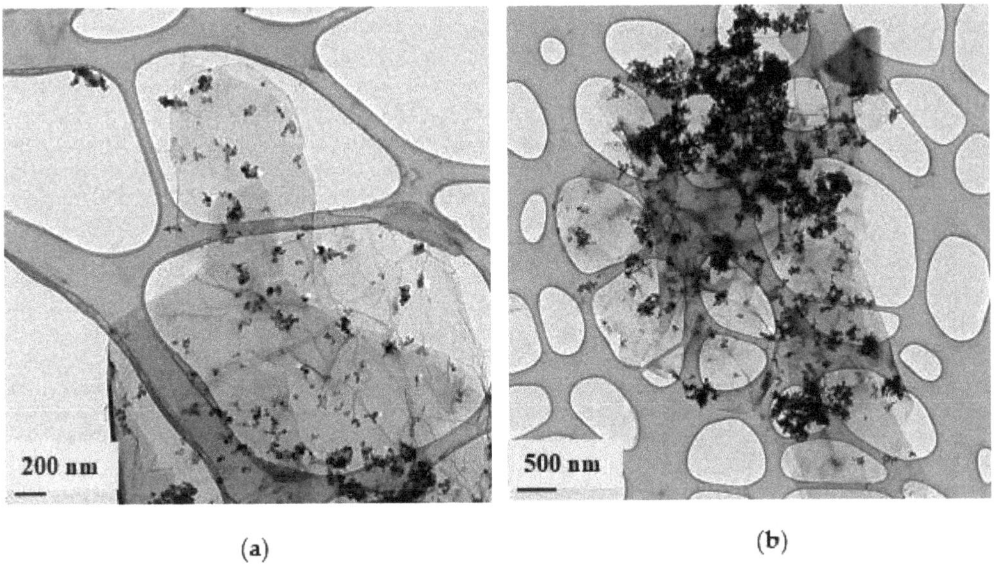

Figure 4. TEM images of the 0.9Pt/TiO_2@rGO samples with NPs:rGO ratios: (**a**) 3:1 and (**b**) 9:1.

2.4. Aerogels Optical Properties

The optical properties of the reduced composites, as well as those of bare TiO_2 and rGO, were analyzed by UV-VIS diffuse reflectance spectroscopy to investigate the samples' photoresponse (Figure 5a). TiO_2 was active in the UV zone and exhibited an abrupt absorption edge around 400 nm, while rGO displayed a continuous absorption in the visible range. For the reduced composites, a broad background absorption in the visible range was observed as a consequence of rGO black characteristics, more notable in the sample with the lowest percentage of NPs, e.g., xPt/TiO_2@rGO (3:1). The contribution of the NPs to the absorption in the UV zone can be clearly appreciated in the composites' spectra, although with a red shift in the adsorption edge that was slight, to 450 nm, for the (9:1) composite and pronounced, to 750 nm, when the amount of rGO was increased in the (3:1) sample. This shift indicates an increased photoresponse in the visible range of the composite aerogels with respect to net TiO_2 NPs. To study the indirect optical band gap of the photocatalyst, a Tauc plot was determined, calculated from the UV-VIS absorption spectrum (see the detailed description in the SI) (Figure 5b) [45,46]. For the bare TiO_2, a band gap energy of 3.1 eV was estimated from the x-axis intercept of the extrapolated line fitted to the linear region of the plot. The effect of the Pt content was analyzed for the xPt/TiO_2@rGO (3:1) composites with an x range of 0.9–0.1 wt%. The three measured bandgap values (at Pt contents of 0.9, 0.5, and 0.1 wt%) were of ca. 1.5 eV, thus demonstrating the low influence of this parameter. Contrarily, enhancing the ratio of NPs:rGO in the 0.9Pt/TiO_2@rGO sample from 3:1 to 9:1 results in an increase in the bandgap from 1.5 to 2.8 eV, with a concomitant increase in the transition energy of the photoexcited electrons. This phenomenon is assigned to the generation of impurity energy levels above the valance band in the NPs upon their incorporation onto the rGO surface. Thus, for the excitation of the charge carriers, less energy is required [47]. It is worth mentioning that the shift in the absorption edge and the decrease in the band gap energy are both more notable in the studied compounds than in similar published systems, reporting shift values of only 0.1–0.4 eV [17,26]. This important result is explained by the formation of a large number of Ti–O–C bonds in the xPt/TiO_2@rGO samples, established between the surface of the TiO_2 and the rGO flakes [48], and occurring during the reduction and elimination of water from the pre-settle hydrogen bond interactions Ti–OH...OH–GO in the xPt/TiO_2@GO intermediates. Although the low band gap energy in the (3:1) aerogel suggests a high photoresponse, this feature does not necessarily mark out the best photocatalyst system, since other factors should be taken into account. Importantly, using a high amount of rGO can evoke some activity loss, as the dark flakes can shield some active NPs, such that not all catalytic units are exposed to the light. Hence, to design the right catalyst, a compromise must be attained between the percentage of components (NPs:rGO) in the composite; on the one hand, to increase the band gap through the decrease in the number of NPs and, on the other hand, to reduce darkness via a decrease in the proportion of rGO.

Photoluminescence experiments were carried out to study the recombination rate of the photogenerated e^-/h^+ pairs in the aerogel catalyst. One of the main drawbacks described for the use of TiO_2 semiconductors in photocatalytic processes is the fast recombination of the photogenerated species [7]. This behavior is clearly evidenced in the photoluminescence spectrum with an intense emission after the photoexcitation of the bare NPs under UV light at 320 nm (Figure 5c). A wide luminescence band is observed for TiO_2, with a maximum at 410 nm (close to the band gap energy of TiO_2), which is followed by a less intense signal at 468 nm. The spectra of the reduced xPt/TiO_2@rGO (3:1) and (9:1) composites did not have the same pattern as that of net TiO_2, being indeed similar to that of rGO. Thus, the photoluminescence intensity was diminished in the composites with respect to bare NPs, which is a usual behavior originated by the e^- acceptor and transport features given by the rGO support, resulting in suppressed charge recombination and less intense light emission. However, the composites with the lowest percentage of rGO showed the weakest photoluminescence intensity values, indicating that there is an optimal rGO content regarding the recombination rate. For the composite with the largest number of

NPs, quantitatively more photoelectrons can be generated, thus resulting in a high number of potential recombinations and increased photoluminescence intensity. The deposited Pt on the TiO_2 surface has been described to act as an electron sink, by trapping the electrons and further transferring them to the rGO support [25]. Comparing the applied Pt ratios, e.g., 0.1, 0.5, and 0.9 wt%, the lower the Pt loading, the weaker the photoluminescence response (Figure 5d). This result suggests that Pt can also act as a recombination center [49].

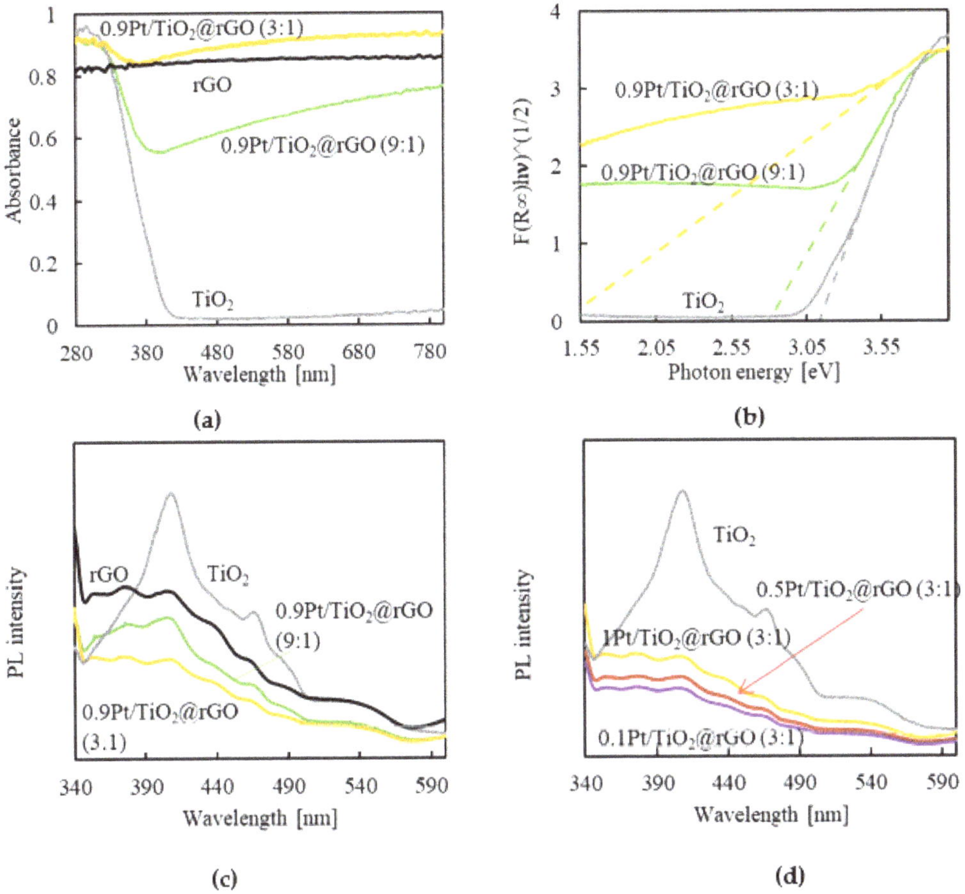

Figure 5. Optical characterization of bare NPs and reduced aerogel composites: (**a**) UV-VIS diffuse reflectance spectra, (**b**) band gap energy determined from the Tauc plot (the linear part of the plot is extrapolated to the *x*-axis), (**c**) photoluminescence spectra at an excitation wavelength of 320 nm, and (**d**) samples of xPt/TiO_2@rGO (3:1) with different Pt ratios.

2.5. Photocatalytic Hydrogen Production

Studies of photocatalytic H_2 production were performed to evaluate the new catalytic system. For that, the applied conditions were first extensively examined and optimized. The most favorable conditions for the use of the new catalyst in a particular catalytic process would depend on both the character of the material (composition and structure) and the setup used. In this study, the analysis is focused on the optimization of the synthesized xPt/TiO_2@rGO composite aerogel in regard to its photocatalytic activity in irradiated aqueous methanol solutions. The applied setup for the photocatalytic reaction is schematized in Figure 6. The studied parameters were catalyst reduction degree (from any to mild reduction), catalyst architecture (one-piece monolith or smashed aerogel), methanol concentration in the aqueous solution (from 0.01 to 1 v%), the added amount of catalyst to the reactor (from 0.03 to 2 $g_{NPs}L^{-1}$), Pt percentage in the catalyst (from 0 to 1 wt% in the Pt/TiO_2 NPs), and NP ratio with respect to rGO (3:1 and 9:1).

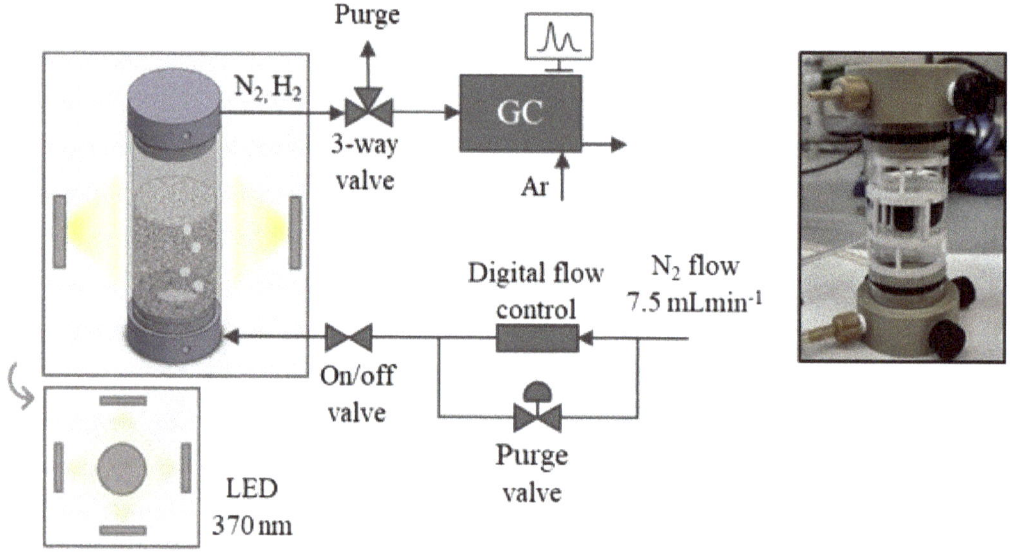

Figure 6. Scheme of the catalyst reactor setup and picture of the vessel involving the hand-made support holding the one-piece aerogel monolith.

In a typical experiment performed with the sample 0.9Pt/TiO_2@rGO (3:1), the aerogel was recovered after the photocatalytic reaction and analyzed in regard to composition. The sample maintained the ratio 3:1 for NPs:rGO, thus indicating the lack of NP leaching, which opens the door for the recyclability of the material.

2.5.1. Aerogel Reduction Degree

The straight use of GO monolithic aerogels with highly hydrophilic character in polar solvents causes the destruction of the macroscopic structure, provoked by strong electrostatic interactions with the solvent once it is immersed into the liquid. To avoid this drawback, the reduction in the GO phase to rGO was the applied solution in this work. The reduction step must be precisely controlled, since excessive reduction leads to highly hydrophobic aerogels that can suffer from low wettability when soaked in polar liquids, such as the water/methanol reaction medium used in this work. Hence, reaching an appropriate reduction degree of the GO composite aerogel is crucial to design an efficient catalyst in which the aqueous solution must easily travel inside the 3D structure [7]. In this work, a soft thermal treatment was applied for the reduction of the intermediate aerogel NPs@GO to NPs@rGO, e.g., 300 °C under a N_2 atmosphere. A reduced aerogel with amphiphilic properties, involving graphitic hydrophobic regions and remaining hydrophilic oxygenated groups (hydroxyl and carboxyl), was thus synthesized. The decrease in hydrophilicity after reduction was depicted by water contact angle measurements, showing an increase in the contact angle for the reduced composite (58.9°) in comparison to the non-reduced (21.8°) (Figure S5). The relatively still high wettability found for the reduced composite is the consequence of the residual oxygenated groups and the involvement of the more hydrophilic TiO_2 NPs, as well.

For the analyzed catalytic process, the necessity of reducing the GO support to rGO was established in a preliminary experiment in which the H_2 evolution with time was compared for similar samples either non-reduced (1Pt/TiO_2@GO (2:1)) or reduced (0.9Pt/TiO_2@rGO (3:1)) under similar conditions in the catalytic reactor (0.5 v/v methanol/water solution, 0.5 $g_{NPs}L^{-1}$, and smashed aerogel). The obtained results indicate that the reduction step was necessary to improve the efficiency of the catalyst (Figure 7a).

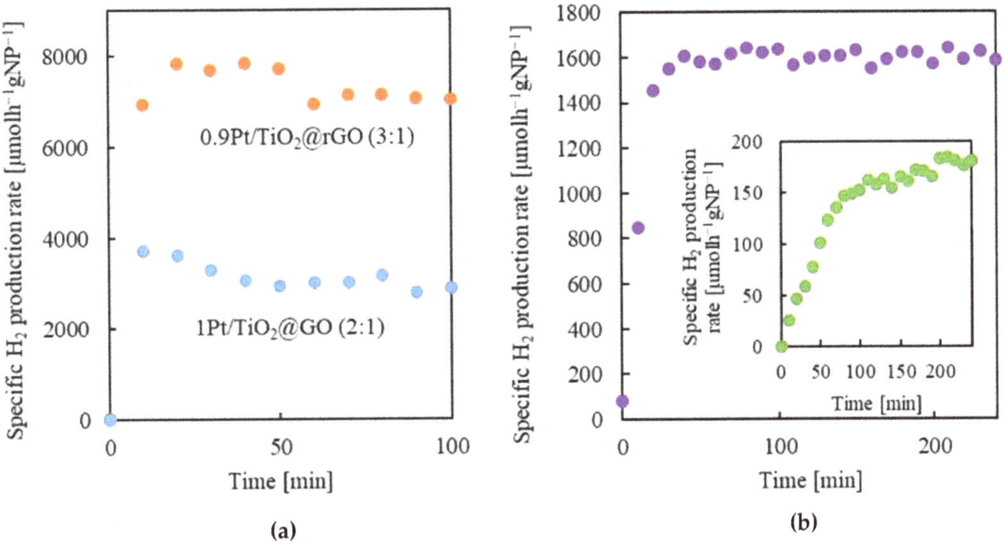

Figure 7. H_2 production rate at different experimental conditions: (**a**) using the non-reduced 1Pt/TiO_2@GO (2:1) and reduced 0.9Pt/TiO_2@rGO (3:1) aerogels (reaction conditions: 0.5 v/v methanol/water, 0.5 $g_{NPs}L^{-1}$, smashed aerogel), and (**b**) using the 0.9Pt/TiO_2@rGO aerogel as a one-piece monolith (green) and smashed (purple) (reaction conditions: 0.5 v/v methanol/water, 2$g_{NPs}L^{-1}$).

Indeed, the H_2 production in steady-state conditions was more than doubled for the reduced aerogel, increasing from 3280 $\mu mol_{H2}h^{-1}g_{NPs}{}^{-1}$ in the 1Pt/TiO_2@GO (2:1) intermediate to 7070 $\mu mol_{H2}h^{-1}g_{NPs}{}^{-1}$ in the 0.9Pt/TiO_2@rGO (3:1) aerogel. The reasons for the positive effect on the H_2 production rate of the GO reduction are two fold. On one hand, upon the removal of the oxygenated groups on the GO surface, the graphitic structure is partially restored, although in small spots due to defect generation [42,43]. Nevertheless, rGO would own more electrical pathways than GO, enhancing the conductivity of the matrix that plays a key role in transferring the photogenerated e^-, thus preventing recombination and improving H_2 production efficiency [50]. On the other hand, the more hydrophobic reduced structure favors the adsorption of the methanol sacrificial agent over water and maintains it close to the Pt/TiO_2 NPs to boost h^+ consumption, which was beneficial for production [26].

2.5.2. Aerogel Architecture

Fabricating GO-based composite aerogels using $scCO_2$ makes possible the creation of 3D monolithic meso/macroporous architectures. Preliminary tests were performed using a set-up designed for the straight use of the monoliths (Figure 6). Four of these cylinders were simultaneously used with a total weight of ca. 40 mg, which represents a catalytic NPs concentration of 2 $g_{NPs}L^{-1}$ in the reactor filled with 14 mL of a 0.5 v/v methanol/water solution. Under these conditions, a specific H_2 production of 180 $\mu molh^{-1}g_{NP}{}^{-1}$ was reached at the steady state for the sample 0.9Pt/TiO_2@rGO (3:1) (Figure 7b). Compared to the literature, this value is similar to those given in some of the published works (e.g., 100–400 $\mu molh^{-1}g_{NP}{}^{-1}$ using TiO_2/Pt/rGO composites [25]), although it is considered to be in the low range of H_2 production [26,27,29,30]. Definitely, the most important drawback of using monolithic one-piece aerogels in the catalytic experiment is relays in the small amount of sample exposed to light, so that during irradiation most of the Pt/TiO_2 NPs in the interior part of the monolith remains inactive. Moreover, a long reaction time was needed to reach the steady state, in the order of 200 min, which was related to the slow diffusion of reactants in the monolithic aerogels with low permeability due to certain densification originated by shrinkage and high tortuosity. Diffusion is further hindered in non-stirred setups such as the one used in this work for the one-piece monoliths. Agitation was not used to avoid turbulence damage to the integrity of the monoliths during the measurements. Hence, in spite of the great prospect of applying monolithic aerogels to diminish the loss of active sites during recycling [7], the utilization of one-piece aerogel photocatalysts in liquid media appears to still face significant challenges [51].

To improve the catalytic activity of the aerogels, a second set of experiments was performed by first dispersing the reduced monoliths in small pieces in the aqueous methanol solvent, using for that soft and short sonication. The aim of the ultrasonic treatment was not to re-exfoliate the rGO flakes but to break the monoliths into pieces. The DLS characterization of the obtained dispersion gave a bimodal pattern with peaks at ca. 10 and 20 μm hydrodynamic sizes, representative of the size of the broken pieces of aerogel (Figure S6). After this treatment, the H_2 production rate for the sample $0.9Pt/TiO_2$@rGO (3:1) increased to 1600 $\mu molh^{-1}g_{NPs}{}^{-1}$, with the particularity that this high rate was achieved after only 40 min (Figure 7b). The shortening in the required time to reach equilibrium is related to an increase in the catalyst permeability occurring for the small pieces with shorter throughout distances than the one-piece monoliths. Hence, permeability, even having similar absolute values for the smashed and one-piece aerogels is not the key parameter limiting the catalytic activity of the micrometric samples. The setup involving small pieces has the advantage that measurements can be performed under stirring, thus minimizing drawbacks related to reagent diffusion, adsorption, and desorption. Moreover, the small pieces of aerogel are continuously moving in the turbulences created by agitation, thus giving more chance for the catalytic NPs for being irradiated by light. All these factors lead to an enhancement in the H_2 production rate.

2.5.3. Methanol Concentration

Concerning methanol sacrificial agent concentration in the aqueous solution, diverse, even contradictory results have been published on TiO_2-based systems used for photocatalytic H_2 production. Actually, some of them conclude that methanol contributes less than its stoichiometric ratio to the overall H_2 formation [52], while others confirm that the overall reaction can be described as the photoconversion of exclusively methanol [53]. In fact, it is expected that an increase in the methanol concentration in the aqueous solution results in enhanced H_2 production, whether it comes either from the water/sacrificial agent or the sacrificial agent exclusively, due to the more effective scavenging of the photogenerated h^+ by the alcohol. Water is known to play an important role in the complete oxidation of alcohol to CO_2, making its presence necessary [54]. For instance, water has the ability to fasten the essential desorption of the reaction products from the catalyst surface, thus enhancing the reaction rate. As a consequence, after a certain increase in methanol concentration, the decrease in H_2 production generally occurs due to the hindered adsorption of water on the catalyst surface already occupied by alcohol molecules. To optimize this parameter for the developed catalyst, methanol was applied in increased concentrations in the aqueous solution, from 10 to 100 v% (0.01 to 1 v/v). Experiments were performed with the smashed aerogel of sample $0.9Pt/TiO_2$@rGO (3:1) and a catalyst concentration of 2 $g_{NPs}L^{-1}$. The measured flow rates of the produced H_2 at the steady state, e.g., at 60 min for each methanol concentration, are shown in Figure 8a. Initially, the increase in the methanol-to-water ratio favored the H_2 production rate up to a maximum reached at a concentration of ca. 0.5 v/v. Thereafter, a further increase in the alcohol concentration results in a smooth decrease in the H_2 evolution. Commonly, a behavior of a sharp decline in H_2 generation has been observed for similar catalysts [55,56]. The lack of severe decrease, caused by excess methanol on the catalytic efficiency of the studied composite, is here related to the swelling characteristics of the used rGO support. Although the photocatalytic reaction is induced by the photogenerated h^+ and e^- on the Pt/TiO_2 NPs, the swelling of the rGO support is influenced by the polarity of the solvent, which affects the interaction between the dispersed active sites on the rGO surface and the reactants. Applying methanol in high concentrations would enhance the swelling of the aerogel pieces since methanol interaction with the hydrophobic graphitic regions in the rGO support would be stronger than for water [57]. The swelled structure would allow better accessibility for the reactants to the attached NPs. Hence, the above-mentioned adverse effects of the excess of methanol are somewhat compensated by aerogel swelling, and the H_2 production is maintained at a relatively high level in all the studied ranges of alcohol concentration.

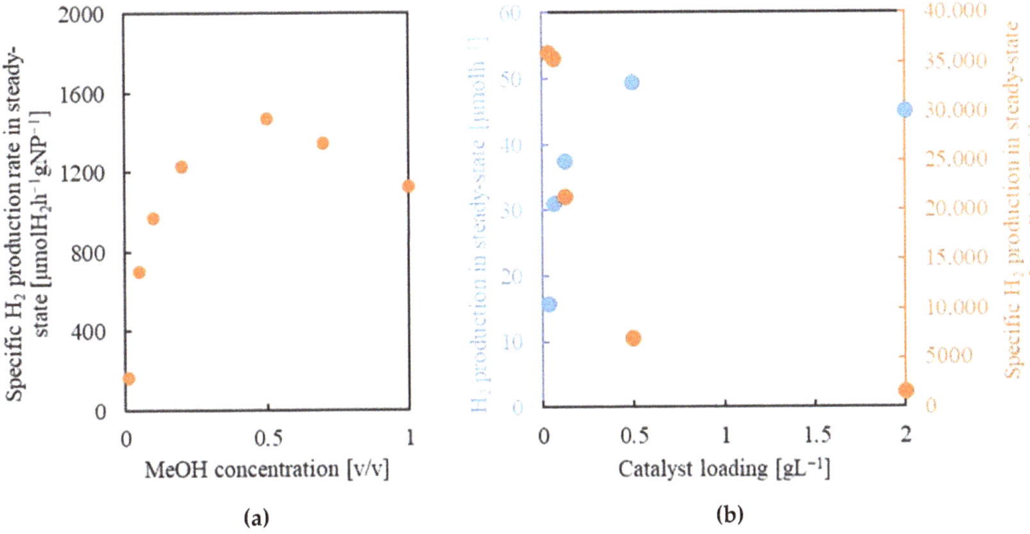

Figure 8. H_2 production rate in steady-state conditions (60 min) for $0.9Pt/TiO_2$@rGO (3:1) smashed aerogel: (**a**) influence of methanol concentration (2 $g_{NPs}L^{-1}$), and (**b**) influence of NP concentration (0.5 v/v methanol/water), in which results are expressed as the non-specific (μmol_{H2} h^{-1}, blue) and specific (μmol_{H2} $h^{-1}g_{NPs}{}^{-1}$, orange) production rates.

2.5.4. Catalyst Concentration

The effect of the catalyst loading, referring to the number of NPs, was investigated in the concentration interval of 0.03 to 2 $g_{NPs}L^{-1}$ (equivalent to 0.04–2.67 $g_{aerogel}L^{-1}$) for the $0.9Pt/TiO_2$@rGO (3:1) smashed aerogel dispersed in 0.5 v/v water/methanol solution. The production rate of H_2 was measured at a steady state (60 min) (Figure 8b). In the studied interval of concentration, the H_2 flow rate, expressed as the specific value, e.g., normalized to the catalyst NPs weight ($\mu mol_{H2}h^{-1}g_{NPs}{}^{-1}$), was very high at low catalyst loading (0.03–0.125 $g_{NPs}L^{-1}$), and then substantially decreased at high concentrations (>0.5 $g_{NPs}L^{-1}$). However, this result, which could be taken at the first instance as an indication of the benefits of working at a very low concentration of catalyst, is just a mathematical artifact since, in fact, the total amount of the produced H_2 can be considered as being in the low range. The representation of the catalytic data as a function of the non-specific H_2 production rate ($\mu mol_{H2}h^{-1}$) indicates that the total amount of evolved H_2 sharply increases with the catalyst loading up to a value of ca. 0.5 $g_{NPs}L^{-1}$. Thereafter, H_2 production slightly decreases by increasing catalyst concentration. This decrease is likely due to light blocking by an excess of dark solid catalyst dispersion [58].

2.5.5. Catalyst Composition

Platinum Content

Regarding the catalyst composition, one important parameter for regulating the photocatalytic activity in the reaction of H_2 production is the amount of Pt added to the TiO_2 NPs. The presence of Pt is necessary for suppressing the recombination of the photogenerated e^- and h^+ in the TiO_2 semiconductor, thus enhancing the formation of H_2. Intimate Pt–TiO_2 contact at the interphase is also necessary to maximize the H_2 production efficiency [59]. To analyze this parameter, a series of experiments was performed with smashed xPt/TiO_2@rGO (3:1) aerogels with four different values of Pt content in the NPs: 0, 0.1, 0.5, and 1.0 wt%. In previous works involving Pt/TiO_2/rGO systems, the proportion of the noble metal is also within this range, typically 0.4–1% [17,25,26,29], which facilitates data comparison. Measurements were carried out at the optimal reaction conditions previously established, e.g., 0.5 v/v water/methanol and smashed catalyst with a concentration of 0.5 $g_{NPs}L^{-1}$. Primary tests indicated that without the addition of the noble metal (sample TiO_2@rGO), the H_2 evolution was negligible, with a value of only 60 $\mu mol_{H2}h^{-1}g_{NPs}{}^{-1}$. This result corroborates previous findings pointing to the inactivity of TiO_2 NPs without the use of a co-catalyst [60]. In the studied range of noble metal loading, the H_2 production increases concomitantly with Pt content (Figure 9a). The decline in H_2 evolution was smoother when the Pt content was decreased from 0.9 to 0.5 wt% than from 0.5 to 0.1 wt%. This observation indicates that, although the overall H_2 production was the highest with the sample of 0.9 wt% Pt content in TiO_2, the total amount of noble metal can be halved without losing significant activity. This is an important result since Pt is the most expensive component of the catalyst; therefore, the amount of noble metal incorporated into the composite would be crucial in any industrial process and must be reduced as much as possible. The design of a catalyst involving an important reduction in the use of Pt is currently an important goal targeted by the European Commission [61].

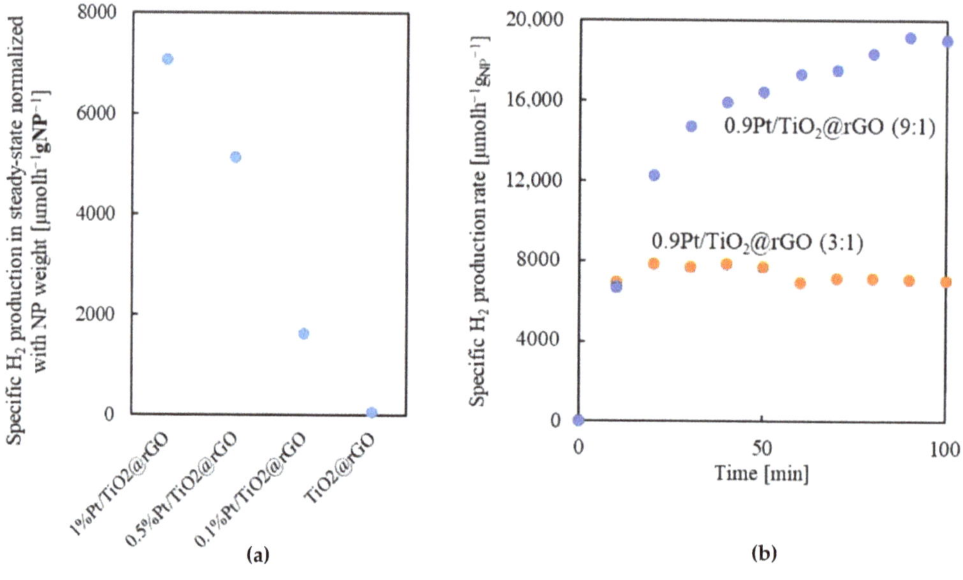

Figure 9. Specific H_2 production rate for smashed aerogel samples: (**a**) xPt/TiO_2@rGO (3:1) with different Pt content, and (**b**) 0.9Pt/TiO_2@rGO with 3:1 and 9:1 NP:rGO ratios. Reaction conditions: 0.5 v/v methanol/water, 0.5 $g_{NPs}L^{-1}$.

NP:rGO Ratio

The influence of modifying the NPs:rGO ratio in the 0.9Pt/TiO_2@rGO composite was investigated by increasing this value from 3:1 to 9:1. This modification resulted in a 2.7-fold enhancement in the H_2 production efficiency, from 7070 to 18,800 $\mu mol_{H2} h^{-1} g_{NPs}{}^{-1}$ at the steady state (Figure 9b). The foremost effect of increasing the NP loading in the composite was to enhance the light-harvesting of the catalyst since, statistically, the probability for the UV radiation to contact active centers rises. On the contrary, the large amount of rGO in the aerogel with the lowest number of NPs shields an important portion of the active centers, hindering overall catalyst activity [62]. Moreover, the steady state in H_2 evolution was reached at 10 and 90 min for the 3:1 and 9:1 samples, respectively. Increasing the rGO percentage in the composite should raise the hydrophobic character of the aerogel, thus improving the accessibility of the methanol vs. water in the pores. Excess amounts of methanol close to the photogenerated holes facilitate the fast h^+ trapping by the alcohol, such that in the initial phase of the reaction, the H_2 production already reaches its limit [57]. Hence, the sample with the lowest proportion of NPs has the advantage of rapidly reaching the steady state, while the sample with the higher proportion revealed the largest H_2 production rate. One point that should be underlined particularly is that the value of the H_2 production rate obtained for the 0.9Pt/TiO_2@rGO (9:1) catalyst exceeds ca. 2–10 times the values published for similar systems using aqueous methanol solutions as reaction media and involving Pt/TiO_2/rGO in the catalyst [25,29].

3. Conclusions

Three-dimensional porous Pt/TiO_2@GO and Pt/TiO_2@rGO composite aerogels were prepared using the one-step low-temperature green supercritical CO_2 method. The produced aerogels are intended for the photocatalytic production of H_2 from aqueous methanol solutions. For this application, optimal working operational conditions resulting in the highest H_2 production rate were settled as a 0.5 $g_{NPs} L^{-1}$ catalyst concentration in a 0.5 v/v methanol/water reaction solution. A two-fold increase in the H_2 production was observed when the GO support was mildly reduced to rGO, an effect assigned to the generation of new electronic pathways upon the partial restoration of the graphene network, and the favored adsorption of the methanol in the reduced structure. The moderate H_2 production rate observed when a one-piece monolith was used (180 $\mu molh^{-1} g_{NP}{}^{-1}$) was significantly improved (ca. 10-fold) when the aerogels were broken into small pieces (1600 $\mu molh^{-1} g_{NP}{}^{-1}$), shortening also the time needed to reach the equilibrium from 200 to only 40 min. This enhancement is the result of the improved light exposure of the active sites and increased reagent and product diffusion. Increasing the NP:rGO ratio from 3:1 to 9:1 caused a 2.7-fold increase in the H_2 evolution due to the reduced amount of shielded, and thus inactive, NPs. Regarding the catalyst composition, low Pt percentages, in the order of 0.9–0.5 wt%, can be used, still giving a high H_2 production rate. In the most favorable conditions, an H_2 production of 18,800 $\mu molh^{-1} g_{NP}{}^{-1}$ was measured for the 0.9Pt/TiO_2@rGO (9:1) aerogel catalyst in aqueous methanol, which is remarkably high compared to the reported similar Pt/TiO_2/rGO systems.

4. Materials and Methods

4.1. Materials

For the preparation of the Pt/TiO_2 NPs, chloroplatinic acid hexahydrate ($H_2PtCl_6.6H_2O$) and TiO_2 NPs (AEROXIDE P25, ca. 20 nm), provided by Alfa Aesar and Evonik, respectively, were used. For the aerogel preparation, a GO water dispersion of 4 $mgmL^{-1}$, supplied by Graphenea Inc. (Spain), was employed. Ethanol and methanol were purchased from Carlo Erba. Liquid CO_2 (99.95 wt%). N_2 and H_2 gasses were delivered by Carburos Metálicos S.A.

4.2. Synthetic Methods

4.2.1. Preparation of NPs of xPt/TiO_2 Composite

NPs of Pt/TiO_2 with different Pt contents were prepared following a reported incipient impregnation deposition and followed by reduction methodology [63]. Briefly, a weighted amount of $H_2PtCl_6.6H_2O$, e.g., 2.7, 13, or 26 mg, was dissolved in 2.5 mL of ultrapure water and used to obtain Pt/TiO_2 composites with 0.1, 0.5, and 1.0 wt% Pt contents, respectively. Each Pt solution was added dropwise to 1 g of TiO_2 NPs at a rate of 8.33 $\mu Lmin^{-1}$, achieved by using a peristaltic pump, while the resulting slurry was continuously sonicated. After that, the deposited suspension was kept under sonication for 2 h. The dense dispersion was dried at 100 °C overnight in an air oven. The recovered powder was treated at 200 °C in a tubular oven, increasing the temperature with a heating ramp of 10 °C min^{-1}, first under N_2 for 1 h and then reduced under a H_2 flow of 50 mL min^{-1} for 3 h. A grey powder was obtained and named as xPt/TiO_2, where x indicates the added Pt weight content in percentage.

4.2.2. Preparation of xPt/TiO_2@rGO Composite Aerogels

For the composite aerogels, a suspension of GO in ethanol with an adjusted concentration of 3.5 mg mL^{-1} was first prepared from water dispersion following a reported protocol [36]. Weighted amounts of bare TiO_2 and composed xPt/TiO_2 NPs were dispersed by sonication in aliquots of 1 mL of the GO-ethanol suspension to obtain weight ratios of 2:1 and 6:1 for NPs:GO. The suspensions were added to assay tubes of ca. 1 cm diameter and 2 mL volume and placed in a 200 mL high-pressure reactor (TharProcess). The aerogels were prepared by drying the suspensions with $scCO_2$ in the batch mode, keeping the autoclave at 200 bar and 45 °C for 48 h (Figure 1a) [36]. Finally, the CO_2 was slowly released from the reactor under isothermal conditions. The xPt/TiO_2@GO samples were recovered as one-piece cylindrical monoliths. The reduction of the recovered 3D GO aerogels to rGO was carried out in a tubular oven at 300 °C under N_2 flow. To reach the target temperature, a heating ramp of 5 °C min^{-1} was used, upholding the temperature for 20 min after reaching 100 and 200 °C, and then maintaining it for 2 h at 300 °C. The reduced xPt/TiO_2@rGO aerogels were also recovered as one-piece monoliths and used either as-synthesized or smashed into small pieces.

4.3. Characterization

The platinum content in the xPt/TiO_2@rGO aerogels was quantified by inductively coupled plasma mass spectrometry (ICP-MS, Agilent 7700x) after digesting the samples in hydrochloric, nitric, and hydrofluoric acids (3:1:0.5 v/v). The structural characterization of the prepared NPs and the reduced composite aerogels was performed by powder X-ray diffraction (PXRD) in a Siemens D5000, using the Cu Kα incident radiation with a step scan of 0.02° in the 2θ 5–40° range. The size of the NPs was estimated using PXRD data and the Scherrer equation. Surface functional groups were studied by Fourier transform infrared (FTIR) spectroscopy (Jasco 4700 Spectrophotometer), after the dispersion of the samples in potassium bromide (KBr). Raman spectra were recorded to ascertain the reduction in GO by using an excitation wavelength of 532 nm. The morphology of the composite aerogels and the size of the NPs, as well as their degree of dispersion on the rGO platelets, were investigated by scanning (SEM, Quanta FEI 200) and transmission (TEM, JEOL 1210) electron microscopies. The BET (Brunauer, Emmet, Teller) surface area (S_a), the BJH (Barrett, Joyner, and Halenda), and cumulative adsorption pore volume (V_p) were determined by collecting N_2 adsorption/desorption isotherms at 77 K (ASAP 2020, Micromeritics Inc., Norcross, GA, USA), after degassing the samples at 393 K for 20 h. For the smashed aerogels, dynamic light scattering (DLS Coulter LS230) was used to study the hydrodynamic size of the aerogel broken pieces dispersed in methanol/water (0.5 v/v). The wettability of the reduced and non-reduced composites was investigated by water contact angle measurement (Biolin Sci. Attension Theta Lite) after preparing a compressed pellet with the monoliths. The optical properties of the aerogels were investigated by UV-VIS diffuse reflectance (Jasco V-560) and photoluminescence (Jasco FP-8300) spectroscopies using the smashed solid samples. For the UV experiments, barium sulfate ($BaSO_4$) powder was used as blank.

4.4. Photocatalytic H_2 Production

The photocatalytic activity of the synthesized composite aerogels for H_2 production was tested by immersing the reduced monoliths, either as recovered in one piece or ultrasonically smashed, in an aqueous methanol solution. For the one-piece samples, a specifically designed basket composed of poly-lactic acid (PLA) polymer was fabricated with 3D printing to hold four monoliths and prevent them from floating during the reaction. The basket was settled inside a cylindrical glass reactor of 20 mL that was filled with 14 mL of a methanol/water mixture (Figure 6). For the tests with the smashed aerogels, four monoliths were added to the methanol/water solution, gentile sonicated for 5 min in an ultrasonic bath, and poured into the 20 mL reactor vessel without the basket but under mechanical stirring (100 rpm). The methanol percentage in the methanol/water mixture was varied from 0.01 to 1 v/v. Different aerogel catalyst concentrations, from 0.04–2.67 $g_{aerogel}L^{-1}$ (equivalent to 0.03 to 2 $g_{NPs}L^{-1}$) were also tested. Before starting the catalytic reaction, the system was purged with N_2, and a stream of this inert gas was continued at a rate of 7.5 mL min^{-1} during the entire experiment. The reaction mixture was irradiated with a four visible-LED system placed at a 4 cm distance from the cylindrical reactor wall. The emission spectrum is shown in the SI (Figure S7). The average nominal irradiance of each LED was 45.0 mW cm^{-2}, determined by using a UV–vis spectroradiometer (OceanOptics USB2000+). The H_2 production rate was analyzed online every 10 min in the headspace of the reactor by using a gas chromatograph (Inficon 3000 MicroGC). The results are expressed referring to either the rate of H_2 production [$\mu mol_{H2}h^{-1}$] or the rate per mass of the NPs [$\mu mol_{H2}h^{-1}g_{NPs}{}^{-1}$]. Steady-state H_2 production data recorded after 60 min were calculated for all the analyses performed with the smashed monoliths.

Supplementary Materials: The following supporting information can be downloaded at: https://www.mdpi.com/article/10.3390/gels8110719/s1, Figure S1: XRD spectra of the non-reduced 1Pt/TiO_2@GO and reduced 0.9Pt/TiO_2@rGO (3:1) composites, Figure S2: TEM images of: (a) bare TiO_2, and (b) 0.9Pt/TiO_2 NPs obtained by subjecting 1Pt/TiO_2 NPs to the reduction treatment; Figure S3: N_2 adsorption/desorption isotherms for the 0.9Pt/TiO_2@rGO sample compared to the precursors GO and 1Pt/TiO_2 NPs.; Figure S4: BJH volumetric pore size distribution calculated from the adsorption branch of the isotherm; Figure S5: Contact angle measurement of: (a) 0.9Pt/TiO_2@rGO and 1Pt/TiO_2@GO composites, measured with a water droplet; Figure S6: DLS analysis of the 1Pt/TiO_2-rGO (3:1) aerogel dispersed in an aqueous methanol solution (0.5 v/v); Determination of the band gap energy of the photocatalysts; Figure S7: Emission spectrum of the used light source.

Author Contributions: Conceptualization, M.K., C.G.S., A.M.L.-P., J.L.F. and C.D.; methodology, M.K. and C.G.S.; investigation, M.K., C.G.S. and A.M.L.-P.; resources, A.M.L.-P., J.L.F. and C.D.; data curation, M.K., C.G.S., A.M.L.-P. and C.D.; writing—original draft preparation, M.K. and C.D.; writing—review and editing, M.K., C.G.S., J.L.F. and C.D.; supervision, C.G.S. and A.M.L.-P.; project administration and funding acquisition, A.M.L.-P., J.L.F. and C.D. All authors have read and agreed to the published version of the manuscript.

Funding: This work was supported by the Spanish Ministry of Science and Innovation MICINN through the Severo Ochoa Program for Centers of Excellence (CEX2019–000917-S) and the Spanish National Plan of Research with project PID2020–115631GB-I00. We would like to thank the scientific collaboration under LA/P/0045/2020 (ALiCE), UIDB/50020/2020, and UIDP/50020/2020 (LSRE-LCM), financed by national funds through FCT/MCTES (PIDDAC). Márta Kubovics acknowledges the financial support from the European Union's Horizon 2020 research and innovation program under the Marie Sklodowska-Curie Cofund grant (agreement no MSCA-COFUND-DP/0320-754397) and the Short Term Scientific Mission (STSM) from the Greenering network by the COST Association. This work has been performed in the framework of the doctoral program "Chemistry" of the Universitat Autònoma de Barcelona by Márta Kubovics.

Institutional Review Board Statement: Not applicable.

Informed Consent Statement: Not applicable.

Data Availability Statement: Not applicable.

Conflicts of Interest: The authors declare no conflict of interest.

References

1. Mohd Shah, N.R.A.; Mohamad Yunus, N.N.; Wong, W.Y.; Arifin, K.; Jeffery Minggu, L. Current progress on 3D graphene-based photocatalysts: From synthesis to photocatalytic hydrogen production. *Int. J. Hydrog. Energy* **2021**, *46*, 9324–9340. [CrossRef]
2. Nishiyama, H.; Yamada, T.; Nakabayashi, M.; Maehara, Y.; Yamaguchi, M.; Kuromiya, Y.; Nagatsuma, Y.; Tokudome, H.; Akiyama, S.; Watanabe, T.; et al. Photocatalytic solar hydrogen production from water on a 100-m2 scale. *Nature* **2021**, *598*, 304–307. [CrossRef] [PubMed]
3. Guo, S.; Li, X.; Li, J.; Wei, B. Boosting photocatalytic hydrogen production from water by photothermally induced biphase systems. *Nat. Commun.* **2021**, *12*, 1–10. [CrossRef] [PubMed]
4. Christoforidis, K.C.; Fornasiero, P. Photocatalytic Hydrogen Production: A Rift into the Future Energy Supply. *ChemCatChem* **2017**, *9*, 1523–1544. [CrossRef]
5. Zhu, S.; Wang, D. Photocatalysis: Basic principles, diverse forms of implementations and emerging scientific opportunities. *Adv. Energy Mater.* **2017**, *7*, 1–24. [CrossRef]
6. Martha, S.; Chandra Sahoo, P.; Parida, K.M. An overview on visible light responsive metal oxide based photocatalysts for hydrogen energy production. *RSC Adv.* **2015**, *5*, 61535–61553. [CrossRef]
7. Kuang, P.; Sayed, M.; Fan, J.; Cheng, B.; Yu, J. 3D Graphene-Based H2-Production Photocatalyst and Electrocatalyst. *Adv. Energy Mater.* **2020**, *10*, 1–53. [CrossRef]
8. Chen, X.; Shen, S.; Guo, L.; Mao, S.S. Semiconductor-based photocatalytic hydrogen generation. *Chem. Rev.* **2010**, *110*, 6503–6570. [CrossRef]
9. Ola, O.; Maroto-Valer, M.M. Review of material design and reactor engineering on TiO_2 photocatalysis for CO_2 reduction. *J. Photochem. Photobiol. C Photochem. Rev.* **2015**, *24*, 16–42. [CrossRef]
10. Eidsvåg, H.; Bentouba, S.; Vajeeston, P.; Yohi, S.; Velauthapillai, D. TiO_2 as a photocatalyst for water splitting—An experimental and theoretical review. *Molecules* **2021**, *26*, 1687. [CrossRef]
11. Blengini, G.A.; Latunussa, C.E.L.; Eynard, U. *Study on the EU's List of Critical Raw Materials—Critical Raw Materials Factsheets;* Final Report; European Comission: Brussels, Belgium, 2020. [CrossRef]
12. Communication from the Commission to the European Parliament, the Council, the European Economic and Social Committee and the Committee of the Regions, Critical Raw Materials Resilience: Charting a Path Towards Greater Security and Sustainability. Available online: https://eur-lex.europa.eu/legal-content/EN/TXT/?uri=CELEX%3A52020DC0474 (accessed on 7 September 2022).
13. Yeh, T.F.; Cihlář, J.; Chang, C.Y.; Cheng, C.; Teng, H. Roles of graphene oxide in photocatalytic water splitting. *Mater. Today* **2013**, *16*, 78–84. [CrossRef]
14. Solís-Fernández, P.; Bissett, M.; Ago, H. Synthesis, structure and applications of graphene-based 2D heterostructures. *Chem. Soc. Rev.* **2017**, *46*, 4572–4613. [CrossRef] [PubMed]
15. Zhou, X.; Zhang, X.; Wang, Y.; Wu, Z. 2D Graphene-TiO_2 Composite and Its Photocatalytic Application in Water Pollutants. *Front. Energy Res.* **2021**, *8*, 1–10. [CrossRef]
16. Zhang, S.; Li, B.; Wang, X.; Zhao, G.; Hu, B.; Lu, Z.; Wen, T.; Chen, J.; Wang, X. Recent developments of two-dimensional graphene-based composites in visible-light photocatalysis for eliminating persistent organic pollutants from wastewater. *Chem. Eng. J.* **2020**, *390*, 124642. [CrossRef]
17. Wang, P.; Zhan, S.; Xia, Y.; Ma, S.; Zhou, Q.; Li, Y. The fundamental role and mechanism of reduced graphene oxide in rGO/Pt-TiO_2 nanocomposite for high-performance photocatalytic water splitting. *Appl. Catal. B* **2017**, *207*, 335–346. [CrossRef]
18. Roy, S.S.; Cheruvathoor Poulose, A.; Bakandritsos, A.; Varma, R.S.; Otyepka, M. 2D graphene derivatives as heterogeneous catalysts to produce biofuels via esterification and trans-esterification reactions. *Appl. Mater. Today* **2021**, *23*, 101053. [CrossRef]
19. Gusmão, R.; Veselý, M.; Sofer, Z. Recent Developments on the Single Atom Supported at 2D Materials beyond Graphene as Catalysts. *ACS Catal.* **2020**, *10*, 9634–9648. [CrossRef]
20. Mao, S.; Lu, G.; Chen, J. Three-dimensional graphene-based composites for energy applications. *Nanoscale* **2015**, *7*, 6924–6943. [CrossRef]
21. Li, X.; Yu, J.; Jaroniec, M. Hierarchical photocatalysts. *Chem. Soc. Rev.* **2016**, *45*, 2603–2636. [CrossRef]
22. Bakbolat, B.; Daulbayev, C.; Sultanov, F.; Beissenov, R.; Umirzakov, A.; Mereke, A.; Bekbaev, A.; Chuprakov, I. Recent developments of TiO_2-based photocatalysis in the hydrogen evolution and photodegradation: A review. *Nanomaterials* **2020**, *10*, 1790. [CrossRef]
23. Navalon, S.; Dhakshinamoorthy, A.; Alvaro, M.; Garcia, H. Metal nanoparticles supported on two-dimensional graphenes as heterogeneous catalysts. *Coord. Chem. Rev.* **2016**, *312*, 99–148. [CrossRef]
24. Yam, K.M.; Guo, N.; Jiang, Z.; Li, S.; Zhang, C. Graphene-based heterogeneous catalysis: Role of graphene. *Catalysts* **2020**, *10*, 53. [CrossRef]

25. Rivero, M.J.; Iglesias, O.; Ribao, P.; Ortiz, I. Kinetic performance of TiO_2/Pt/reduced graphene oxide composites in the photocatalytic hydrogen production. *Int. J. Hydrog. Energy* **2019**, *44*, 101–109. [CrossRef]
26. Mohan, P.S.; Purkait, M.K.; Chang, C.T. Experimental evaluation of Pt/TiO_2/rGO as an efficient HER catalyst via artificial photosynthesis under UVB & visible irradiation. *Int. J. Hydrog. Energy* **2020**, *45*, 17174–17190. [CrossRef]
27. Zeng, P.; Zhang, Q.; Zhang, X.; Peng, T. Graphite oxide-TiO_2 nanocomposite and its efficient visible-light-driven photocatalytic hydrogen production. *J. Alloys Compd.* **2012**, *516*, 85–90. [CrossRef]
28. Wang, Z.; Yin, Y.; Williams, T.; Wang, H.; Sun, C.; Zhang, X. Metal link: A strategy to combine graphene and titanium dioxide for enhanced hydrogen production. *Int. J. Hydrog. Energy* **2016**, *41*, 22034–22042. [CrossRef]
29. Da Silva, R.O.; Heiligtag, F.J.; Karnahl, M.; Junge, H.; Niederberger, M.; Wohlrab, S. Design of multicomponent aerogels and their performance in photocatalytic hydrogen production. *Catal. Today* **2015**, *246*, 101–107. [CrossRef]
30. Lin, C.C.; Wei, T.Y.; Lee, K.T.; Lu, S.Y. Titania and Pt/titania aerogels as superior mesoporous structures for photocatalytic water splitting. *J. Mater. Chem.* **2011**, *21*, 12668–12674. [CrossRef]
31. Borrás, A.; Rosado, A.; Fraile, J.; López-Periago, A.M.; Giner Planas, J.; Yazdi, A.; Domingo, C. Meso/microporous MOF@graphene oxide composite aerogels prepared by generic supercritical CO_2 technology. *Microporous Mesoporous Mater.* **2022**, *335*, 111825. [CrossRef]
32. Borrás, A.; Fraile, J.; Rosado, A.; Marbán, G.; Tobias, G.; López-Periago, A.M.; Domingo, C. Green and Solvent-Free Supercritical CO_2 -Assisted Production of Superparamagnetic Graphene Oxide Aerogels: Application as a Superior Contrast Agent in MRI. *ACS Sustain. Chem. Eng.* **2020**, *8*, 4877–4888. [CrossRef]
33. Rosado, A.; Borrás, A.; Fraile, J.; Navarro, J.A.R.; Suárez-García, F.; Stylianou, K.C.; López-Periago, A.M.; Giner Planas, J.; Domingo, C.; Yazdi, A. HKUST-1 Metal-Organic Framework Nanoparticle/Graphene Oxide Nanocomposite Aerogels for CO_2 and CH4 Adsorption and Separation. *ACS Appl. Nano Mater.* **2021**, *4*, 12712–12725. [CrossRef]
34. Smirnova, I.; Gurikov, P. Aerogel production: Current status, research directions, and future opportunities. *J. Supercrit. Fluids* **2018**, *134*, 228–233. [CrossRef]
35. Veres, P.; López-Periago, A.M.; Lázár, I.; Saurina, J.; Domingo, C. Hybrid aerogel preparations as drug delivery matrices for low water-solubility drugs. *Int. J. Pharm.* **2015**, *496*, 360–370. [CrossRef] [PubMed]
36. Borrás, A.; Goncalves, G.; Marbán, G.; Sandoval, S.; Pinto, S.; Marques, P.A.A.P.; Fraile, J.; Tobais, G.; López-Periago, A.M.; Domingo, C. Preparation and Characterization of Graphene Oxide Aerogels: Exploring the Limits of Supercritical CO_2 Fabrication Methods. *Chem. A Eur. J.* **2018**, *24*, 15903–15911. [CrossRef] [PubMed]
37. Luo, P.; Lin, Y. Further thermal reduction of reduced graphene oxide aerogel with excellent rate performance for supercapacitors. *Appl. Sci.* **2019**, *9*, 2188. [CrossRef]
38. Liu, W.; Speranza, G. Tuning the Oxygen Content of Reduced Graphene Oxide and Effects on Its Properties. *ACS Omega* **2021**, *6*, 6195–6205. [CrossRef]
39. Litke, A.; Frei, H.; Hensen, E.J.M.; Hofmann, J.P. Interfacial charge transfer in Pt-loaded TiO_2 P25 photocatalysts studied by in-situ diffuse reflectance FTIR spectroscopy of adsorbed CO. *J. Photochem. Photobiol. A Chem.* **2018**, *370*, 84–88. [CrossRef]
40. Gillespie, P.N.O.; Martsinovich, N. Origin of Charge Trapping in TiO_2/Reduced Graphene Oxide Photocatalytic Composites: Insights from Theory. *ACS Appl. Mater. Interfaces* **2019**, *11*, 31909–31922. [CrossRef]
41. Saleem, H.; Haneef, M.; Abbasi, H.Y. Synthesis route of reduced graphene oxide via thermal reduction of chemically exfoliated graphene oxide. *Mater. Chem. Phys.* **2018**, *204*, 1–7. [CrossRef]
42. Díez-Betriu, X.; Álvarez-García, S.; Botas, C.; Álvarez, P.; Sánchez-Marcos, J.; Prieto, C.; Menéndez, R.; de Andrés, A. Raman spectroscopy for the study of reduction mechanisms and optimization of conductivity in graphene oxide thin films. *J. Mater. Chem. C Mater.* **2013**, *1*, 6905–6912. [CrossRef]
43. Stankovich, S.; Dikin, A.D.; Piner, R.D.; Kohlhaas, K.A.; Kleinhammes, A.; Jia, Y.; Wu, Y.; Nguyen, S.T.; Ruoff, R.S. Synthesis of graphene-based nanosheets via chemical reduction of exfoliated graphite oxide. *Carbon N. Y.* **2007**, *45*, 1558–1565. [CrossRef]
44. López-Díaz, D.; López Holgado, M.; García-Fierro, J.L.; Velázquez, M.M. Evolution of the Raman Spectrum with the Chemical Composition of Graphene Oxide. *J. Phys. Chem. C* **2017**, *121*, 20489–20497. [CrossRef]
45. Makuła, P.; Pacia, M.; Macyk, W. How To Correctly Determine the Band Gap Energy of Modified Semiconductor Photocatalysts Based on UV-Vis Spectra. *J. Phys. Chem. Lett.* **2018**, *9*, 6814–6817. [CrossRef]
46. Munk, P.; Kubelka, F. A Contribution to the Optics of Pigments. *Z. Tech. Phys.* **1931**, *12*, 593–601.
47. Lan, Z.A.; Zhang, G.; Wang, X. A facile synthesis of Br-modified g-C3N4 semiconductors for photoredox water splitting. *Appl. Catal. B* **2016**, *192*, 116–125. [CrossRef]
48. Zhang, H.; Lv, X.; Li, Y.; Wang, Y.; Li, J. P25-graphene composite as a high performance photocatalyst. *ACS Nano* **2010**, *4*, 380–386. [CrossRef] [PubMed]
49. Lee, J.; Choi, W. Photocatalytic Reactivity of Surface Platinized TiO_2: Substrate Specificity and the Effect of Pt Oxidation State. *J. Phys. Chem.* **2005**, *109*, 7399–7406. [CrossRef]
50. Zeng, W.; Tao, X.; Lin, S.; Lee, C.; Shi, D.; Lam, K.; Huang, B.; Wang, Q.; Zhao, Y. Defect-engineered reduced graphene oxide sheets with high electric conductivity and controlled thermal conductivity for soft and flexible wearable thermoelectric generators. *Nano Energy* **2018**, *54*, 163–174. [CrossRef]
51. Wan, W.; Zhang, R.; Ma, M.; Zhou, Y. Monolithic aerogel photocatalysts: A review. *J. Mater. Chem. A Mater.* **2018**, *6*, 754–775. [CrossRef]

52. Guzman, F.; Chuang, S.S.C.; Yang, C. Role of methanol sacrificing reagent in the photocatalytic evolution of hydrogen. *Ind. Eng. Chem. Res.* **2013**, *52*, 61–65. [CrossRef]
53. Nomikos, G.N.; Panagiotopoulou, P.; Kondarides, D.I.; Verykios, X.E. Kinetic and mechanistic study of the photocatalytic reforming of methanol over Pt/TiO_2 catalyst. *Appl. Catal. B* **2014**, *146*, 249–257. [CrossRef]
54. Shimura, K.; Yoshida, H. Heterogeneous photocatalytic hydrogen production from water and biomass derivatives. *Energy Environ. Sci.* **2011**, *4*, 2467–2481. [CrossRef]
55. Chen, W.T.; Dong, Y.; Yadav, P.; Aughterson, R.D.; Sun-Waterhouse, D.; Waterhouse, G.I.N. Effect of alcohol sacrificial agent on the performance of Cu/TiO_2 photocatalysts for UV-driven hydrogen production. *Appl. Catal. A Gen.* **2020**, *602*, 117703. [CrossRef]
56. Chen, W.T.; Chan, A.; Sun-Waterhouse, D.; Llorca, J.; Idriss, H.; Waterhouse, G.I.N. Performance comparison of Ni/TiO_2 and Au/TiO_2 photocatalysts for H2 production in different alcohol-water mixtures. *J. Catal.* **2018**, *367*, 27–42. [CrossRef]
57. Pendolino, F.; Capurso, G.; Maddalena, A.; lo Russo, S. The structural change of graphene oxide in a methanol dispersion. *RSC Adv.* **2014**, *4*, 32914–32917. [CrossRef]
58. Curcó, J.; Giménez, D.; Addardak, A.; Cervera-March, S.; Esplugas, S. Effects of radiation absorption and catalyst concentration on the photocatalytic degradation of pollutants. *Catal. Today* **2002**, *76*, 177–188. [CrossRef]
59. Jiang, X.; Fu, X.; Zhang, L.; Meng, S.; Chen, S. Photocatalytic reforming of glycerol for H2 evolution on Pt/TiO_2: Fundamental understanding the effect of co-catalyst Pt and the Pt deposition route. *J. Mater. Chem. A Mater.* **2015**, *3*, 2271–2282. [CrossRef]
60. Yang, J.; Wang, D.; Han, H.; Li, C. Roles of cocatalysts in photocatalysis and photoelectrocatalysis. *Acc. Chem. Res.* **2013**, *46*, 1900–1909. [CrossRef]
61. Communication from the Commission to the European Parliament, the Council, the European Economic and Social Committee and the Committee of the Regions, The European Green Deal. Available online: https://eur-lex.europa.eu/legal-content/EN/TXT/?uri=COM%3A2019%3A640%3AFIN (accessed on 7 September 2022).
62. Lu, Y.; Ma, B.; Yang, Y.; Huang, E.; Ge, Z.; Zhang, T.; Zhang, S.; Li, L.; Guan, N.; Ma, Y.; et al. High activity of hot electrons from bulk 3D graphene materials for efficient photocatalytic hydrogen production. *Nano Res.* **2017**, *10*, 1662–1672. [CrossRef]
63. Naffati, N.; Sampaio, M.J.; Da Silva, E.S.; Nsib, M.F.; Arfaoui, Y.; Houas, A.; Faria, J.L.; Silva, C.G. Carbon-nanotube/TiO_2 materials synthesized by a one-pot oxidation/hydrothermal route for the photocatalytic production of hydrogen from biomass derivatives. *Mater. Sci. Semicond. Process.* **2020**, *115*, 105098. [CrossRef]

gels

MDPI

Communication

Reduction of PVA Aerogel Flammability by Incorporation of an Alkaline Catalyst

Zhi-Han Cheng [1], Mo-Lin Guo [1], Xiao-Yi Chen [1], Ting Wang [2], Yu-Zhong Wang [2] and David A. Schiraldi [1,*]

1 Department of Macromolecular Science & Engineering, Case Western Reserve University, Cleveland, OH 44106, USA; zxc357@case.edu (Z.-H.C.); mxg501@case.edu (M.-L.G.); xxc293@case.edu (X.-Y.C.)

2 College of Chemistry, Sichuan University, Chengdu 610064, China; wangting1993@mail.xhu.edu.cn (T.W.); yzwang@scu.edu.cn (Y.-Z.W.)

* Correspondence: das44@case.edu

Abstract: Sodium hydroxide was used as a base catalyst to reduce the flammability of poly(vinyl alcohol) (PVA) aerogels. The base-modified aerogels exhibited significantly enhanced compressive moduli, likely resulting in decreased gallery spacing and increased numbers of "struts" in their structures. The onset of decomposition temperature decreased for the PVA aerogels in the presence of the base, which appears to hinder the polymer pyrolysis process, leading instead to the facile formation of dense char. Cone calorimetry testing showed a dramatic decrease in heat release when the base was added. The results indicate that an unexpected base-catalyzed dehydration occurs at fire temperatures, which is the opposite of the chemistry normally observed under typical synthesis conditions.

Keywords: aerogel; flammability; base; alkali; char

Citation: Cheng, Z.-H.; Guo, M.-L.; Chen, X.-Y.; Wang, T.; Wang, Y.-Z.; Schiraldi, D.A. Reduction of PVA Aerogel Flammability by Incorporation of an Alkaline Catalyst. *Gels* **2021**, *7*, 57. https://doi.org/10.3390/gels7020057

Academic Editor: István Lázár

Received: 25 March 2021
Accepted: 4 May 2021
Published: 8 May 2021

1. Introduction

Aerogels are one of the lowest density families of known materials, first reported by Kistler in 1931 [1], who used silicon alkoxides as sol-gel precursors to silica, which were then carefully dried to avoid capillary collapse. The name "aerogel" means a wet gel whose solvent has been exchanged with air; these materials possess the unique properties of low densities (ranging from 0.005 to 0.1 g/cm^3), high porosities, high specific surface areas and low thermal conductivities [2–5]. As these material properties are valuable in designing consumer and industrial products, there has been a steady increase in the use of aerogels in thermal insulation, liquid absorbents, energy storage, and catalysis [6–10]. Inorganic aerogels tend to be brittle, which has led to the evaluation of polymer-based aerogels, which are materials that can exhibit polymer foam-like properties [11,12]. Poly(vinyl alcohol) (PVA) is an ideal candidate for fabricating such polymer aerogels because of its good chemical stability, low toxicity and favorable mechanical properties [13]. As a water soluble polymer with abundant hydroxyl groups, PVA aerogels can be prepared using an environmentally friendly, freeze drying method [14–17].

PVA aerogels are potentially promising candidates for the replacement of traditional polymer foams in insulation, packaging and building areas; these polymer foams suffer from their inherent flammability [3,17]. More than 1.3 million fires were reported in 2017 in the United States alone, resulting in an estimated 3400 civilian deaths and $23 billion in property loss, hence the importance of materials flammability cannot be over emphasized [18]. Flame retardants can be incorporated into polymer systems, but often suffer from toxicity and/or mechanical properties issues; additives such as polybrominated diphenyl ethers (PBDEs) and polybrominated biphenyls (PBBs) are often effective in limiting polymer flammabilities, but are under increasing scrutiny for toxicological reasons [19–21]. The few known flame retardants that could potentially resolve the toxicity/mechanical properties problems require high loadings, increasing final product densities and costs [22], limiting their commercial attractiveness [23,24]. The thermal degradation of PVA has been addressed by some authors, and to a limited extent the flammability of this polymer has also been explored [25–28]. These previous works have focused on dense films and molded PVA samples; one of those studies also showed that added NaOH could lower the flammability of the system, though no explanation for the effect was given [25].

In the present work, we report a novel method of increasing the mechanical properties of PVA aerogels, while decreasing their flammabilities, maintaining low densities, and making use of a low cost/toxicity additive. Sodium hydroxide (NaOH) can act as a catalyst at flame temperatures to promote a char forming process without the addition of other flame retardants. Low additive levels of NaOH were found to profoundly alter the aerogel properties, minimizing their impact on product density. To the best of our knowledge, there is no similar system of PVA aerogels that incorporates low levels of NaOH, nor of a systematic study varying that additive's concentration previously reported in the literature.

2. Results and Discussion

The aerogel compositions examined in this study are given in Table 1.

Table 1. Compositions of aerogels.

Sample	PVA (g)	NaOH (g)/(mol/L)	DI Water (g)	pH
P_5	5	0/0	100	5.8
$P_5/S_{0.001}$	5	$0.001/2.5 \times 10^{-4}$	100	6.0
$P_5/S_{0.01}$	5	$0.01/2.5 \times 10^{-3}$	100	8.3
$P_5/S_{0.1}$	5	$0.1/2.5 \times 10^{-2}$	100	11.7
$P_5/S_{0.5}$	5	$0.5/2.5 \times 10^{-2}$	100	12.8

Poly(vinyl alcohol), PVA; sodium hydroxide, NaOH; deionized water, DI water.

2.1. Apparent Density and Mechanical Performance

The apparent densities of the aerogels are shown in Table 2. No significant differences in densities were noted, which is consistent with the percentage of solids used for each of the compositions; the exception to this observation occurred in $P_5/S_{0.5}$ when a large amount of sodium hydroxide was added to the formulation, increasing the aerogel density. No obvious shrinkage was observed after freeze-drying and all samples maintained good shape in the mold during the process. While the mechanical properties of the PVA aerogel were reported to decrease with the addition of inorganic matter (because the poor interfacial adhesion between them) [29], the addition of NaOH actually increased the mechanical properties of PVA aerogels produced in this study. The initial compressive modulus of P5/S0.5 was 1.30 ± 0.30 MPa, which is nearly four times that of the control, potentially because phase separated NaOH could densify the solid "struts" in the aerogel structure

Table 2. Observed aerogel properties.

Sample	Modulus (MPa)	Density (g/cm^3)	Specific Modulus (MPa cm^3/g)
P_5	0.31 ± 0.11	0.065 ± 0.002	4.8 ± 1.6
$P_5/S_{0.001}$	0.36 ± 0.01	0.066 ± 0.002	5.5 ± 0.2
$P_5/S_{0.01}$	0.47 ± 0.08	0.067 ± 0.001	7.1 ± 1.3
$P_5/S_{0.1}$	0.55 ± 0.16	0.066 ± 0.002	8.3 ± 2.0
$P_5/S_{0.5}$	1.30 ± 0.30	0.079 ± 0.001	16.3 ± 3.0

To eliminate the influence of density, the specific compressive modulus was also compared. The specific moduli were calculated by dividing the ultimate compressive modulus values by the sample densities. As can be seen in Table 2, the specific compressive moduli increased from 4.8 ± 1.6 to 16.3 ± 3.0 MPa cm^3/g as the NaOH was increased from

0% to 9.1%. The increases in mechanical properties of the aerogels with increasing NaOH levels differs from the observed mechanical properties of PVA films, which incorporate that same additive [28]. Our previous work with PVA aerogels has shown that their mechanical properties are extremely sensitive to the skeletal density of aerogel "struts", which resemble the cellular walls in foams, as well as the morphology of the aerogels [3,5,11–13]. It has also been demonstrated that increasing levels of hydrogen bonding within PVA aerogels can significantly enhance their mechanicals properties [3]. It is not surprising that even low levels of additives, capable of binding to the polymer via hydrogen bonding, would bring about skeletal densification and morphological changes.

2.2. Morphology

SEM was used to investigate the aerogel morphologies and the relationship between the structure and properties. Figure 1A,B shows typical "house of cards" aerogel structures, which is a lamellar structure caused by ice growth [30,31]. While with the addition of 0.02% NaOH, the lamellar structure was retained, the structure of the individual layers was not as neat as those of unmodified PVA aerogels samples. As the level of NaOH added to the polymer was increased, structural changes were observed—previous work has shown that mechanical properties trend with an aerogel structure [32]. In samples wherein the amount of NaOH was increased to 3.8%, the lamellar structure was interrupted (Figure 1E,F), gallery spaces decreased in size and increased in number; the samples shrank/densified, with a concomitant increase in compressive moduli (typical of such aerogels).

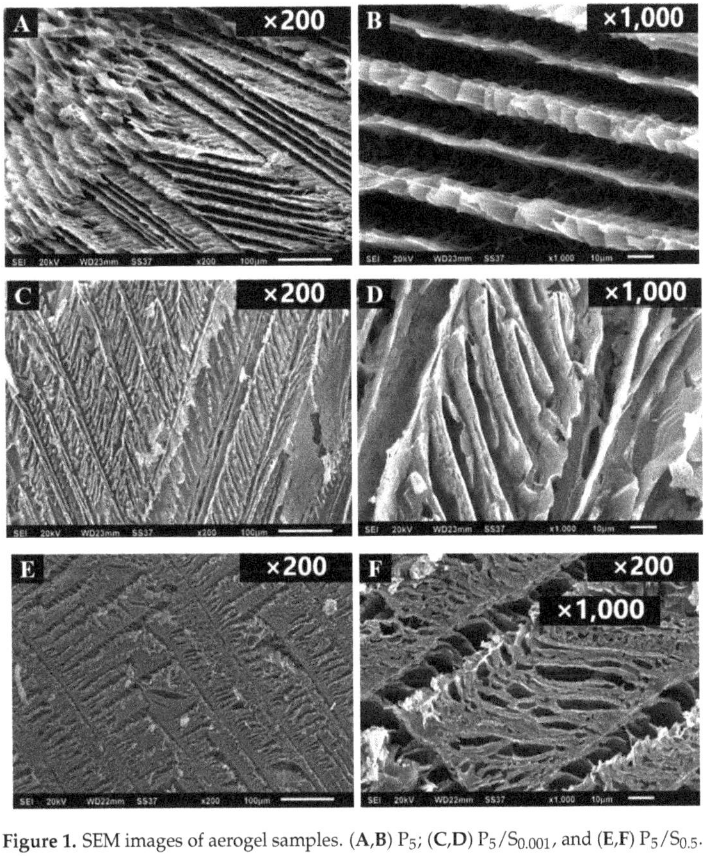

Figure 1. SEM images of aerogel samples. (**A**,**B**) P_5; (**C**,**D**) $P_5/S_{0.001}$, and (**E**,**F**) $P_5/S_{0.5}$. P_X/S_Y defines the polymer and sodium hydroxide concentrations used to produce the aerogels.

2.3. *Thermal Stability & Degradation Mechanism*

The thermal stabilities of the aerogels in this study were investigated by thermogravimetric analysis (TGA; Figure 2) and differential thermogravimetry (DTG; Figure 3). The related data are shown in Table 3, which includes the decomposition temperatures at 5% weight loss ($Td_{5\%}$), at 20% weight loss ($Td_{20\%}$), and at the maximum decomposition rates (Td_{max}), the values at maximum mass decomposition rate (dW/dT) and the char. PVA aerogels could easily capture water from the atmosphere because the existence of abundant hydroxyl groups on its chain. To avoid the influence of moisture, the samples were heated to 100 °C from room temperature at a heating rate of 40 °C/min and then equilibrated at 100 °C for 2 min. After that it was heated to 700 °C at a heating rate of 10 °C/min under nitrogen.

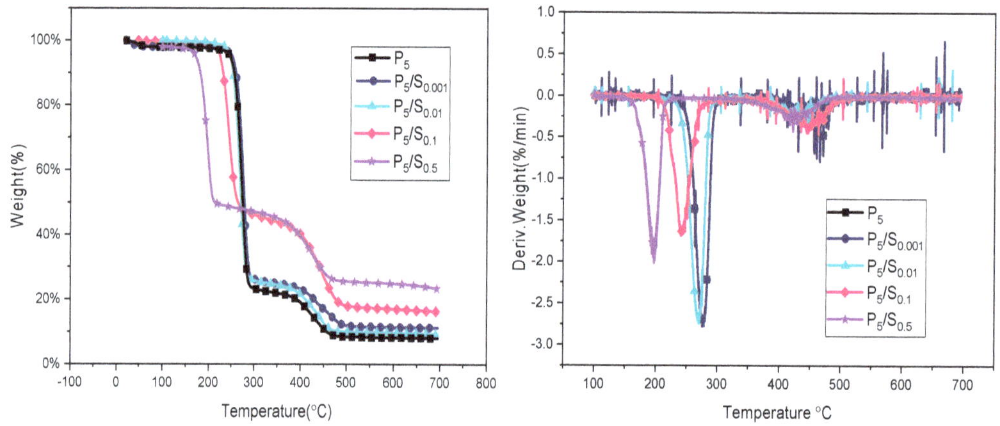

Figure 2. TGA and DTG (differential gravimetry) curves of PVA and PVA/NaOH aerogels at a heating rate of 10 °C/min under nitrogen.

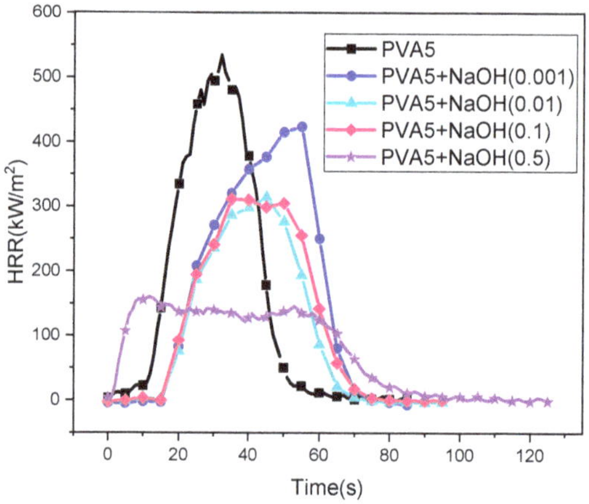

Figure 3. HRR plots of PVA and PVA/NaOH aerogels.

Table 3. TGA (thermogravimetric analysis) data of freeze dried PVA and PVA/NaOH aerogels.

Sample	$Td_{5\%}$ (°C)	$Td_{20\%}$ (°C)	Td_{max} (°C)	Dw/dt (%/°C)	Residue (%)
P_5	246	262	274	2.71	7.9
$P_5/S_{0.001}$	249	266	277	2.68	11.3
$P_5/S_{0.01}$	244	258	270	2.63	9.3
$P_5/S_{0.1}$	224	238	243	1.60	16.4
$P_5/S_{0.5}$	171	189	199	1.98	23.5

As hydrophilic materials rich in surface hydroxyl, PVA aerogels were difficult to remove all physisorbed water by regular drying methods. The weight loss stage before 100 °C is attributed to the loss of this water. The main decomposition step occurred between 150–500 °C. The onset decomposition temperature was evaluated by $Td_{5\%}$. With the addition of NaOH, the onset temperature decreased, which is consistent with previously reported observations [31,33]. We propose that NaOH is acting as a base catalyst at the elevated temperatures associated with burning polymers. In this mechanism, which is speculative at this time, PVA undergoes a rapid chain-stripping elimination of water once it is heated above the decomposition temperature (shown in Scheme 1) [34,35]. The chain stripping reaction (which is known chemistry in the absence of base catalysis) could produce polyenes through dehydration, which is in a competitive relationship with chain scission. Such polyenes are char precursors. With base catalysis, hydroxide may decrease the C-H bond strength, decreasing the onset decomposition temperature of the PVA in the aerogels, producing higher levels of polyenes and ultimately char was produced rather than chain scission products. Catalysis of the mechanistic steps shown in Scheme 1 are not normally associated with alkaline materials (rather with acids); the normal mechanisms of organic chemistry are not necessarily in play at the elevated temperatures associated with fire events—activation energies for reactions can be overcome when reaction temperatures are increased by hundreds of degrees. Recent innovations in high temperature alkaline catalysis (sodium hydroxide and calcium carbonate) used in biodiesel refining [36,37], as well as in the Guerbet Coupling reaction [38], suggest that the hypothetical mechanism shown in Scheme 1 has a prior art basis. In the most generally-accepted mechanism for the Guerbet Coupling of ethanol to butanol (or of butanol to 2-ethylhexanol), the high temperature dehydration of an alcohol is a key step, just as it is in the chain stripping mechanism. Hence, new chemistries may need to be considered when examining the mechanisms of flame retardation of organic polymers.

As can be seen in Table 3, the onset temperature for PVA aerogels decreased from 246 °C to 171 °C with an increasing base catalyst. A change in the degradation mechanism, enhancing chain stripping at the expense of chain scissions, was similarly observed with increasing levels of the base catalyst (Figures 2 and 3). With the greater extent of chain stripping in the presence of base, more char (less weight loss) resulted. In the proposed char-forming pathway, the polyenes undergo a Diels-Alder or intramolecular cyclization reaction [34,35], producing substituted cyclohexenes and cyclohexadienes, respectively, which can then aromatize to substituted aromatics. The fusion of aromatics results in the final, observed char products [39]. With the increase in char yield from 7.9 to as much as 23.5% (at 0.5% NaOH), approximately logarithmic relations between NaOH concentration and char yield are observed, with the data point at 0.001 being the only outlier; by increasing the addition levels of the base, the maximum mass-decomposition rates (dW/dT) dropped, as expected from 2.71 to under 2.0. Previously reported work by Arora and coworkers [28], examining the effects of as much as 5 wt% NaOH to PVA films, proposes that dehydration of alcohol groups in PVA likely commence at decreased temperatures when the alkaline agent is added—this is consistent with our proposed alkaline catalysis in PVA aerogels. The previous workers observed an increase in limiting oxygen index (LOI) from 20.5 to 27.2 with the incorporation of 4.5% NaOH. These authors also suggested that water generated by alcohol dehydrations could provide a diluent effect in reducing flammability. We cannot rule out such a mechanism, but also propose the graphitization/char formation shown in Scheme 1 to play an important role in the observed flame retardation.

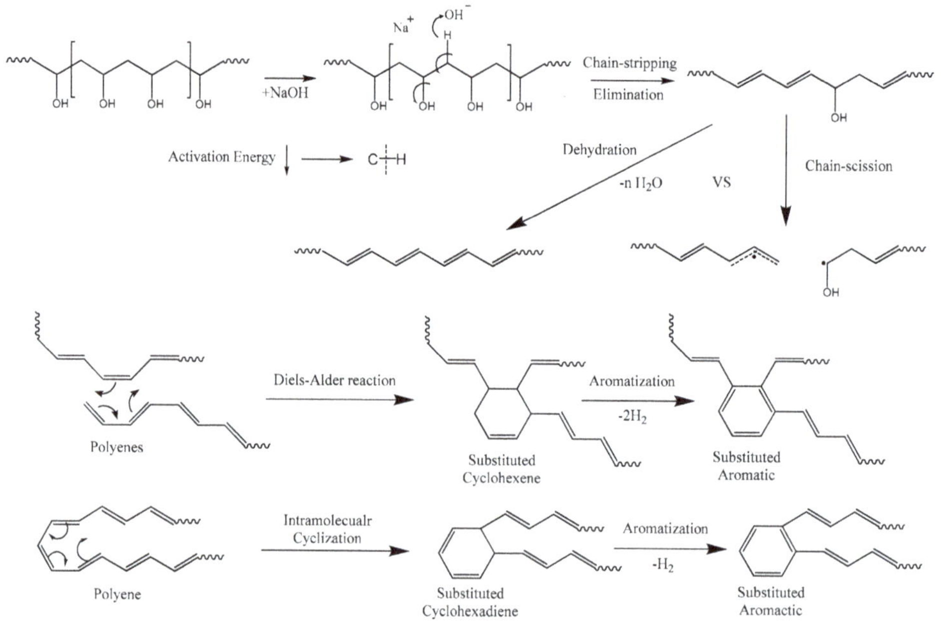

Scheme 1. Thermal pyrolysis process of PVA with the addition of NaOH.

2.4. Combustion Behavior

The combustion behavior of the aerogels was studied using cone calorimetry. Cone calorimetry is widely used in fire studies, it could provide plentiful data, including the time to ignition (TTI), peak of heat release (PHRR), time to peak of heat release (TTPHRR) and total heat release (THR); the results are shown in Table 4.

Table 4. Cone Calorimetry Data for PVA and PVA/NaOH Aerogels.

Sample	Weight (g)	TTI (s)	PHRR (kW/m^2)	TTPHRR (s)	THR (MJ/m^2)	THR/Mass ($MJ/(m^2\ g)$)
P_5	5.3	8	533 ± 35	32	12.9	2.4 ± 0.0
$P_5/S_{0.001}$	6.6	10	424 ± 37	55	14.0	2.1 ± 0.0
$P_5/S_{0.01}$	4.8	11	314 ± 13	45	9.9	2.1 ± 0.0
$P_5/S_{0.1}$	5.3	10	311 ± 31	35	11.2	2.1 ± 0.0
$P_5/S_{0.5}$	5.9	5	160 ± 20	13	9.5	1.6 ± 0.0

The samples were ignited under the heat flux of 50 kW/m^2. While all the samples were ignited in a short time, their overall flammability's were significantly reduced with incorporating NaOH. Figure 3 shows the HRR plots of PVA/NaOH aerogels and the control sample. The HRR curve of pristine PVA samples exhibited a sharp peak, with a PHRR of 533 kW/m^2; with the addition of a relatively small amount of NaOH, the curve showed a similar, relatively broad peak, and PHRR was decreased to about 310 kw/m^2; When the amount of NaOH was "sufficient", the HRR curve showed a very broad peaks with the lowest PHRR. The shape of $P_5/S_{0.5}$ aerogel was a typical HRR curve of thick charring materials [40]. The PHRR of $P_5/S_{0.5}$ was 160 kw/m^2, which was 30% of the number for pure PVA aerogel. The results indicate the addition of NaOH could decrease the fire risk of PVA based aerogel material in a big fire test. No dripping was observed for any of the samples during combustion testing.

Total heat release (THR) data are listed in Table 4. THRs of aerogel samples were all lower than that of the control samples, except for $P_5/S_{0.001}$, which likely contained insufficient base catalyst to produce an observable effect. To eliminate the influence of mass differences and to better illustrate the combustion behavior of the samples, THR/mass was compared. The $P_5/S_{0.5}$ aerogel exhibited the lowest THR/mass value, which was 1.6 MJ/(m^2 g), much lower than the results of the control PVA aerogels. The THR/mass values of P5/S0.001, P5/S0.01, and P5/S0.1 samples were similar, which is consistent with the PHRR results, indicating similar flammabilities.

The data above all showed that the addition of NaOH decreases the flammability of the PVA aerogels. The samples containing NaOH could form char to protect the unignited matrix polymer. The samples with relatively small amounts of NaOH in the PVA system showed slightly lower flammabilities compared to the pristine PVA aerogels, but the char forming process was insufficient to protect the unignited polymer. With a sufficient level of NaOH incorporated into the aerogel system, the samples could form denser char rapidly to fully cover the unignited part. The TGA char yields, given in Table 3, support the increasing levels of char with increased NaOH addition levels (8–11% char at/under 0.01% NaOH; 16% char at 0.1% NaOH; 24% char at 0.5%); these controlled values are consistent with the qualitative observations from cone calorimetry experiments. The mechanism of graphitization of PVA under flame conditions is consistent with the literature of PVA degradation, with the largely undescribed addition of alkaline catalysis. This hampered heat transfer and isolated the unignited matrix from the ignition source. Figure 4 illustrates the chars remaining after cone calorimetry testing. The control samples burned totally with little char forming. $P_5/S_{0.01}$ aerogel formed slightly more char than the pristine PVA aerogel, while $P_5/S_{0.5}$ formed a much denser char and was sustained in a good shape, even after the test. This result is in accordance with our proposed mechanism of base catalyzed char formation.

(A) (B) (C)

Figure 4. Images of aerogel samples after cone calorimetry test. (**A**) P_5; (**B**) $P_5/S_{0.01}$; (**C**) $P_5/S_{0.5}$.

To better understand the catalytic function of NaOH in PVA aerogel during combustion, the residues were studied using Raman spectroscopy (Figure 5). The spectra of chars with and without NaOH all exhibit two broad peaks with intensity maxima at 1580–1600 cm^{-1} (G band) and 1350 cm^{-1} (D band) [41,42]. The G band corresponds to the stretching vibration mode with E2 g symmetry in the aromatic layers of the graphite crystalline, whereas the D band is due to the disordered graphite [41,43,44]. The graphitization degree of the chars were estimated by the ratio of the intensity of the D and G bands (I_D/I_G) [41,45]. The I_D/I_G of pristine PVA(0.76) is lower than that of the sample prepared with NaOH(0.80). While lower I_D/I_G means better graphitization degree of the char and the better flame retardancy, it was more important that the graphitized carbons could realize an effective aggregate or crosslinking. The amount of char plays the key role in providing protection from the ignition source [43,45]. Both spectra exhibited a shoulder at ~1200 cm^{-1} in the D band, assigned as D_4, and attributed to the presence of polyene sp^2-sp^3 bonds or C–C and C=C stretching vibrations of polyene-like structures [46–49]. The charring residue of PVA/S had a higher D_4 intensity than was seen with the char residue of pristine PVA aerogels, indicating more polyenes were produced during the combustion tests in the presence of the base, again consistent with the proposal that NaOH was playing the role of catalyst promoting polyenes formation. The char structure enabled by the NaOH could hinder the transfer of combustible volatiles and heat, restricting the further decomposition of the polymer matrix.

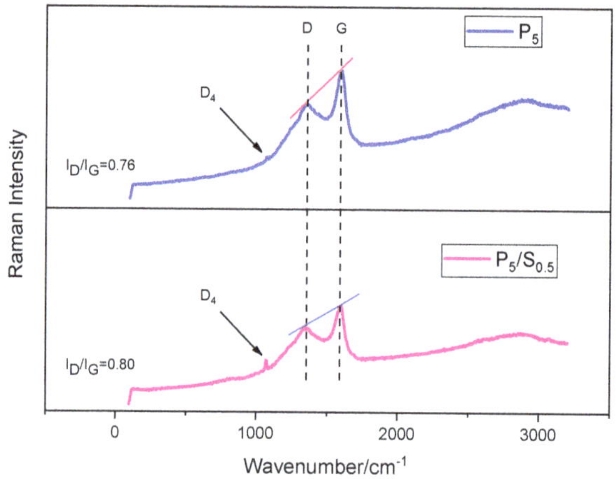

Figure 5. Raman spectra of char residues of P_5 and $P_5/S_{0.5}$.

A potential concern to the addition of sodium hydroxide to the PVA aerogels prepared in this study is that this additive is both hydroscopic and water soluble. One could reasonably worry that properties could change with time and with environmental aging. To the former point, samples were allowed to age under laboratory conditions and appear not to suffer from any water uptake, perhaps due to hydrogen bonding between the polymer and additive, and environmental equilibration native to the PVA polymer itself. We have not carried out any experiments wherein the polymer is exposed to significant washing with water—this is a point that should be addressed with any polymer system that makes use of such basic catalysis.

3. Conclusions

PVA aerogels incorporating differing levels of NaOH were fabricated through a simple and environmentally friendly freeze-drying process. The corresponding apparent densities, compressive properties, morphologies, thermal stabilities, combustion behaviors, and Raman spectroscopy were investigated. The moduli of PVA aerogels increased with increasing NaOH concentration, likely due to higher overall levels of solids and densification of polymeric "struts" and morphological changes in the aerogels. The specific moduli increased by more than three times with the addition 9.1% NaOH. The microstructure of the aerogels showed that the gallery spaces decreased in size and increased in number with increasing base levels. While the thermal degradation onset temperature was lower with the addition of NaOH, the decomposition process was shunted to a pathway wherein higher char yields were obtained once a threshold level of the base was present. Combustion tests showed the addition of NaOH formed much higher char yields once the threshold level of NaOH was met; that char does protect the unignited polymer matrix, leading to overall lower levels of combustion. The PHRR decreased from 533 kw/m^2 to 160 kw/m^2 and THR/Mass decreased from 2.4 MJ/(m^2 g) to 1.6 MJ/(m^2 g) in the presence of the base. The results of this study suggest that NaOH can act as catalyst, promoting polyenes formation in PVA, therefore forming significant char. With a small quantity of base catalyst, the PVA aerogel is mechanically stronger and much less flammable, maintaining, the most important advantage of aerogel, low density without using any char forming agents or halogens compounds in order to decrease flammability.

4. Materials and Methods

Poly(vinyl alcohol) (PVA; Mw 13,000–23,000, 98% hydrolyzed, SIGMA-Aldrich, St. Louis, MO, USA) and sodium hydroxide pellets (ACS grade, >98%; Innovating Science, Avon, NY, USA) were used without further purification. Deionized (DI) water was obtained using a Barnstead RoPure low-pressure, reverse-osmosis system (Lake Balboa, CA, USA).

4.1. Preparation of Aerogels

The aerogel compositions are shown in Table 1. PVA solutions were prepared by stirring PVA powder with DI water for six hours at 90 °C. Sodium hydroxide was mixed with DI water under magnetic stirring at 40 °C for 30 min before addition to the PVA solutions; the solutions were mixed to the desired PVA concentrations at 50 °C until they were homogeneous suspensions (the final compositions are given in Table 1). The suspensions were then cast into a 100 mm × 100 mm × 10 mm rectangular mold (for cone calorimetry testing) or poured into 12.5 mL polystyrene vials (for compression testing) and frozen in a solid carbon dioxide/ethanol bath (−70 °C). These samples were freeze dried using a VirTis Advantage EL-85 lyophilizer, with the shelf temperature set to 25 °C, and the pressure set to under 10 μbar. The products were named as P_x/S_y, where P represents PVA and S represents sodium hydroxide; the subscripts indicate their content per 100 g water. P5 were prepared by same method and fabricated as the general controls. The samples were stored in a desiccator after fabrication and were dried in a vacuum oven at 50 °C for 20 min before characterization.

4.2. Characterization

The pH values were measured using a pH meter (Hanna Instruments, Smithfield, RI, USA). The meter was calibrated by two standard buffers (pH 4.01, 7.01) before every test; each composition was tested 3 times, with average values given in Table 1.

The apparent densities were calculated by measuring the mass and dimensions using a Mettler Toledo (Columbus, OH, USA) AB204-S analytical balance and an electronic digital caliper (RCBS); each composition was tested with 5 samples.

Compression testing was conducted on Instron model 5500 universal testing machine (Instron, Norwood, MA, USA), fitted with a 1 KN load cell and 10 mm/min compression rate. The modulus values were attained from the slope of the area under the linear portion of the stress-strain curves, using specimens measuring 20 mm in both height and diameter. Each composition was tested with a minimum of five cylindrical samples, and the results were averaged.

Thermal properties were measured by thermogravimetric analysis (TGA), using a TGA Q500 (TA Instruments, New Castle, DE, USA). The specimens were placed in a platinum pan, equilibrated at 100 °C for 3 min, then heated to 700 °C at a heating rate of 10 °C/min. The process was conducted under a nitrogen flow rate of 40 mL/min. Each composition was replicated 3 times.

The morphological microstructure of the aerogels was investigated using a JEOL JSM 5900LV scanning electron microscopy (SEM)(JEOL, Tokyo, Japan) with an accelerating voltage of 20 kV. Samples were frozen in liquid nitrogen for 20 min, then cryofractured before imaging.

The combustion behavior of the aerogels was measured using a cone calorimeter (Fire Testing Technology, East Grinstead, UK) in accordance with the ASTM E 1354 standard, with the limitation noted below. The heat flux was set to 50 kW/m^2. The rectangular samples (100 mm × 100 mm × 10 mm) were wrapped with aluminum foil before testing. Note that smoke generation measurements were not made due to equipment limitations.

Raman spectroscopy was conducted on an inVia laser Raman spectrometer (Renishaw, Gloucestershire, UK) at room temperature with excitation provided by a 514 nm laser line.

Author Contributions: Conceptualization, Z.-H.C. and D.A.S..; methodology, Z.-H.C., M.-L.G., X.-Y.C., T.W., T.W. and D.A.S.; validation, D.A.S.; formal analysis, Z.-H.C., Y.-Z.W., D.A.S.; investigation, Z.-H.C., M.-L.G., X.-Y.C.; resources, D.A.S.; data curation, Z.-H.C.; writing—original draft preparation, Z.-H.C.; writing—review and editing, D.A.S.; visualization, Z.-H.C.; supervision, Y.-Z.W. and D.A.S.; project administration, D.A.S.; funding acquisition, D.A.S. All authors have read and agreed to the published version of the manuscript.

Funding: This research received no external funding.

Data Availability Statement: Data is contained within the article.

Conflicts of Interest: The authors declare no conflict of interest.

References

1. Kistler, S.S. Coherent Expanded Aerogels and Jellies. *Nature* **1931**, *127*, 741. [CrossRef]
2. Aaltonen, O.; Jauhiainen, O. The Preparation of Lignocellulosic Aerogels from Ionic Liquid Solutions. *Carbohydr. Polym.* **2009**, *75*, 125–129. [CrossRef]
3. Wang, Y.-T.; Zhao, H.-B.; Degracia, K.; Han, L.-X.; Sun, H.; Sun, M.; Wang, Y.-Z.; Schiraldi, D.A. Green Approach to Improving the Strength and Flame Retardancy of Poly(Vinyl Alcohol)/Clay Aerogels: Incorporating Biobased Gelatin. *ACS Appl. Mater. Interfaces* **2017**, *9*, 42258–42265. [CrossRef]
4. Long, H.; Harley-Trochimczyk, A.; Pham, T.; Tang, Z.; Shi, T.; Zettl, A.; Carraro, C.; Worsley, M.A.; Maboudian, R. High Surface Area MoS_2/Graphene Hybrid Aerogel for Ultrasensitive NO_2 Detection. *Adv. Funct. Mater.* **2016**, *26*, 5158–5165. [CrossRef]
5. Chen, H.-B.; Schiraldi, D.A. Flammability of Polymer/Clay Aerogel Composites: An Overview. *Polym. Rev.* **2019**, *59*, 1–24. [CrossRef]
6. Feng, J.; Le, D.; Nguyen, S.T.; Tan Chin Nien, V.; Jewell, D.; Duong, H.M. Silica-cellulose Hybrid Aerogels for Thermal and Acoustic Insulation Applications. *Colloids Surfaces A Physicochem. Eng. Asp.* **2016**, *506*, 298–305. [CrossRef]
7. Feng, J.; Nguyen, S.T.; Fan, Z.; Duong, H.M. Advanced Fabrication and Oil Absorption Properties of Super-Hydrophobic Recycled Cellulose Aerogels. *Chem. Eng. J.* **2015**, *270*, 168–175. [CrossRef]
8. Cui, J.; Xi, Y.; Chen, S.; Li, D.; She, X.; Sun, J.; Han, W.; Yang, D.; Guo, S. Prolifera-Green-Tide as Sustainable Source for Carbonaceous Aerogels with Hierarchical Pore to Achieve Multiple Energy Storage. *Adv. Funct. Mater.* **2016**, *26*, 8487–8495. [CrossRef]
9. Xia, W.; Qu, C.; Liang, Z.; Zhao, B.; Dai, S.; Qiu, B.; Jiao, Y.; Zhang, Q.; Huang, X.; Guo, W.; et al. High-Performance Energy Storage and Conversion Materials Derived from a Single Metal–Organic Framework/Graphene Aerogel Composite. *Nano Lett.* **2017**, *17*, 2788–2795. [CrossRef]
10. Pajonk, G.M. Aerogel Catalysts. *Appl. Catal.* **1991**, *72*, 217–266. [CrossRef]
11. Finlay, K.; Gawryla, M.D.; Schiraldi, D.A. Biologically Based Fiber-Reinforced/Clay Aerogel Composites. *J. Ind. Eng. Chem. Res.* **2008**, *47*, 615–619. [CrossRef]
12. Gawryla, M.D.; Nezamzadeh, M.; Schiraldi, D.A. Foam-like Materials Produced from Abundant Natural Resources. *Green Chem.* **2008**, *10*, 1078. [CrossRef]
13. Chen, H.-B.; Wang, Y.-Z.; Schiraldi, D.A. Preparation and Flammability of Poly(Vinyl Alcohol) Composite Aerogels. *ACS Appl. Mater. Interfaces* **2014**, *6*, 6790–6796. [CrossRef]
14. Zheng, Q.; Cai, Z.; Gong, S. Green Synthesis of Polyvinyl Alcohol (PVA)–Cellulose Nanofibril (CNF) Hybrid Aerogels and Their Use as Superabsorbents. *J. Mater. Chem. A* **2014**, *2*, 3110–3118. [CrossRef]
15. De France, K.J.; Hoare, T.; Cranston, E.D. Review of Hydrogels and Aerogels Containing Nanocellulose. *Chem. Mater.* **2017**, *29*, 4609–4631. [CrossRef]
16. Liu, A.; Medina, L.; Berglund, L.A. High-Strength Nanocomposite Aerogels of Ternary Composition: Poly(Vinyl Alcohol), Clay, and Cellulose Nanofibrils. *ACS Appl. Mater. Interfaces* **2017**, *9*, 6453–6461. [CrossRef]
17. Wicklein, B.; Kocjan, A.; Salazar-Alvarez, G.; Carosio, F.; Camino, G.; Antonietti, M.; Bergström, L. Thermally Insulating and Fire-Retardant Lightweight Anisotropic Foams Based on Nanocellulose and Graphene Oxide. *Nat. Nanotechnol.* **2015**, *10*, 277–283. [CrossRef] [PubMed]
18. *Fire Loss in the United States During 2017*; NFPA National Fire Protectio n Association: Quincy, MA, USA, 2018.
19. Darnerud, P.O. Toxic Effects of Brominated Flame Retardants in Man and in Wildlife. *Environ. Int.* **2003**, *29*, 841–853. [CrossRef]
20. Rahman, F.; Langford, K.H.; Scrimshaw, M.D.; Lester, J.N. Polybrominated Diphenyl Ether (PBDE) Flame Retardants. *Sci. Total Environ.* **2001**, *275*, 1–17. [CrossRef]
21. Safe, S.; Hutzinger, O. Polychlorinated Biphenyls (PCBs) and Polybrominated Biphenyls (PBBs): Biochemistry, Toxicology, and Mechanism of Action. *CRC Crit. Rev. Toxicol.* **1984**, *13*, 319–395. [CrossRef] [PubMed]

22. Chen, H.-B.; Wang, Y.-Z.; Sánchez-Soto, M.; Schiraldi, D.A. Low Flammability, Foam-like Materials Based on Ammonium Alginate and Sodium Montmorillonite Clay. *Polymers* **2012**, *53*, 5825–5831. [CrossRef]
23. Troitzsch, J.H. Overview of Flame Retardants. *Chem. Today* **1998**, *16*, 18–24.
24. Morgan, A.B.; Wilkie, C.A. *Flame Retardant Polymer Nanocomposites*; Wiley-Interscience: Hoboken, NJ, USA, 2007.
25. Holland, B.J.; Hay, J.N. The thermal degradation of poly(vinyl alcohol). *Polymer* **2001**, *42*, 6775–6783. [CrossRef]
26. Yang, H.; Xu, S.; Jiang, L.; Dan, Y. Thermal Decomposition Behavior of Poly (Vinyl Alcohol) with Different Hydroxyl Content. *J. Macromol. Sci. Part B Phys.* **2012**, *51*, 464–480. [CrossRef]
27. Zaikov, G.E.; Lomakin, S.M. New aspects of ecologically friendly polymer flame retardant systems. *Polym. Degrad. Stab.* **1996**, *54*, 223–233. [CrossRef]
28. Arora, S.; Kumar, M.; Kumar, M. Thermal and flammability studies of poly(vinyl alcohol) composites filled with sodium hydroxide. *J. Appl. Polym. Sci.* **2013**, *127*, 3877–3884. [CrossRef]
29. Wang, L.; Sánchez-Soto, M.; Maspoch, M.L. Polymer/Clay Aerogel Composites with Flame Retardant Agents: Mechanical, Thermal and Fire Behavior. *Mater. Des.* **2013**, *52*, 609–614. [CrossRef]
30. Johnson, J.R.; Spikowski, J.; Schiraldi, D.A. Mineralization of Clay/Polymer Aerogels: A Bioinspired Approach to Composite Reinforcement. *ACS Appl. Mater. Interfaces* **2009**, *1*, 1305–1309. [CrossRef] [PubMed]
31. Chen, H.-B.; Liu, B.; Huang, W.; Wang, J.-S.; Zeng, G.; Wu, W.-H.; Schiraldi, D.A. Fabrication and Properties of Irradiation-Cross-Linked Poly(vinyl alcohol)/Clay Aerogel Composites. *ACS Appl. Mater. Interfaces* **2014**, *6*, 16227–16236. [CrossRef]
32. Wang, Y.; Gawryla, M.D.; Schiraldi, D.A. Effects of Freezing Conditions on the Morphology and Mechanical Properties of Clay and Polymer/Clay Aerogels. *J. Appl. Polym. Sci.* **2013**, *129*, 1637–1641. [CrossRef]
33. Cheng, Z.; DeGracia, K.; Schiraldi, D.; Cheng, Z.; DeGracia, K.; Schiraldi, D.A. Sustainable, Low Flammability, Mechanically-Strong Poly(Vinyl Alcohol) Aerogels. *Polymers (Basel)* **2018**, *10*, 1102. [CrossRef]
34. Gilman, J.W.; Vanderhardt, D.L.; Kashiwagi, T. Thermal Decomposition Chemistry of Poly (Vinyl Alcohol): Char Characterization and Reactions with Bismaleimides. In *Fire and Polymers II*; A.C.S.: Washington, DC, USA, 1995; pp. 161–185.
35. Pandey, S.; Pandey, S.K.; Parashar, V.; Mehrotra, G.K.; Pandey, A.C. Ag/PVA Nanocomposites: Optical and Thermal Dimensions. *J. Mater. Chem.* **2011**, *21*, 17154. [CrossRef]
36. Mittelbach, M. Advances in biodiesel catalysts and processing technologies. In *Advances in Biodiesel Production*; Luque, R., Melero, J.A., Eds.; Woodhead Publishing: Cambridge, UK, 2012; pp. 133–153.
37. Cheirsilp, B.; Srinuanpan, S.; Mandik, Y.I. Efficient Harvesting of Microalgal biomass and Direct Conversion of Microalgal Lipids into Biodiesel. In *Microalgae Cultivation for Biofuels Production*; Yousuf, A., Ed.; Academic Press: Cambridge, MA, USA, 2020; pp. 83–96.
38. Hanspal, S. The Guerbet Coupling of Ethanol into Butanol over Calcium Hydroxyapatite Catalysts. Ph.D. Thesis, University of Virginia, Charlottesville, VA, USA, May 2016.
39. Grand, A.; Wilkie, C. *Fire Retardancy of Polymeric Materials*; CRC Press: Boca Raton, FL, USA, 2000.
40. Schartel, B.; Hull, T.R. Development of Fire-Retarded Materials—Interpretation of Cone Calorimeter Data. *Fire Mater.* **2007**, *31*, 327–354. [CrossRef]
41. Ferrari, A.C.; Robertson, J. Interpretation of Raman Spectra of Disordered and Amorphous Carbon. *Phys. Rev. B* **2000**, *61*, 14095–14107. [CrossRef]
42. Ferrari, A.C.; Meyer, J.C.; Scardaci, V.; Casiraghi, C.; Lazzeri, M.; Mauri, F.; Piscanec, S.; Jiang, D.; Novoselov, K.S.; Roth, S.; et al. Raman Spectrum of Graphene and Graphene Layers. *Phys. Rev. Lett.* **2006**, *97*, 187401. [CrossRef]
43. Huang, Y.; Yang, Y.; Ma, J.; Yang, J. Preparation of Ferric Phosphonate/Phosphinate and Their Special Action on Flame Retardancy of Epoxy Resin. *J. Appl. Polym. Sci.* **2018**, *135*, 46206. [CrossRef]
44. Kaufman, J.H.; Metin, S.; Saperstein, D.D. Symmetry Breaking in Nitrogen-Doped Amorphous Carbon: Infrared Observation of the Raman-Active *G* and *D* Bands. *Phys. Rev. B* **1989**, *39*, 13053–13060. [CrossRef]
45. Song, S.; Ma, J.; Cao, K.; Chang, G.; Huang, Y.; Yang, J. Synthesis of a Novel Dicyclic Silicon-/Phosphorus Hybrid and Its Performance on Flame Retardancy of Epoxy Resin. *Polym. Degrad. Stab.* **2014**, *99*, 43–52. [CrossRef]
46. Dippel, B.; Jander, H.; Chemical, J.H.-P.C. NIR FT Raman Spectroscopic Study of Flame Soot. *Phys. Chem. Chem. Phys.* **1999**, *1*, 4707–4712. [CrossRef]
47. Dippel, B.; Heinzenberg, J. Soot Characterization in Atmospheric Particles from Different Sources by NIR FT. Raman Spectroscopy. *J. Aerosol Sci.* **1999**, *30*, S907–S908. [CrossRef]
48. Sadezky, A.; Muckenhuber, H.; Grothe, H.; Niessner, R.; Pöschl, U. Raman Microspectroscopy of Soot and Related Carbonaceous Materials: Spectral Analysis and Structural Information. *Carbon N. Y.* **2005**, *43*, 1731–1742. [CrossRef]
49. Coccato, A.; Jehlicka, J.; Moens, L.; Vandenabeele, P. Raman Spectroscopy for the Investigation of Carbon-Based Black Pigments. *J. Raman Spectrosc.* **2015**, *46*, 1003–1015. [CrossRef]

MDPI
St. Alban-Anlage 66
4052 Basel
Switzerland
www.mdpi.com

Gels Editorial Office
E-mail: gels@mdpi.com
www.mdpi.com/journal/gels

www.ingramcontent.com/pod-product-compliance
Lightning Source LLC
LaVergne TN
LVHW070702100526
838202LV00013B/1013